Agronomy and Crop Science

Agronomy and Crop Science

Edited by **Jamie Hanks**

Callisto Reference

New York

Published by Callisto Reference,
106 Park Avenue, Suite 200,
New York, NY 10016, USA
www.callistoreference.com

Agronomy and Crop Science
Edited by Jamie Hanks

International Standard Book Number: 978-1-63239-657-0 (Hardback)

Printed in the United States of America.

Contents

Preface

The main aim of this book is to educate learners and enhance their research focus by presenting diverse topics covering this vast field. This is an advanced book which compiles significant studies by distinguished experts. This book addresses successive solutions to the challenges arising in the area of application, along with it; the book provides scope for future developments.

Agronomy is the field of science that takes into account a more holistic and integrated view of the agriculture and all the important fields related to it. It encompasses soil classification, crop rotation, irrigation and drainage, plant physiology, plant breeding, soil fertility, weed control, insect and pest control. Crop science on the other hand focuses on the effects of drought, water use efficiency, effect of temperatures on crops, mineral deficiency and toxicity stress and to reduce them. These are overlaping fields as they both concentrate on crops. This book attempts to understand the multiple branches that fall under the disciplines of agronomy and crop science and how such concepts have practical applications. The various studies that are constantly contributing towards advancing technologies and evolution of these fields are examined in detail. From theories to research to practical applications, case studies related to all contemporary topics of relevance to this field have been included in this book. It will help the readers in keeping pace with the rapid changes in this field. It will serve as a reference to a broad spectrum of readers.

It was a great honour to edit this book, though there were challenges, as it involved a lot of communication and networking between me and the editorial team. However, the end result was this all-inclusive book covering diverse themes in the field.

Finally, it is important to acknowledge the efforts of the contributors for their excellent chapters, through which a wide variety of issues have been addressed. I would also like to thank my colleagues for their valuable feedback during the making of this book.

Editor

Preface

1

A Draft Genome Sequence for *Ensete ventricosum*, the Drought-Tolerant "Tree Against Hunger"

James Harrison [1], Karen A. Moore [1], Konrad Paszkiewicz [1], Thomas Jones [1], Murray R. Grant [1], Daniel Ambacheew [2], Sadik Muzemil [2] and David J. Studholme [1,*]

[1] College of Life and Environmental Sciences, University of Exeter, Geoffrey Pope Building, Stocker Road, Exeter EX4 4QD, UK; E-Mails: jh288@exeter.ac.uk (J.H.); K.A.Moore@exeter.ac.uk (K.A.M.); k.h.paszkiewicz@exeter.ac.uk (K.P.); tj234@exeter.ac.uk (T.J.); m.r.grant@exeter.ac.uk (M.R.G.)

[2] Southern Agricultural Research Institution (SARI). P.O. Box. 06, Hawassa, Ethiopia; E-Mails: ethiodan@gmail.com (D.A.); croprch@sari.gov.et (S.M.)

* Author to whom correspondence should be addressed; E-Mail: d.j.studholme@exeter.ac.uk

---

Abstract: We present a draft genome sequence for enset (*Ensete ventricosum*) available via the Sequence Read Archive (accession number SRX202265) and GenBank (accession number AMZH01. Enset feeds 15 million people in Ethiopia, but is arguably the least studied African crop. Our sequence data suggest a genome size of approximately 547 megabases, similar to the 523-megabase genome of the closely related banana (*Musa acuminata)*. At least 1.8% of the annotated *M. acuminata* genes are not conserved in *E. ventricosum*. Furthermore, enset contains genes not present in banana, including reverse transcriptases and virus-like sequences as well as a homolog of the RPP8-like resistance gene. We hope that availability of genome-wide sequence data will stimulate and accelerate research on this important but neglected crop.

Keywords: enset; Ethiopia; drought-tolerance; Musaceae

---

1. Introduction

Enset (*Ensete ventricosum*) is one of the most important crop plants grown in Ethiopia, where it makes a major contribution of to the food security of the country, feeding at least 15 million people. It buffers food deficit during dry spells and recurrent drought and has been dubbed as the "tree against hunger" [1]. Enset is a multi-purpose crop, with all parts of the plant being utilized for human food, animal forage, medicine, or ornamental uses [2]. Furthermore, it has the capacity for high yield, can be stored for long periods, can be harvested at any time of the year and at any stage over a period of several years [3], thereby offering advantages over seasonal crops.

The genus *Ensete* falls within the botanical family Musaceae, which also includes bananas and plantains (genus *Musa*). Enset is susceptible to some of the same diseases that threaten banana, including bacterial wilt caused by *Xanthomonas campestris* pathovar *musacearum* [4]. Unlike banana, the main edible parts of the enset plant are the starchy corm and pseudostem. The genome of enset is diploid with $n = 9$ [5], while the recently published doubled-haploid banana genome sequence has $n = 11$ [6].

There are many clones and landraces of enset in Ethiopia [1,3]. A collection of more than 600 clones and landraces from major enset growing areas of Ethiopia has been assembled and conserved *ex situ* by the Southern Agricultural Research Institute at Areka and some of these differ in important agronomic characteristics and tolerance to disease [7]. Some attempts at molecular characterization of enset clones or landraces have been made using amplified fragment length polymorphism AFLP [8,9] and random amplified polymorphic DNA RAPD techniques [10,11], revealing the existence of genetic diversity and, therefore, the potential for improvement by breeding, if suitable markers were available. However, despite its importance and value, enset has been relatively neglected by scientific research and is arguably the least-studied African crop. There is an urgent need for efficient improvement of this crop. Our aim was to help accelerate enset research and crop improvement by providing draft genome sequence data and identifying single-nucleotide polymorphisms (SNPs) that might serve as molecular markers for marker-assisted breeding. We also aimed to investigate genetic similarity between enset and banana thus to assess the usefulness of banana genomic resources for application to enset.

2. Results and Discussion

2.1. Whole-Genome Sequencing

We generated 40.4 gigabases of whole-genome shotgun sequence data from the enset genome consisting of 202 million pairs of 100-nucleotide Illumina sequence reads. The sequence reads are freely available from the Sequence Read Archive under accession number SRX202265. Our approach was similar to that of Davey and colleagues [12] who recently re-sequenced the banana B genome (*M. balbisiana*) using 281 million pairs of 100-nucleotide Illumina sequence reads. Their attempt at *de novo* assembly yielded a highly fragmented genome assembly consisting of a large number of short contigs. However, they were able to gain insights into the B genome by aligning their sequence reads against the previously sequenced A genome (*M. acuminata*) and calling a consensus alignment [12]. Likewise, we used both *de novo* sequence assembly (that is, without using a reference genome sequence) and an approach based upon alignment of reads against the banana A-genome reference sequence as described in the sections below. Our aligned enset genomic sequence reads covered 47% of the *M. acuminata* reference genome sequence (247 out of 523 Mb). This is less than the coverage by Davey and colleagues' alignment of *M. balbisiana* reads against the same reference genome, which covered 341 out of 523 Mb (65%), perhaps not surprisingly given the larger evolutionary distance between enset and the *Musa* species.

To check for contamination, we aligned our enset genomic sequence reads against all of the 2735 available complete prokaryotic genomes [13] using the Burrows-Wheeler Aligner BWA [14]. We found that 8.27% of our sequence reads were alignable against prokaryotic bacterial sequences. The genome sequences showing the greatest coverage were *Pseudomonas fluorescens* SBW25 [15] and *Methylobacterium radiotolerans* JCM 2831 ([16], GenBank: CP001001) with sequence reads covering 30.6% and 33.5% of the lengths of their genomes, respectively. These prokaryotic sequences possibly originate from endophytes and/or epiphytes associated with the plant even though we attempted to clean and sterilize the surface of the plant material by wiping with ethanol. We note that in the study by Davey and colleagues [12] there was also some bacterial sequence present in the *M. balbisiana* genomic re-sequencing data: 3.03% of Davey's data aligned to the prokaryotic genome sequences, with coverage of 94.3% of the *Propionibacterium acnes* 266 [17] chromosome, and 60.8% of the *Serratia marcescens* WW4 [18] chromosome. Therefore, it seems that bacterial contamination of plant genome sequence data is not unique to our study. We also note that the depth of coverage of any single bacterial genome by "plant" genomic reads is very low: no more than 2.03× for the *P. fluorescens* and *M. radiotolerans* genomes and no more than 9.1× for the *P. acnes* and *S. marcescens* genomes mentioned above, and, therefore, not enough to be effectively assembled *de novo*.

2.2. Estimation of the Enset Genome Length

Based on alignment against enset nuclear DNA sequences available in the GenBank database (Table 1), we estimate the depth of coverage as 67.67×. Given that we generated a total of 37.05 gigabases of sequence data (after removing prokaryote-matching reads) this would indicate a genome size of approximately 547 megabases. This is close to the haploid genome size of 523 megabases for the closely related *M. acuminata* [6].

2.3. Conservation of Protein-Coding Sequences between Enset and Banana

To identify which banana protein-coding genes are conserved in enset, we aligned our enset shotgun sequence reads against the 36,542 *M. acuminata* coding sequences identified by D'Hont and colleagues [6] using BWA [19]. The advantage of this approach is that it is not confounded by incomplete assembly of or gene prediction in the enset data. The frequency distribution for breadth of coverage across these 36,542 sequences is shown in Figure 1. The breadths of coverage follow a bi-modal distribution with peaks close to zero and close to 100% coverage. The peak close to zero corresponds to banana genes that are either absent from the enset genome or else they are so divergent that the corresponding enset sequences fail to align. There are 662 (1.8%) banana protein-coding sequences that have zero coverage by the aligned enset data and are, therefore, absent, or very divergent, in enset. The Supplementary Data includes a spreadsheet indicating the breadths of coverage of each *M. acuminata* gene.

Table 1. Depths of coverage of previously published enset nuclear DNA sequences. The median depth of coverage is 67.67 times.

GenBank accession number and description	Depth
HM118700.1 TCP-1-eta subunit gene	80.71
HM118740.1 mRNA capping enzyme large subunit family protein gene	79.26
HM118605.1 electron transport protein gene	79.06
HM118577.1 ATP:citrate lyase gene	75.76
HM118779.1 succinoaminoimidazole-carboximide ribonucleotide synthetase family	74.08
HM118753.1 methylcrotonyl-CoA carboxylase beta chain-like gene	72.01
HM118766.1 annexin-like protein gene	71.61
HM118805.1 initiation factor 2B family protein gene	68.05
HM118660.1 zeaxanthin epoxidase gene	67.67
HM118646.1 CASP protein-like gene, partial sequence	65.98
HM118632.1 endoribonuclease dicer protein-like gene, partial sequence	65.39
HM118673.1 Na/H antiporter gene	65.16
HM118591.1 stomatal cytokinesis defective protein gene	64.52
HM118819.1 DNA polymerase delta catalytic subunit gene	63.05
HM118713.1 NAD+ synthase domain protein gene	61.95
HM118619.1 non-phototropic hypocotyl 3-like gene, partial sequence	61.72
HM118686.1 DUF89 family protein gene	57.14

Figure 1. Frequency distribution for breadth of coverage on 36,542 banana gene sequences by enset whole-genome shotgun sequence reads aligned against the banana genome using BWA.

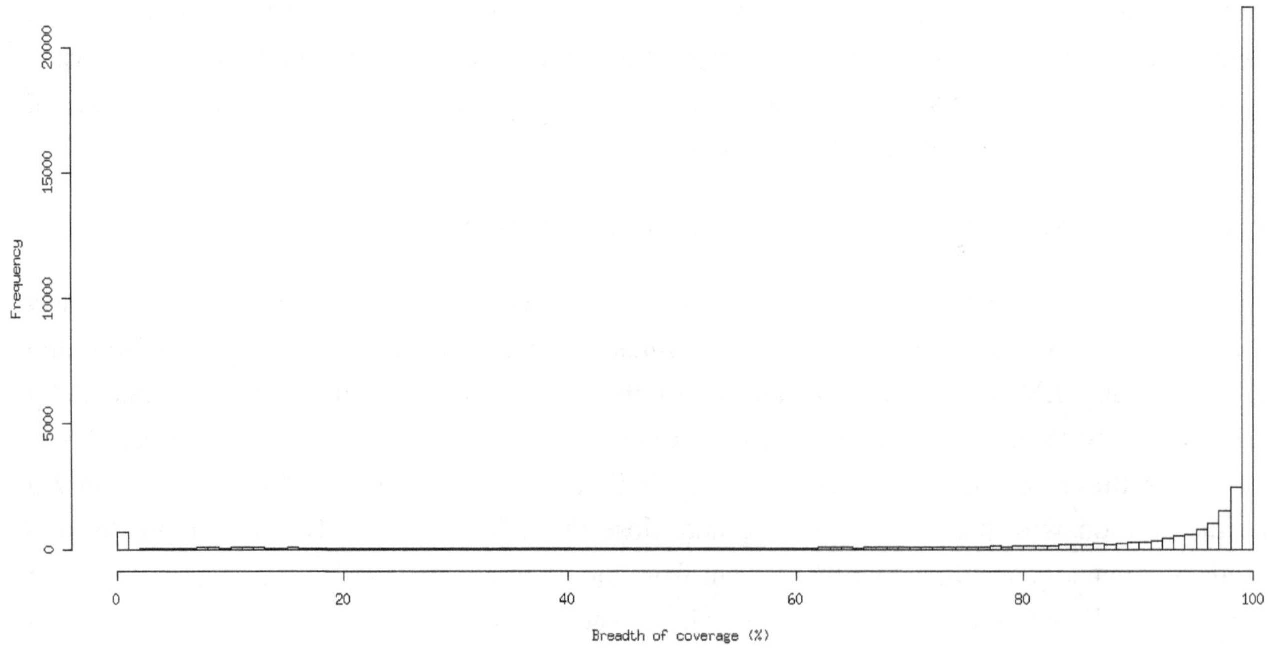

2.4. Heterozygosity and Single-Nucleotide Polymorphisms (SNPs)

Single-nucleotide polymorphisms (SNPs) can be valuable markers for crop improvement [20] but have not previously been reported for enset. Given the very fragmented nature of our *de novo* assembly of the enset genome, we followed the example of Davey and colleagues [12] by performing SNP calling against the high-quality reference genome sequence of *M. acuminata* [6]. To do the alignment, we used BWA [14] and only considered sequence reads that uniquely align to a single genomic location. By aligning the enset shotgun sequence reads against this banana genome sequence, we were able to identify 30,287 sites at which there was an approximately 50:50 ratio between the two most frequent aligned nucleotides (where the most abundant base accounts for between 49% and 51% of the aligned bases and where coverage is at least 10×). These sites are distributed over the whole genome (see Figure 2) and occur on average every 17.3 kb. If we are less stringent and include all sites where the frequency of the most abundant base is between 48% and 52%, then the number of heterozygous sites increases to 76,416, a density of one site per 6.8 kb of banana genome. See Figure 3 for an example of such a locus, containing three heterozygous sites. See the Supplementary Data for a list of these heterozygous sites. The rationale for using the banana genome as a reference sequence for identifying heterozygous SNPs is that the banana reference genome sequence is much more contiguous and better annotated than the enset *de novo* genome sequence. However, one limitation of this approach is that it will fail to identify heterozygous sites that fall within enset-specific sequences. We found that alignment between enset genomic sequences reads and the banana reference genome sequence covered only 47% of the banana genome and occurred much more frequently in genes rather than intergenic regions, as also observed by Davey and colleagues [12] for alignment of *M. balbisiana* genomic reads against the same reference genome. To circumvent this limitation, we also generated lists of heterozygous sites called on the enset *de novo* assembly; these can be found in the Supplementary Data.

2.5. De Novo *Assembly of the Enset Genome Sequence*

Although alignment of raw sequence reads against the banana reference genome sequence is useful for identifying SNPs and sequences conserved between both plant species, we required a *de novo* assembly of the enset data in order to examine gene order and to identify enset sequences that are not present in the banana genome. Our assembly had a total length of 459.5 megabases. This represents 84% of the estimated enset genome-size of 547 megabases and is 97.3% of the length of the recently published banana genome assembly of 472.2 megabases [6]. Given that our estimate of the enset genome size based on sequence coverage is very approximate and assuming that the enset genome is of similar size to the banana genome, then this suggests that our *de novo* assembly represents nearly complete coverage of the enset genome.

Figure 2. Positions on the banana genome that display heterozygosity in enset. The horizontal axis indicates position on the chromosome and the vertical axis indicates the frequency of the most common base (A, C, G, or T). Only those sites are shown at which there is at least 10× coverage and at which the frequency of the most abundant base is between 49% and 51% inclusive.

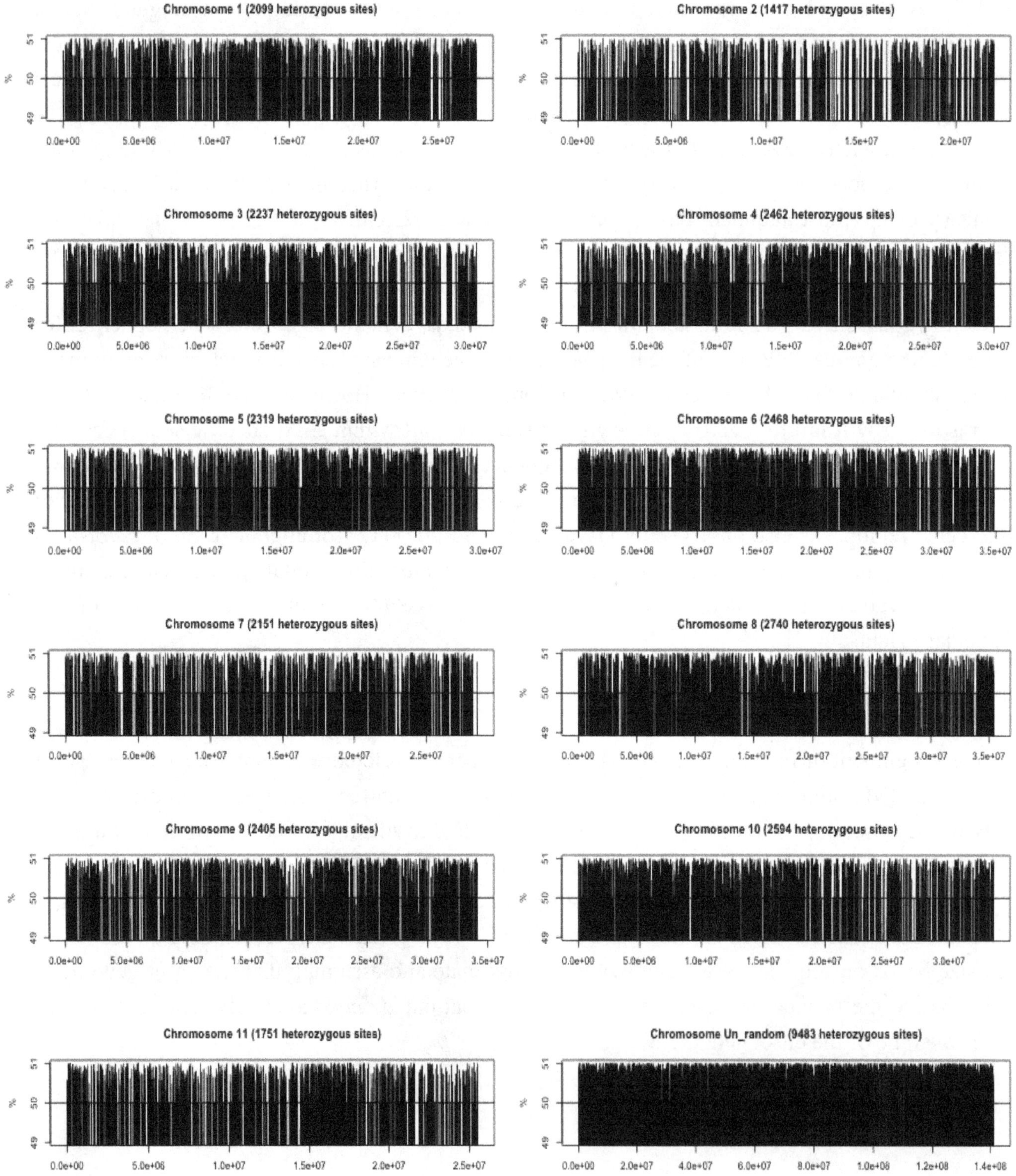

Figure 3. Example of a protein-coding gene that is heterozygous in enset. We aligned enset genomic sequence reads against the banana genome using BWA. The figure shows a 40-nucleotide region of the alignment falling within a protein-coding gene (GSMUA_Achr1T20250_001), encoding a predicted acyl-transferase. This region includes three single-nucleotide polymorphisms, at which the enset genome sequence is heterozygous with approximately 50:50 frequencies for two haplotypes (C…C…T and G…T…C).

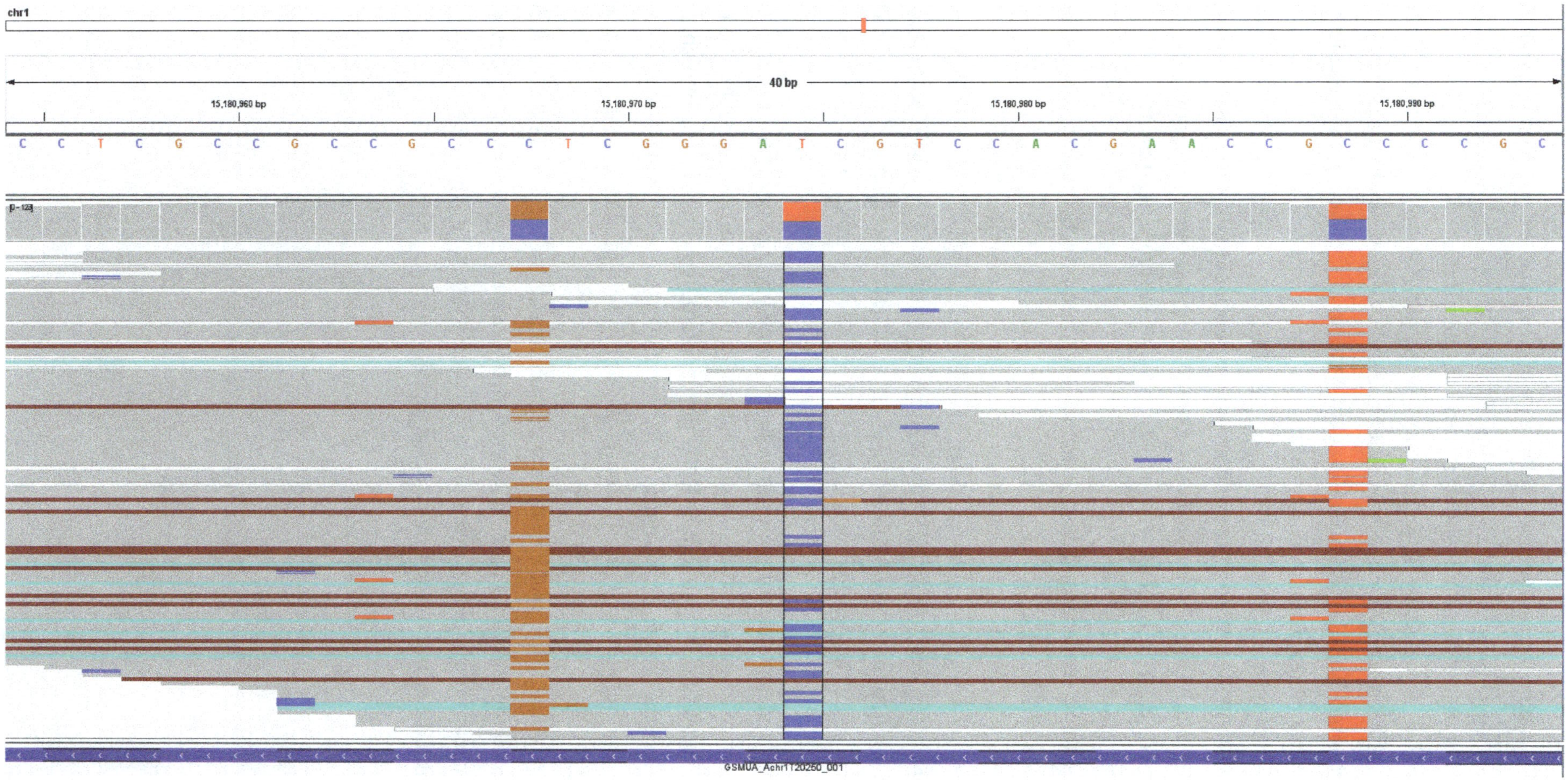

The enset genome sequence assembly is available via the GenBank database under accession number AMZH01. Due to restrictions on the numbers of contigs and supercontigs that GenBank can accept within a whole-genome shotgun project, GenBank only includes the enset contigs and super-contigs that are at least five kilobases in length. The full assembly, including contigs and super-contigs of between 200 and 5000 nucleotides, is available via Figshare [21]. Approximately 70% of the enset genome assembly is alignable against the banana genome sequence and average nucleotide sequence identity is 89.90% over the alignable sequence, as judged by the *dnadiff* tool in the MUMmer [22] software package.

Given that about 8% of our genomic sequence reads actually originated from prokaryotes rather than from the plant, we checked our *de novo* assembly for prokaryotic sequences by performing Basic Local Alignment Search Tool nucleotide (BLASTN) searches against the 2735 available complete prokaryotic genomes [13]. A total of 81,795 bp (0.018%) of the enset *de novo* assembly matched prokaryotic genome sequences. These sequences were removed from the data submitted to GenBank (accession AMZH01).

We performed a preliminary annotation of the enset genome assemblies using FGENESH [23] to predict protein-coding genes; summary statistics are given in Table 2 and the protein sequences, their genomic coordinates, results of BLASTP searches against the *M. acuminata* proteome, and the results of functional prediction using PfamScan [24] are available via Figshare [21] (the file was too large to be included in the Supplementary Data). Of 42,749 predicted proteins, 9967 did not have any significant sequence similarity to the banana proteome detectable by BLASTP. It should be noted that due to the fragmented nature of the draft *de novo* assembly, the number of predicted genes is likely to be significantly over-estimated as some gene models are split between multiple contigs. We used RfamScan [25] to identify non-coding RNA genes, including microRNAs, which are listed in Table 3, and we used RepeatMasker [26] to search for matches to repeat sequences (Table 4), as described in the Experimental Section. Overall, the enset assembly was predicted to have a greater repeat-content (32.65%) than the banana A genome (20.31%).

Gene order was highly conserved between banana and enset, at least over the scale of tens of kilobases, as exemplified in Figure 4, which shows an alignment of the longest enset super-contig against banana chromosome 5. However, we did identify some differences in gene-content between the two genomes as described in the following sections.

Table 2. Assembly statistics.

	Complete assembly	Subset of assembly submitted to GenBank (AMZH00000000.1)
Number of scaffolds	123,779	14,787
N_{50} scaffold length	11,149	13,657
NG_{50} scaffold length (bp)	9,954	n.a. *
Shortest scaffold (bp)	200	5,000
Longest scaffold (bp)	105,416	103,995
Sum of scaffold lengths (bp)	458,655,998	172,241,963
Mean scaffold length (bp)	3,705	15,952
Median scaffold length (bp)	1,056	13,404
Number of contigs	259,028	19,109
N_{50} contig length (bp)		8,724
NG_{50} contig length (bp)	2,428	n.a. *
Shortest contig (bp)	201	5,000
Longest contig (bp)	56,178	56,178
Sum of contig lengths (bp)	390,884,093	163,735,150
Mean contig length (bp)	1,509	8,568
Median contig length (bp)	555	7,448
Number of gene models	42,749	23,423
Mean length of predicted protein (aa)	311.64	353.84
G + C (%)	38.95	39.14

* NG_{50} lengths [27] were calculated on the basis of an estimated genome length of 50 Mb. The total length of the scaffolds submitted to GenBank (under accession AMZH00000000.1) was less than 50% of this estimated length (7.54 Mb *versus* 25 Mb); therefore, it is not possible to calculate NG_{50} length for this dataset.

Table 3. Predicted non-coding RNAs in the enset genome assembly predicted by Rfam version 11.

GenBank accession number	Scaffold name	Start and end positions	Strand	Rfam ID (and accession number)	Rfam scan E value
KB218331.1	scf_22030_17941	4842–4920	+	Intron_gpII (RF00029)	$2.89e^{-04}$
KB218832.1	scf_22030_39767	2365–2435	−	Intron_gpII (RF00029)	$3.47e^{-08}$
KB218412.1	scf_22030_21016	944–1028	+	mir-156 (RF00073)	$7.66e^{-17}$
KB220497.1	scf_22030_77035	4888–4971	−	mir-156 (RF00073)	$1.34e^{-17}$
KB220497.1	scf_22030_77035	4888–4971	+	mir-156 (RF00073)	$4.11e^{-09}$
KB220618.1	scf_22030_78211	2918–3003	−	mir-156 (RF00073)	$1.57e^{-14}$
KB220618.1	scf_22030_78211	2918–3003	+	mir-156 (RF00073)	$8.68e^{-09}$
KB220859.1	scf_22030_80462	10702–10791	+	mir-156 (RF00073)	$1.65e^{-17}$
KB220859.1	scf_22030_80462	10702–10791	−	mir-156 (RF00073)	$7.33e^{-09}$
KB220860.1	scf_22030_80478	14044–14147	+	mir-156 (RF00073)	$3.70e^{-17}$
KB220947.1	scf_22030_81257	2331–2413	+	mir-156 (RF00073)	$2.41e^{-16}$
KB220073.1	scf_22030_72447	11922–12159	−	MIR159 (RF00638)	$1.44e^{-35}$
KB220073.1	scf_22030_72447	11924–12161	+	MIR159 (RF00638)	$9.81e^{-22}$

Table 3. *Cont.*

GenBank accession number	Scaffold name	Start and end positions	Strand	Rfam ID (and accession number)	Rfam scan E value
KB220655.1	scf_22030_78562	4140–4330	−	MIR159 (RF00638)	$1.15e^{-37}$
KB220655.1	scf_22030_78562	4142–4332	+	MIR159 (RF00638)	$2.01e^{-21}$
KB218508.1	scf_22030_25031	13232–13319	+	mir-160 (RF00247)	$3.76e^{-23}$
KB218508.1	scf_22030_25031	13231–13319	−	mir-160 (RF00247)	$1.52e^{-09}$
KB219059.1	scf_22030_50116	8622–8711	+	mir-160 (RF00247)	$3.16e^{-23}$
KB219059.1	scf_22030_50116	8622–8711	−	mir-160 (RF00247)	$1.35e^{-11}$
KB218046.1	scf_22030_5366	30669–30758	−	mir-160 (RF00247)	$7.21e^{-21}$
KB218046.1	scf_22030_5366	30669–30756	+	mir-160 (RF00247)	$3.20e^{-08}$
KB219346.1	scf_22030_59171	24014–24101	+	mir-160 (RF00247)	$1.18e^{-20}$
KB219346.1	scf_22030_59171	24014–24101	−	mir-160 (RF00247)	$6.30e^{-09}$
KB218895.1	scf_22030_42834	6184–6270	−	MIR164 (RF00647)	$5.38e^{-19}$
KB218895.1	scf_22030_42834	6184–6270	+	MIR164 (RF00647)	$3.11e^{-12}$
KB219508.1	scf_22030_63187	11271–11378	+	MIR164 (RF00647)	$1.12e^{-18}$
KB219508.1	scf_22030_63187	11271–11378	−	MIR164 (RF00647)	$1.02e^{-12}$
KB218104.1	scf_22030_8363	10326–10443	−	MIR164 (RF00647)	$6.46e^{-23}$
KB218104.1	scf_22030_8363	10326–10443	+	MIR164 (RF00647)	$6.71e^{-16}$
KB217991.1	scf_22030_2485	3315–3401	−	mir-166 (RF00075)	$5.93e^{-21}$
KB217991.1	scf_22030_2485	3315–3401	+	mir-166 (RF00075)	$2.53e^{-10}$
KB218022.1	scf_22030_4161	21528–21639	+	mir-166 (RF00075)	$3.99e^{-20}$
KB218022.1	scf_22030_4161	21528–21639	−	mir-166 (RF00075)	$1.31e^{-10}$
KB219071.1	scf_22030_50479	2432–2530	−	mir-166 (RF00075)	$2.04e^{-22}$
KB219071.1	scf_22030_50479	2432–2530	+	mir-166 (RF00075)	$1.27e^{-12}$
KB219643.1	scf_22030_65797	40153–40244	−	mir-166 (RF00075)	$2.40e^{-22}$
KB219643.1	scf_22030_65797	40153–40244	+	mir-166 (RF00075)	$9.30e^{-12}$
KB220445.1	scf_22030_76496	6198–6315	−	mir-166 (RF00075)	$2.47e^{-23}$
KB220445.1	scf_22030_76496	6198–6315	+	mir-166 (RF00075)	$5.31e^{-12}$
KB220707.1	scf_22030_79012	6213–6322	−	mir-166 (RF00075)	$2.17e^{-24}$
KB220707.1	scf_22030_79012	6213–6322	+	mir-166 (RF00075)	$8.47e^{-13}$
KB221155.1	scf_22030_81490	17577–17697	+	mir-166 (RF00075)	$6.47e^{-17}$
KB221155.1	scf_22030_81490	17577–17697	−	mir-166 (RF00075)	$4.00e^{-08}$
KB218667.1	scf_22030_31606	22038–22152	+	MIR167_1 (RF00640)	$6.27e^{-22}$
KB218667.1	scf_22030_31606	22039–22153	−	MIR167_1 (RF00640)	$4.21e^{-16}$
KB218973.1	scf_22030_46697	19560–19671	+	MIR167_1 (RF00640)	$2.76e^{-17}$
KB218973.1	scf_22030_46697	19561–19672	−	MIR167_1 (RF00640)	$9.11e^{-14}$
KB220367.1	scf_22030_75599	1–83	+	MIR167_1 (RF00640)	$1.83e^{-11}$
KB220367.1	scf_22030_75599	1–81	−	MIR167_1 (RF00640)	$5.81e^{-09}$
KB220896.1	scf_22030_80878	14228–14335	+	MIR168 (RF00677)	$1.12e^{-22}$
KB220896.1	scf_22030_80878	14227–14333	−	MIR168 (RF00677)	$2.28e^{-14}$
KB218337.1	scf_22030_18159	17587–17690	−	MIR169_2 (RF00645)	$1.07e^{-26}$
KB218337.1	scf_22030_18159	13143–13246	−	MIR169_2 (RF00645)	$2.24e^{-21}$
KB218337.1	scf_22030_18159	12902–12993	−	MIR169_2 (RF00645)	$3.40e^{-21}$
KB218337.1	scf_22030_18159	17589–17692	+	MIR169_2 (RF00645)	$2.10e^{-15}$
KB218337.1	scf_22030_18159	12904–12995	+	MIR169_2 (RF00645)	$2.36e^{-15}$
KB220127.1	scf_22030_72989	786–899	−	MIR169_2 (RF00645)	$9.28e^{-18}$

Table 3. *Cont.*

GenBank accession number	Scaffold name	Start and end positions	Strand	Rfam ID (and accession number)	Rfam scan E value
KB220321.1	scf_22030_74988	935–1052	+	MIR169_2 (RF00645)	$7.84e^{-18}$
KB220321.1	scf_22030_74988	933–1050	−	MIR169_2 (RF00645)	$9.12e^{-11}$
KB218337.1	scf_22030_18159	17584–17696	−	MIR169_5 (RF00865)	$3.86e^{-08}$
KB218337.1	scf_22030_18159	17583–17695	+	MIR169_5 (RF00865)	$5.88e^{-08}$
KB220127.1	scf_22030_72989	780–906	+	MIR169_5 (RF00865)	$1.94e^{-19}$
KB220127.1	scf_22030_72989	781–907	−	MIR169_5 (RF00865)	$1.46e^{-06}$
KB220321.1	scf_22030_74988	928–1058	−	MIR169_5 (RF00865)	$7.73e^{-20}$
KB220321.1	scf_22030_74988	927–1057	+	MIR169_5 (RF00865)	$9.15e^{-06}$
KB220807.1	scf_22030_80059	3863–3990	+	MIR169_5 (RF00865)	$4.61e^{-11}$
KB218810.1	scf_22030_38865	27461–27559	+	MIR171_1 (RF00643)	$1.79e^{-16}$
KB218810.1	scf_22030_38865	27459–27557	−	MIR171_1 (RF00643)	$8.90e^{-14}$
KB220711.1	scf_22030_79061	2105–2214	+	MIR171_1 (RF00643)	$2.74e^{-19}$
KB220711.1	scf_22030_79061	2103–2212	−	MIR171_1 (RF00643)	$4.15e^{-13}$
KB219420.1	scf_22030_61010	2619–2748	−	mir-172 (RF00452)	$2.11e^{-19}$
KB219420.1	scf_22030_61010	2619–2748	+	mir-172 (RF00452)	$1.03e^{-15}$
KB218089.1	scf_22030_7511	28886–28982	−	mir-287 (RF00788)	$3.04e^{-04}$
KB218983.1	scf_22030_47118	10649–10756	−	MIR390 (RF00689)	$1.99e^{-21}$
KB218983.1	scf_22030_47118	10649–10756	+	MIR390 (RF00689)	$1.75e^{-14}$
KB219488.1	scf_22030_62701	16710–16837	+	MIR390 (RF00689)	$3.68e^{-23}$
KB219488.1	scf_22030_62701	16710–16837	−	MIR390 (RF00689)	$8.85e^{-12}$
KB218810.1	scf_22030_38865	36369–36475	+	MIR394 (RF00688)	$9.23e^{-14}$
KB219360.1	scf_22030_59359	18185–18287	−	mir-395 (RF00451)	$5.48e^{-14}$
KB219360.1	scf_22030_59359	18185–18287	+	mir-395 (RF00451)	$6.44e^{-11}$
KB219922.1	scf_22030_70572	3837–3927	+	MIR396 (RF00648)	$1.03e^{-20}$
KB219922.1	scf_22030_70572	1415–1528	+	MIR396 (RF00648)	$1.35e^{-17}$
KB219922.1	scf_22030_70572	3836–3926	−	MIR396 (RF00648)	$2.37e^{-15}$
KB219922.1	scf_22030_70572	1414–1527	−	MIR396 (RF00648)	$2.41e^{-13}$
KB219961.1	scf_22030_71131	9924–10008	−	MIR396 (RF00648)	$1.30e^{-15}$
KB219961.1	scf_22030_71131	9925–10009	+	MIR396 (RF00648)	$3.38e^{-12}$
KB220512.1	scf_22030_77233	7423–7504	+	MIR396 (RF00648)	$1.50e^{-20}$
KB220512.1	scf_22030_77233	7422–7503	−	MIR396 (RF00648)	$6.96e^{-17}$
KB221106.1	scf_22030_81441	12748–12911	+	MIR408 (RF00690)	$2.85e^{-09}$
KB219476.1	scf_22030_62392	5876–5979	+	MIR535 (RF00714)	$4.25e^{-19}$
KB219838.1	scf_22030_69379	8499–8600	+	MIR535 (RF00714)	$1.44e^{-23}$
KB219838.1	scf_22030_69379	8497–8598	−	MIR535 (RF00714)	$1.83e^{-17}$
KB220694.1	scf_22030_78899	5550–5652	−	MIR535 (RF00714)	$3.74e^{-18}$
KB220154.1	scf_22030_73255	538–819	+	Plant_SRP (RF01855)	$1.43e^{-24}$
KB220490.1	scf_22030_76954	17439–17650	+	Plant_U3 (RF01847)	$2.04e^{-36}$
KB219898.1	scf_22030_70290	25811–25954	+	snoF1_F2 (RF00482)	$1.49e^{-19}$
KB218033.1	scf_22030_4706	9374–9436	−	snoJ33 (RF00315)	$4.02e^{-07}$
KB219471.1	scf_22030_62284	16444–16526	−	snoJ33 (RF00315)	$5.63e^{-09}$
KB219426.1	scf_22030_61169	69226–69316	−	snoR11 (RF00349)	$1.31e^{-17}$
KB219685.1	scf_22030_66563	26216–26343	−	snoR111 (RF01228)	$1.27e^{-14}$
KB220857.1	scf_22030_80459	12071–12174	−	snoR113 (RF01420)	$4.15e^{-20}$

Table 3. *Cont.*

GenBank accession number	Scaffold name	Start and end positions	Strand	Rfam ID (and accession number)	Rfam scan E value
KB218307.1	scf_22030_16452	15390–15476	−	snoR118 (RF01424)	$1.15e^{-15}$
KB218657.1	scf_22030_31300	24736–24824	+	snoR14 (RF01280)	$8.40e^{-14}$
KB218015.1	scf_22030_3847	11974–12060	−	snoR16 (RF00296)	$1.39e^{-18}$
KB218015.1	scf_22030_3847	12491–12577	−	snoR16 (RF00296)	$1.11e^{-17}$
KB220504.1	scf_22030_77091	17217–17303	−	snoR16 (RF00296)	$4.81e^{-19}$
KB220504.1	scf_22030_77091	16789–16875	−	snoR16 (RF00296)	$9.43e^{-19}$
KB220539.1	scf_22030_77514	2858–2933	+	snoR160 (RF00203)	$1.40e^{-15}$
KB219378.1	scf_22030_59710	15789–15866	+	snoR28 (RF00355)	$4.91e^{-22}$
KB218307.1	scf_22030_16452	15543–15617	−	snoR66 (RF00202)	$2.49e^{-16}$
KB219947.1	scf_22030_70993	16528–16659	+	snoR80 (RF01224)	$2.92e^{-20}$
KB220353.1	scf_22030_75402	20181–20308	−	snoR86 (RF00303)	$1.06e^{-24}$
KB219338.1	scf_22030_58993	16769–16872	−	snoR97 (RF01215)	$1.30e^{-18}$
KB219443.1	scf_22030_61493	32748–32838	−	SNORD15 (RF00067)	$2.00e^{-09}$
KB219661.1	scf_22030_66054	15711–15796	−	SNORD25 (RF00054)	$5.96e^{-22}$
KB219661.1	scf_22030_66054	15482–15566	−	SNORD25 (RF00054)	$5.50e^{-21}$
KB219661.1	scf_22030_66054	14874–14958	−	SNORD25 (RF00054)	$2.14e^{-20}$
KB219661.1	scf_22030_66054	15075–15159	−	SNORD25 (RF00054)	$9.04e^{-17}$
KB219898.1	scf_22030_70290	25498–25585	+	SNORD33 (RF00133)	$5.82e^{-16}$
KB218015.1	scf_22030_3847	12999–13097	−	SNORD43 (RF00221)	$7.53e^{-11}$
KB220504.1	scf_22030_77091	17701–17798	−	SNORD43 (RF00221)	$6.80e^{-12}$
KB220504.1	scf_22030_77091	17915–18012	−	SNORD43 (RF00221)	$9.20e^{-11}$
KB219898.1	scf_22030_70290	25347–25436	+	snoU31b (RF01285)	$4.66e^{-17}$
KB220870.1	scf_22030_80641	5915–5999	+	snoU36a (RF01302)	$5.82e^{-21}$
KB219426.1	scf_22030_61169	68869–68977	−	snoZ152 (RF00350)	$2.58e^{-16}$
KB219947.1	scf_22030_70993	16107–16211	+	snoZ157 (RF00333)	$1.58e^{-18}$
KB219898.1	scf_22030_70290	25690–25775	+	snoZ196 (RF00134)	$2.75e^{-14}$
KB220870.1	scf_22030_80641	6066–6159	+	snoZ223 (RF00135)	$1.98e^{-19}$
KB218327.1	scf_22030_17743	7560–7631	+	snoZ266 (RF00332)	$8.06e^{-09}$
KB219338.1	scf_22030_58993	17401–17516	−	snoZ278 (RF00201)	$1.76e^{-16}$
KB219338.1	scf_22030_58993	17113–17226	−	snoZ278 (RF00201)	$9.06e^{-13}$
KB219250.1	scf_22030_57131	12714–12875	−	U1 (RF00003)	$9.36e^{-39}$
KB219770.1	scf_22030_68191	6294–6455	+	U1 (RF00003)	$3.43e^{-41}$
KB220529.1	scf_22030_77416	6949–7110	+	U1 (RF00003)	$5.34e^{-36}$
KB220746.1	scf_22030_79451	5096–5256	+	U1 (RF00003)	$2.21e^{-27}$
KB218084.1	scf_22030_7289	6288–6438	−	U12 (RF00007)	$1.92e^{-27}$
KB219620.1	scf_22030_65416	19689–19820	−	U2 (RF00004)	$2.10e^{-17}$
KB220509.1	scf_22030_77120	23424–23564	−	U4 (RF00015)	$1.19e^{-08}$
KB218936.1	scf_22030_44766	5102–5143	+	U5 (RF00020)	$2.13e^{-09}$
KB218979.1	scf_22030_47021	19677–19800	+	U5 (RF00020)	$4.89e^{-10}$
KB218084.1	scf_22030_7289	12644–12761	−	U5 (RF00020)	$4.29e^{-18}$
KB220567.1	scf_22030_77768	17710–17830	+	U5 (RF00020)	$3.52e^{-11}$
KB217934.1	scf_22030_16	16123–16225	−	U6 (RF00026)	$1.54e^{-10}$
KB218759.1	scf_22030_36539	4240–4337	+	U6 (RF00026)	$2.72e^{-11}$

Figure 4. BLASTN alignment of an enset supercontig (GenBank: KB219804) against banana chromosome 5, displayed using the Artemis Comparison Tool (ACT).

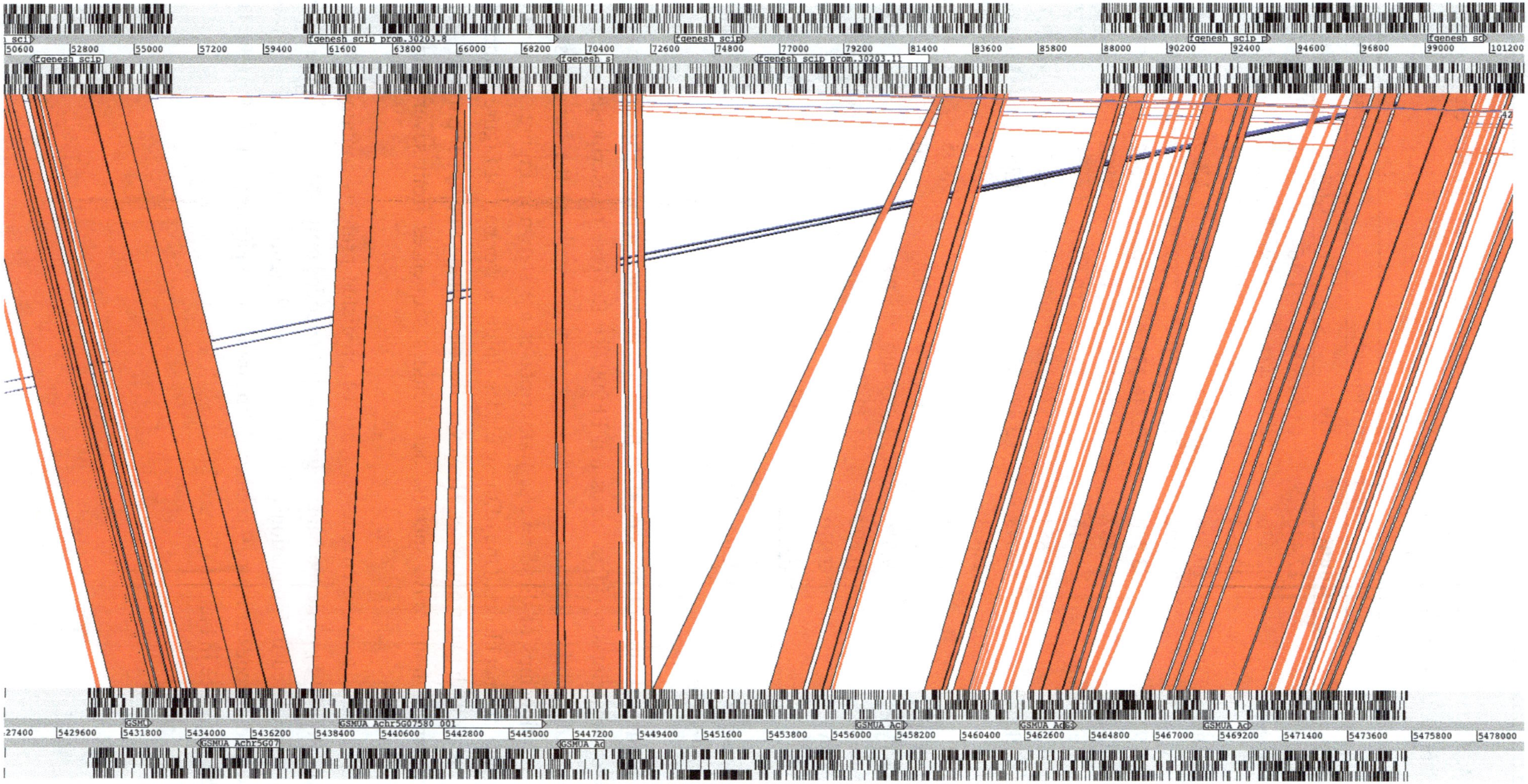

Table 4. Overview and classification of the repeats present in the enset genome and comparison with those in the *M. acuminata* genome.

	Ensete Ventricosum			Musa Acuminata		
Class	**Count**	**Bp**	**%**	**Count**	**Bp**	**%**
Ty1/Copia	17,446	6,064,590	1.36	5,053	2,476,355	0.75
Copia/Angela	102,430	39,177,431	8.78	15,025	10,764,293	3.24
Copia/SIRE1Maximus	102,464	27,386,896	6.14	37,446	26,594,658	8.01
Copia/Tnt1	10,144	4,915,981	1.10	2,869	3,300,009	0.99
Ty3/Gypsy	24,694	11,556,851	2.59	5,047	4,552,048	1.37
Gypsy/CRM	3,740	2,246,235	0.50	542	534,904	0.16
Gypsy/Galadriel	12,452	6,626,137	1.49	1,874	2,210,611	0.67
Gypsy/Galadriel-lineage	16	734	0.00	5	237	0.00
Gypsy/Reina	65,858	23,579,479	5.29	6,170	4,243,784	1.28
Gypsy/Tekay	14,043	5,490,598	1.23	4,351	3,031,464	0.91
LINE	5,833	1,346,085	0.30	1,745	552,483	0.17
RE	31,224	4,967,551	1.11	9,005	2,824,122	0.85
Satellite/Type1	178	69,579	0.02	20	30,828	0.01
Satellite/Type2	9,516	3,563,409	0.80	18	29,902	0.01
clDNA	6,590	1,126,726	0.25	2,652	430,368	0.13
DNA/hAT	2,910	783,511	0.18	1,916	637,668	0.19
Total	409,538	138,901,793	31.14	93,738	62,213,734	19.74

2.6. Enset—Specific Genes Include Reverse Transcriptases, Viral Sequences, and a Putative Disease-Resistance Gene

Among the enset genes not conserved in the *M. acuminata* genome [6], are several predicted to encode reverse transcriptases (Pfam accession PF00078). Reverse transcriptases are characteristic of several classes of mobile elements, including retroviruses, such as the banana streak virus. The phylogenetic relationships of these reverse transcriptases are shown in Figure 5, which indicates that they fall into two distinct clades. One of these clades (in the lower part of Figure 5) includes two genes from banana along with two from enset. However, the other clade (the upper part of Figure 5) includes no known sequences from *Musa* species, but includes sequences from several other monocot and dicot plants.

Similarly, the enset genome encodes at least 14 predicted proteins containing the integrase core domain (Pfam: PF00665) while the banana genome [6] encodes only one (see Figure 6). The integrase core domain is involved in integration of a copy of a viral genome into the host chromosome. The enset genome also encodes at least 19 predicted retrotransposon gag proteins (Pfam: PF03732) with no closely related sequence in banana (Figure 7).

Figure 5. Maximum-Likelihood phylogenetic tree for enset reverse transcriptase-domain proteins. Protein sequences from *E. ventricosum* are indicated by circles. The sequences from *M. acuminata* are indicated by diamonds. Bootstrap values of greater than 50% are indicated as numbers on the branches.

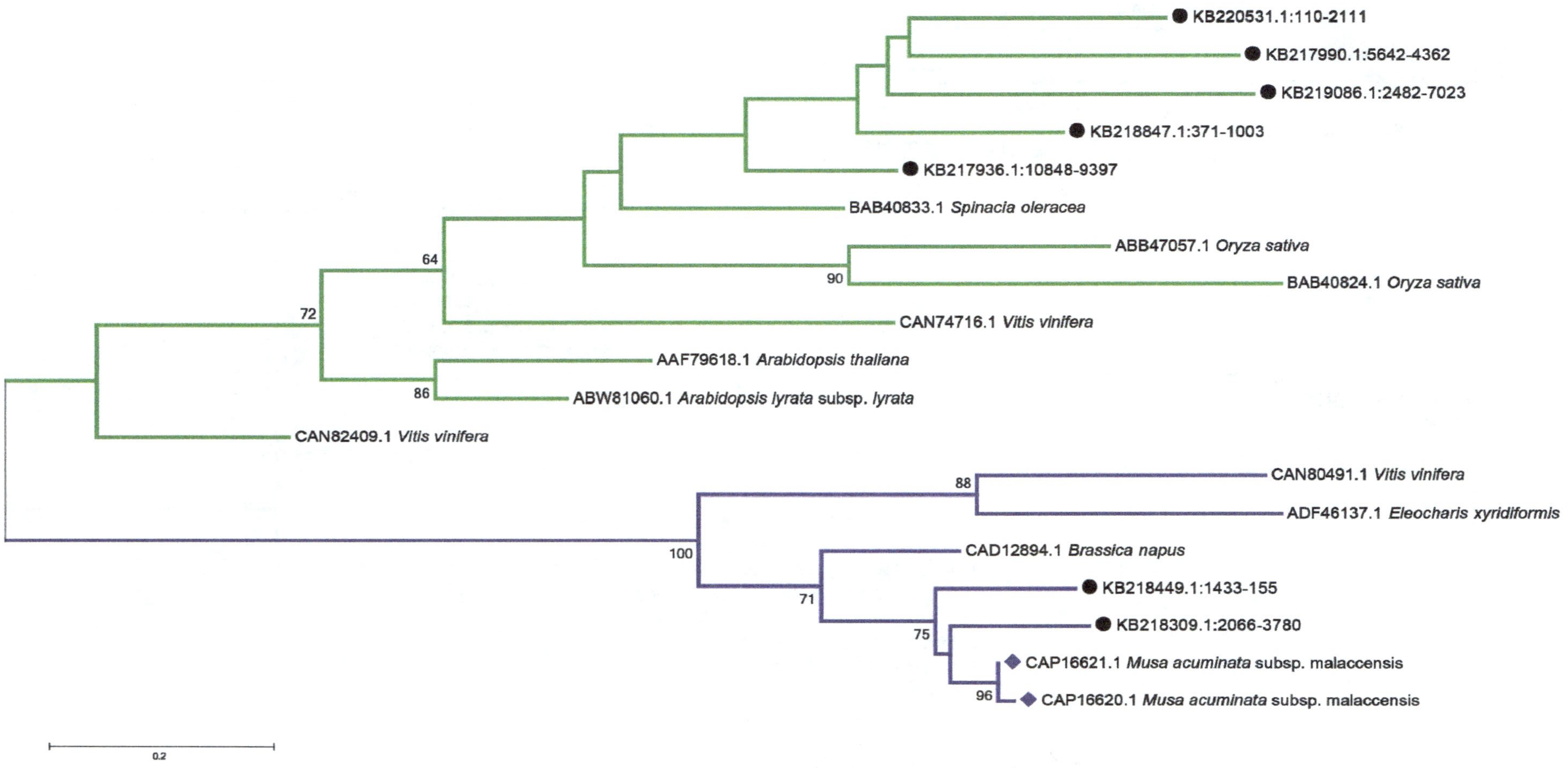

Figure 6. Maximum-Likelihood phylogenetic tree for enset integrase core-domain proteins. Protein sequences from *E. ventricosum* are indicated by circles. Bootstrap values of greater than 50% are indicated as numbers on the branches.

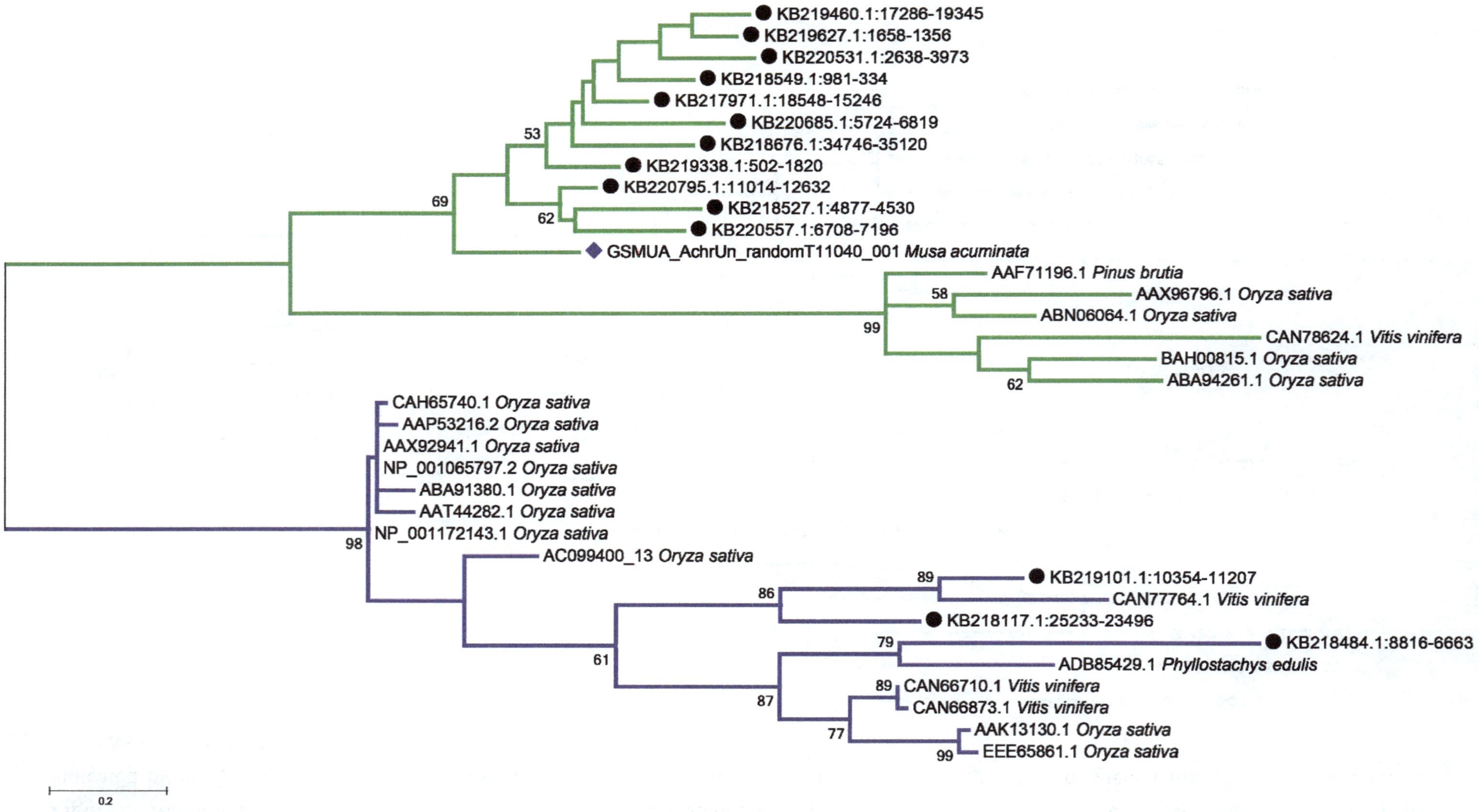

Figure 7. Maximum-Likelihood phylogenetic tree for enset integrase core-domain proteins. Proteins sequences from *E. ventricosum* are indicated by circles. The sequence from *M. acuminata* is indicated by a diamond. Bootstrap values of greater than 50% are indicated as numbers on the branches.

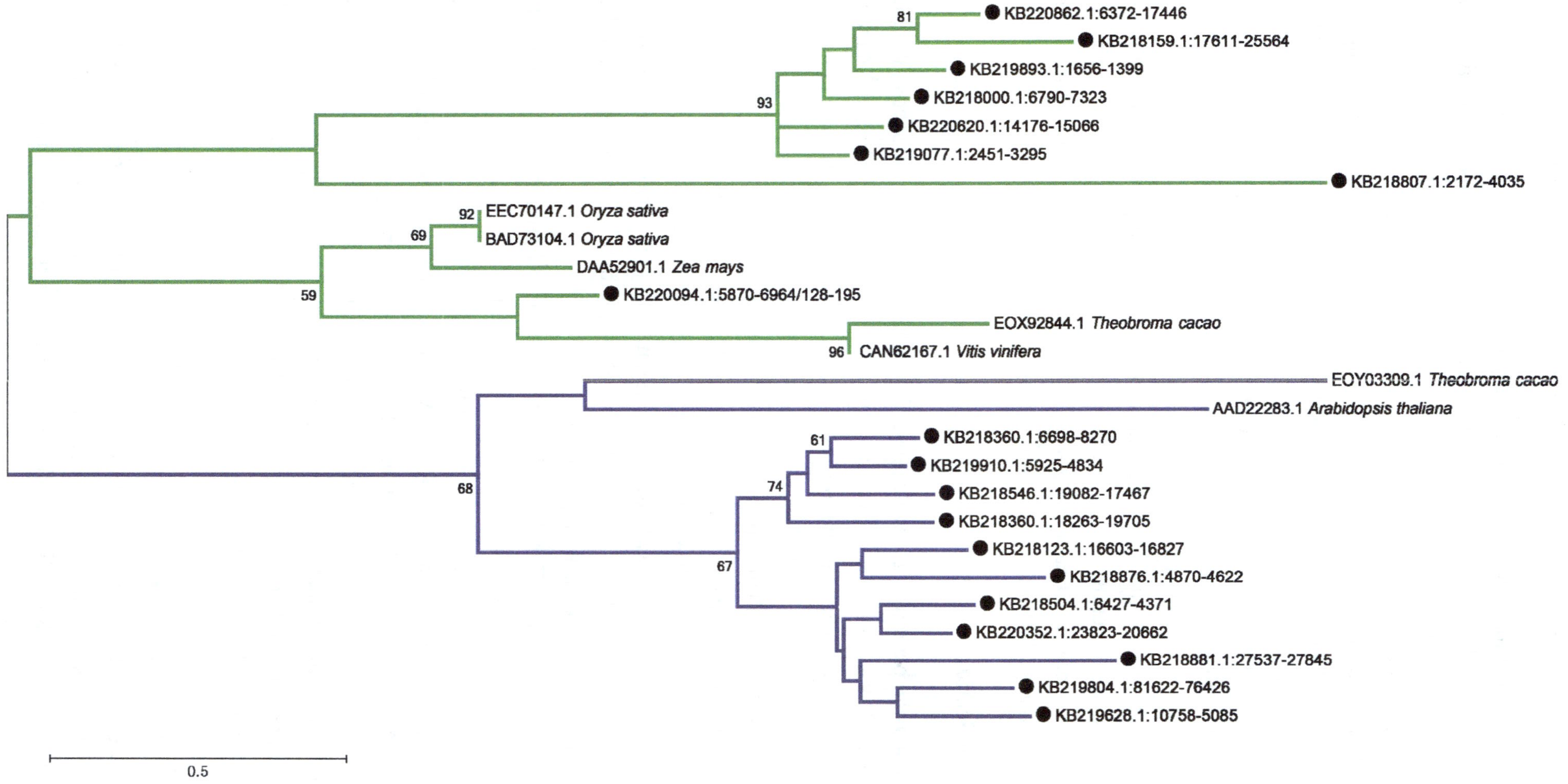

It has been shown that the genomes of some *Musa* species contain endogeneous retroviruses that are integrated into the host chromosome [28]. The genome of *E. ventricosum* contains several sequences that resemble retrovirus sequences and therefore may represent endogeneous integrated viruses. Specifically, a *M. balbisiana* sequence containing eBSOLV (endogeneous *Obino l'Ewai virus*) sequence (GenBank: HE983609 [28]) is highly conserved in *E. ventricosum*, though this sequence is absent from the *M. acuminata* genome [6]. Similarly, *E. ventricosum* contains sequences with 86% nucleotide identity to a 2.25-kb fragment of banana streak UA virus (GenBank: AEC49874) and 79% identity to a 1.1-kb fragment of the sugarcane bacilliform virus (SCBV) BT20231 (GenBank: FJ439799 [29]). It is not clear whether any of these virus sequences represent viruses that can become infectious as they can in *Musa* species [28].

Other enset proteins not found in the banana genome include a protein (GenBank: KB218027) that shares 42% amino-acid identity with *Arabidopsis thaliana* protein At1g53350, annotated as an RPP8-like resistance protein. Examples such as this are candidates for future studies on disease resistance in enset and perhaps even for introgression into banana.

3. Experimental Section

The *E. ventricosum* plant was grown from seed purchased from Jungle Seeds (Wallington, UK). We extracted genomic DNA using the DNAEasy Plant Minikit supplied by Qiagen (Manchester, UK). We sequenced genomic DNA using an Illumina HiSeq 2500, according to the manufacturer's instructions. We used a single lane of an eight-lane flowcell and generated 202 million pairs of 100-nucleotide reads with a mean insert-length of approximately 350 nucleotides.

For alignment of sequence reads against reference sequences, we used BWA version 0.7.5a-r405 [14] and visualized BWA alignments using the Integrative Genomics Viewer IGV [30]. For *de novo* assembly we used SOAPdenovo version 1.05 [31]. Prior to assembly, we removed all sequence reads that contained "N"s. Calculations of N_{50} and NG_{50} were based on the definitions of these two statistics stated by Assemblathon [27].

We used BLAST [32] and MUMMER [22] for pairwise alignments of assembled sequences and reference sequences and visualized BLAST alignments using the Artemis Comparison Tool (ACT) [33]. We used MEGA5 [22] for phylogenetic analysis.

To identify repeat sequences, we used RepeatMasker version open-4.0.1 [26,34,35] in default mode run with RMBLAST version 2.2.27+ against the customized library of *M. acuminata* repeats (1903 sequences) from Hřibová and colleagues [36,37]. This is the same library of banana-specific repeats used in the *M. balbisiana* genome re-sequencing project [12].

For *ab initio* gene prediction from our de novo genome assembly, we used FGENESH v.3.1.1 [22] with parameters tuned for 'monocot plant'.

4. Conclusions

Here we present the first genome-wide sequencing study of enset (*Ensete ventricosum*). We have identified more than 1000 candidate SNPs, and by using less stringent criteria, many more candidates could be identified. These data will be useful as a reference sequence for future "omics studies" on this neglected crop. Armed with this initial draft genome sequence, we can now extend our studies to genotypic variation among different Ethiopian varieties of enset, both cultivated and wild.

Acknowledgments

This work was funded in part by the Wellcome Trust Biomedical Informatics Hub at the University of Exeter. We are grateful to Eva Hřibová for making available the library of banana repeat sequences.

Conflicts of Interest

The authors declare no conflict of interest.

References and Notes

1. Brandt, S.A.; Spring, A.; Hiebsch, C.; McCabe, J.T.; Tabogie, E.; Diro, M.; Wolde-Michael, G.; Yntiso, G.; Shigeta, M.; Tesfaye, S. *The "Tree Against Hunger" Enset-Based Agricultural Systems in Ethiopia*; American Association for the Advancement of Science: Washington, DC, USA, 1997; pp. 1–58.
2. Pijls, L.T.J.; Timmer, A.A.M.; Wolde-Gebriel, Z.; West, C.E. Cultivation, preparation and consumption of ensete *(Ensete ventricosum)* in Ethiopia. *J. Sci. Food Agric.* **1995**, *67*, 1–11.
3. Asfaw, B.T. *Studies on Landraces Diversity* in vivo *and* in vitro *Regeneration of Enset: (*Enset ventricosum *Welw.)*; Köster: Milan, Lombardy, Italy, 2002; p. 127.
4. Biruma, M.; Pillay, M.; Tripathi, L.; Blomme, G.; Abele, S.; Mwangi, M.; Bandyopadhyay, R.; Muchunguzi, P.; Kassim, S.; Nyine, M.; *et al*. Banana *Xanthomonas* wilt: A review of the disease, management strategies and future research directions. *Afr. J. Biotechnol.* **2007**, *6*, 953–962.
5. Cheesman, E. Classification of the bananas: The genus ensete horan. *Kew Bull.* **1947**, *2*, 97–106.
6. D'Hont, A.; Denoeud, F.; Aury, J.-M.J.; Baurens, F.-C.F.; D'Hont, A.; Carreel, F.; Garsmeur, O.; Noel, B.; Bocs, S.; Droc, G.; *et al.* The banana (*Musa acuminata*) genome and the evolution of monocotyledonous plants. *Nature* **2012**, *488*, 213–217.
7. Ethiopian Institute of Agricultural Research (EIAR). Enset Research and Development Experiences in Ethiopia. In Proceedings of Enset National Workshop, Wolkite, Ethiopia, 19–20 August 2010; Yesuf, M., Hunduma, T., Eds.; Ethiopian Institute of Agricultural Research (EIAR): Addis Ababa, Ethiopia, 2012.
8. Tsegaye, A. On Indigenous Production, Genetic Diversity and Crop Ecology of Enset (*Ensete ventricosum* (Welw.) Cheesman). Ph.D. Thesis, Wageningen University, Wageningen, The Netherlands, 22 April 2002; p. 198.
9. Negash, A.; Niehof, A. The significance of enset culture and biodiversity for rural household food and livelihood security in southwestern Ethiopia. *Agric. Human Values* **2004**, *21*, 61–71.
10. Birmeta, G.; Nybom, H.; Bekele, E. RAPD analysis of genetic diversity among clones of the Ethiopian crop plant *Ensete ventricosum*. *Euphytica* **2002**, *124*, 315–325.
11. Birmeta, G.; Nybom, H.; Bekele, E. Distinction between wild and cultivated enset (*Ensete ventricosum*) gene pools in Ethiopia using RAPD markers. *Hereditas* **2004**, *140*, 139–148.

12. Davey, M.W.; Gudimella, R.; Harikrishna, J.A.; Sin, L.W.; Khalid, N.; Keulemans, J. A draft *Musa balbisiana* genome sequence for molecular genetics in polyploid, inter- and intra-specific *Musa* hybrids. *BMC Genomics* **2013**, *14*, doi:10.1186/1471-2164-14-683.
13. National Center for Biotechnology Information. Available online: ftp://ftp.ncbi.nlm.nih.gov/genomes/Bacteria/ (accessed on 22 December 2013).
14. Li, H.; Durbin, R. Fast and accurate short read alignment with Burrows-Wheeler transform. *Bioinformatics* **2009**, *25*, 1754–1760.
15. Silby, M.W.; Cerdeño-Tárraga, A.M.; Vernikos, G.S.; Giddens, S.R.; Jackson, R.W.; Preston, G.M.; Zhang, X.-X.; Moon, C.D.; Gehrig, S.M.; Godfrey, S.A.C.; *et al.* Genomic and genetic analyses of diversity and plant interactions of *Pseudomonas fluorescens*. *Genome Biol.* **2009**, *10*, R51.
16. Copeland, A.; Lucas, S.; Lapidus, A.; Glavina del Rio, T.; Dalin, E.; Tice, H.; Bruce, D.; Goodwin, L.; Pitluck, S.; Kiss, H.; *et al.* US DOE Joint Genome Institute, Walnut Creek, CA, USA. Unpublished work, 2008.
17. Brzuszkiewicz, E.; Weiner, J.; Wollherr, A.; Thürmer, A.; Hüpeden, J.; Lomholt, H.B.; Kilian, M.; Gottschalk, G.; Daniel, R.; Mollenkopf, H.-J.; Meyer, T.F.; Brüggemann, H. Comparative genomics and transcriptomics of *Propionibacterium acnes*. *PLoS One* **2011**, *6*, e21581.
18. Chung, W.-C.; Chen, L.-L.; Lo, W.-S.; Kuo, P.-A.; Tu, J.; Kuo, C.-H. Complete genome sequence of *Serratia marcescens* WW4. *Genome Announc.* **2013**, *1*, e0012613.
19. Li, H.; Durbin, R. Fast and accurate long-read alignment with Burrows-Wheeler transform. *Bioinformatics* **2010**, *26*, 589–595.
20. Mammadov, J.; Aggarwal, R.; Buyyarapu, R.; Kumpatla, S. SNP markers and their impact on plant breeding. *Int. J. Plant Genomics* **2012**, *2012*, 728398.
21. Studholme, D. *Ensete ventricosum* Genome Sequence. Available online: http://figshare.com/articles/Ensete_ventricosum_genome_sequence/894306 (accessed on 6 January 2014).
22. Kurtz, S.; Phillippy, A.; Delcher, A.L.; Smoot, M.; Shumway, M.; Antonescu, C.; Salzberg, S.L. Versatile and open software for comparing large genomes. *Genome Biol.* **2004**, *5*, R12.
23. Solovyev, V. Statistical Approaches in Eukaryotic Gene Prediction. In *Handbook of Statistical Genetics*; John Wiley & Sons, Ltd.: Chichester, West Sussex, UK, 2004; pp. 97–159.
24. Finn, R.D.; Bateman, A.; Clements, J.; Coggill, P.; Eberhardt, R.Y.; Eddy, S.R.; Heger, A.; Hetherington, K.; Holm, L.; Mistry, J.; *et al.* Pfam: The protein families database. *Nucleic Acids Res.* **2013**, *42*, D222–D230.
25. Gardner, P.P.; Daub, J.; Tate, J.; Moore, B.L.; Osuch, I.H.; Griffiths-Jones, S.; Finn, R.D.; Nawrocki, E.P.; Kolbe, D.L.; Eddy, S.R.; *et al.* Rfam: Wikipedia, clans and the "decimal" release. *Nucleic Acids Res.* **2011**, *39*, D141–D145.
26. Tempel, S.; Repeatmasker, U. Using and understanding RepeatMasker. *Methods Mol. Biol.* **2012**, *859*, 29–51.
27. Earl, D.; Bradnam, K.; St John, J.; Darling, A.; Lin, D.; Fass, J.; Yu, H.O.K.; Buffalo, V.; Zerbino, D.R.; Diekhans, M.; *et al.* Assemblathon 1: A competitive assessment of de novo short read assembly methods. *Genome Res.* **2011**, *21*, 2224–2241.

28. Chabannes, M.; Baurens, F.-C.; Duroy, P.-O.; Bocs, S.; Vernerey, M.-S.; Rodier-Goud, M.; Barbe, V.; Gayral, P.; Iskra-Caruana, M.-L. Three infectious viral species lying in wait in the banana genome. *J. Virol.* **2013**, *87*, 8624–8637.
29. Muller, E.; Dupuy, V.; Blondin, L.; Bauffe, F.; Daugrois, J.-H.; Nathalie, L.; Iskra-Caruana, M.-L. High molecular variability of sugarcane bacilliform viruses in Guadeloupe implying the existence of at least three new species. *Virus Res.* **2011**, *160*, 414–419.
30. Thorvaldsdóttir, H.; Robinson, J.T.; Mesirov, J.P. Integrative Genomics Viewer (IGV): High-Performance genomics data visualization and exploration. *Briefings Bioinforma.* **2013**, *14* , 178–192.
31. Luo, R.; Liu, B.; Xie, Y.; Li, Z.; Huang, W.; Yuan, J.; He, G.; Chen, Y.; Pan, Q.; Liu, Y.; *et al.* SOAPdenovo2: An empirically improved memory-efficient short-read *de novo* assembler. *Gigascience* **2012**, *1*, doi:10.1186/2047-217X-1-18.
32. Altschul, S.F.; Gish, W.; Miller, W.; Myers, E.W.; Lipman, D.J. Basic local alignment search tool. *J. Mol. Biol.* **1990**, *215*, 403–410.
33. Carver, T.J.; Rutherford, K.M.; Berriman, M.; Rajandream, M.-A.; Barrell, B.G.; Parkhill, J. ACT: The Artemis Comparison Tool. *Bioinformatics* **2005**, *21*, 3422–3423.
34. Tarailo-Graovac, M.; Chen, N. Using RepeatMasker to identify repetitive elements in genomic sequences. *Curr. Protoc. Bioinformatics* **2009**, *4*, doi:10.1002/0471250953.bi0410s25.
35. RepeatMasker. Available online: http://www.repeatmasker.org (accessed on 20 December 2013).
36. Hribová, E.; Neumann, P.; Matsumoto, T.; Roux, N.; Macas, J.; Dolezel, J. Repetitive part of the banana (*Musa acuminata*) genome investigated by low-depth 454 sequencing. *BMC Plant Biol.* **2010**, *10*, 204.
37. Institute of Experimental Botany. Available online: http://wwwueb.asuch.cas.cz/Olomouc1/ banana-sequencing-data/BananaREP.tar.gz (accessed on 20 December 2013).

2

Biological Control of the Weed Hemp Sesbania (*Sesbania exaltata*) in Rice (*Oryza sativa*) by the Fungus *Myrothecium verrucaria*

Clyde D. Boyette [1,*], Robert E. Hoagland [2] and Kenneth C. Stetina [1]

1 USDA-ARS, Biological Control of Pests Research Unit, Stoneville, MS, 38776, USA; E-Mail: kenneth.stetina@ars.usda.gov
2 USDA-ARS, Crop Production Systems Research Unit, Stoneville, MS, 38776, USA; E-Mail: bob.hoagland@ars.usda.gov

* Author to whom correspondence should be addressed; E-Mail: doug.boyette@ars.usda.gov

---

Abstract: In greenhouse and field experiments, a mycelial formulation of the fungus *Myrothecium verrucaria* (IMI 361690) containing 0.20% Silwet L-77 surfactant exhibited high bioherbicidal efficacy against the problematic weed hemp sesbania. Infection and mortality levels of 100% of hemp sesbania seedlings occurred within 48 h after fungal application in the greenhouse. In rice field tests conducted over a three year period, *M. verrucaria* at an inoculum concentration of 50 g L^{-1} (dry mycelium equivalent) controlled 95% of ≤20 cm tall hemp sesbania plants. *M. verrucaria* also controlled larger plants (≥60 cm tall) using this high inoculum concentration. This level of weed control, as well as rice yields from plots where weeds were effectively controlled, were similar to those which occurred with the herbicide acifluorfen. These results suggest that a mycelial formulation of *M. verrucaria* has potential as a bioherbicide for controlling hemp sesbania in rice.

Keywords: bioherbicide; biocontrol; hemp sesbania; *Myrothecium verrucaria*; mycelial formulation

---

1. Introduction

Hemp sesbania [*Sesbania exaltata* (Rydb.) ex. A.W. Hill] is an aggressive, annual, nodulating leguminous weed that infests rice (*Oryza sativa* L.), soybean [*Glycine max* (L.) Merr.], cotton (*Gossypium hirsutum* L.) and sunflower (*Helianthus annuus* L.) [1]. This plant is distributed in the U.S. coastal plain of Virginia to Florida to Texas in ditches, on stream banks, fallow fields and waste places [2] and has been rated as one of the 10 most troublesome weeds in Arkansas, Louisiana, and Mississippi [3]. It can attain a height of 3 m at maturity [4], produce abundant seeds (21,000 seeds per plant) [5] and reduce crop yield via shading and competition [6,7]. During rice production, it can interfere with harvesting operations, since its fibrous and woody stem biomass can damage combine blades, thereby lengthening harvest time and increasing harvesting and grain drying costs. Contamination by the black hemp sesbania seeds in harvested rice grain also lowers grain quality and value of the crop [5]. This weed is also problematic in sunflower, resulting in 35% yield reduction when competing with sunflower for the entire growing season [1].

Hemp sesbania is toxic to livestock and humans, and seeds appear to be the most toxic plant part [2]. Cattle can die from ingesting this weed and symptoms include hemorrhagic diarrhea, constipation, reduced respiration and elevated pulse rate. They become prostrate and comatose before death. An opened rumen may reveal sprouted seeds and a hemorrhagic inflammation of the abomasum and intestines [8].

Hemp sesbania can add nitrogen into soils through the nodulation process. Beneficial aspects of a related plant, *Sesbania rostrata* Brem., have been evaluated as a green manure for lowland rice in Sierra Leone [9] and in the Philippines [10], resulting in significant rice yield increases when used in a rotation program with rice.

Single or multiple applications of herbicides to control using hemp sesbania have been intensely evaluated, showing that some herbicide treatments can provide effective control. Table 1 lists the common and chemical names of herbicides cited in this paper. For example, both acifluorfen and fomesafen effectively controlled hemp sesbania [11]. Minimum effective rates of acifluorfen or fomesafen to provide 80 and 100% control of 50- to 60-cm tall hemp sesbania in soybeans were 50 and 140 g ha^{-1}, respectively [12]. Lactofen effectively controlled hemp sesbania without the addition of glyphosate, and acifluorfen and chlorimuron combined with glyphosate reduced hemp sesbania fresh weight nearly two-fold more than glyphosate alone [13].

Rice yields are improved when hemp sesbania is effectively controlled. For example, rice grain yields were higher with herbicides that increased control of hemp sesbania, including bentazon plus acifluorfen, carfentrazone and halosulfuron, compared to yields in plots treated with imazethapyr plus bensulfuron or triclopyr mixtures [14,15].

Hemp sesbania has been reported to have tolerance to glyphosate [16,17,18]. Glyphosate alone controlled hemp sesbania by only 28% and 45% at rates of 0.84 and 1.12 kg active ingredient (ai) ha^{-1}, respectively [11]. Other reports indicate that single applications of glyphosate are not adequate to control hemp sesbania [19,20,21]. Several pre-emergence herbicides at half or full rates followed by glyphosate did not completely control hemp sesbania in soybeans, necessitating an acifluorfen application [22]. However, some herbicides or tank mixtures of certain herbicides with glyphosate increased hemp sesbania control [11].

Table 1. Common and chemical names of herbicides mentioned in text [23].

Common Name	Chemical Name
Acifluorfen	5-[2-chloro-4-(trifluoromethyl)phenoxy]-2-nitrobenzoic acid
Bensulfuron	2-[[[[[(4,6-dimethoxy-2-pyrimidinyl)amino] carbonyl]amino]sulfonyl]methyl] benzoic acid
Bentazon	3-(1-methylethyl)-(1*H*)-2,1,3-benzothiadiazin-4(3*H*)-one 2,2-dioxide
Carfentrazone	α,2-dichloro-5-[4-(difluoromethyl)-4,5-dihydro-3-methyl-5-oxo-1*H*-1,2,4-triazol-1-yl] -4-fluorobenzenepropanoic acid
Chlorimuron	2-[[[[(4-chloro-6-methoxy-2-pyrimidinyl)amino]carbonyl]amino]sulfonyl]benzoic acid
Fomesafen	5-[2-chloro-4-(trifluoromethyl)phenoxy]-*N*-(methylsulfonyl)-2-nitrobenzamide
Glyphosate	*N*-(phosphonomethyl)glycine
Halosulfuron	Methyl 5-[((4,6-dimethoxy-2-pyrimidinyl)amino)carbonylamino-sulfonyl]-3-chloro-1-methyl-1*H*-pyrazole-4-carboxylate
Imazapyr	(±)-2-[4,5-dihydro-4-methyl-4-(1-methylethyl)-5-oxo-1*H*-imidazol-2-yl]-3-pyridinecarboxylic acid
Imazethapyr	2-(4,5-dihydro-4-methyl-4-(1-methylethyl)-5-oxo-1*H*-imidazol-2-yl)-5-ethyl-3-pyridine-carboxylic acid
Lactofen	(±)-2-ethoxy-1-methyl-2-oxoethyl-5-[2-chloro-4-(trifluoro-methyl)phenoxy]-2-nitrobenzoate
Pendimethalin	*N*-(1-ethylpropyl)-3,4-dimethyl-2,6-dinitrobenzenamine
Triclopyr	[(3,5,6-trichloro-2-pyridinyl)oxy]acetic acid

Although herbicides are the most effective and immediate solution to weed control, other solutions, including biological control (e.g., bioherbicides) are becoming available [24]. The use of bioherbicides has been recognized as a potential technological alternative to chemical herbicides in certain situations [25–28]. Global interest exists in the bioherbicide concept, and active research and development projects by commercial entities have been established in the USA, Canada, Europe, Australia, Japan, and other countries [26–29].

Several different fungal pathogens have been shown to possess biological control potential of this weed. For example, spore mixtures of *Alternaria crassa* (Sacc.) Rands, with either plant filtrates or fruit pectin, infected and killed hemp sesbania seedlings [30]. The fungus *Colletotrichum truncatum* (Schw.) Andrus and Moore has shown bioherbicidal potential for controlling hemp sesbania [31–34]. *Colletotrichum gloeosporioides* f. sp. *aeschynomene* (Penz) Sacc. (COLLEGO™, Encore Technologies,

Inc., Minnkota, MN, USA; presently marketed as LockDown® [35]), registered for northern jointvetch [*Aeschynomene virginica* (L.) B.S.P] control [36] effectively controlled hemp sesbania in rice when formulated in an invert emulsion [37,38].

The fungus *Myrothecium verrucaria* (Alb. and Schwein.) Ditmar:Fr. (strain IMI 361690) has been extensively evaluated as a bioherbicide for several weeds. *M. verrucaria* has exhibited bioherbicide activity against various weeds including sicklepod [*Senna obtusifolia* (L.) H.S. Barneby], kudzu [*Pueraria lobata* var. *montana* Willd. (Ohwii.)] [39], and morninglory spp. (*Ipomoea* spp.) [40], as well as hemp sesbania [41,42], when formulated with the surfactant Silwet L-77. Although *M. verrucaria* can potentially control several weed species, its production of toxic macrocyclic trichothecenes (mycotoxins) by fungal spores [43] is an issue that limits its practical usage. Thus, a biologically effective, mycotoxin-free formulation would greatly expand the bioherbicidal potential of this bioherbicide.

Most fungi that have been evaluated as bioherbicides utilize formulations of fungal spores as the active component [26], but highly effective mycelial bioherbicide formulations of several fungal bioherbicides also have been developed [44–47]. Experiments in our laboratory demonstrated that a spore-free liquid culture of *M. verrucaria* comprised of mycelia fragments was void of trichothecenes, and exhibited high bioherbicidal activity against the weed kudzu under field conditions [48].

Due to the high efficacy exhibited by *M. verrucaria* against several weeds, including hemp sesbania, and because *M. verrucaria* spores had no significant effects on rice seedlings [41], we hypothesized that a mycelial formulation of *M. verrucaria* may have potential for controlling hemp sesbania in a rice cropping situation. Thus, the objectives of this study were to determine if various concentrations of a mycelial formulation of *M. verrucaria* could effectively control hemp sesbania at several different stages of growth in rice under field conditions, and to determine the effects of *M. verrucaria* on rice yields.

2. Materials and Methods

2.1. Culture, Storage, and Mass Production

A strain of *M. verrucaria*, originally isolated from diseased sicklepod [41], was used in the present studies. The fungus was sub-cultured on potato dextrose agar (PDA, Difco Laboratories, Detroit, MI, USA) in Petri dishes and stock culture samples were stored on twice-sterilized soil (25% water holding capacity) at 4 °C [49]. Inoculated Petri dishes were inverted and placed on open-mesh wire shelves in an incubator (Precision Scientific Inc., Chicago, IL, USA) at 28 °C. The light intensity at dish level was 200 $\mu Em^{-2}s^{-1}$ photosynthetically active radiation (PAR).

A soy flour-corn meal liquid medium was used to produce *M. verrucaria* inocula, since it yielded the most efficacious and highest mycelial content of this fungus when several media were compared in experiments involving kudzu [48]. *M. verrucaria* starter inoculum (500 mL) was grown in 1-L flasks incubated in rotary shakers (185–200 rpm, 28 °C for 7 days). The above medium was also adopted for scaled-up production in laboratory fermenters (Models MF-214 and CMF-128, New Brunswick Corp., Edison, NJ, USA. Fermentations were conducted at 185–200 rpm and 28 °C for 48 h. Harvested mycelia batches from the flasks were filtered (#40 Whatman filter paper) and oven-dried (80 °C, 24 h) and dry weights were recorded in order to determine mycelia biomass (referred to as dry mycelium equivalents). Efficacy tests on hemp sesbania plants under greenhouse and field conditions utilized the raw fermentation product (liquid, unspent medium, and mycelium). For all experiments, prior to spray application to plants, the fermentation product was homogenized in 3–4 L aliquots with an electric blender (high speed, 3 min, Waring Model CB1043, Springfield, MO, USA). All efficacy tests were repeated twice.

2.2. Laboratory and Greenhouse Protocols

Hemp sesbania seeds were obtained from a local seed company (Azlin Seed Co., Leland, MS, USA). Seeds were surface-sterilized (0.05% NaOCl, 5 min), rinsed with sterile distilled water, and germinated (28 °C) on moistened filter paper in Petri dishes. After germination (~48 h), seeds were planted in a commercial potting mix (Jiffy-mix; Jiffy Products of America, Batavia, IL, USA) contained in peat strips, with each strip containing 12 plants. The potting mixture was supplemented with a controlled-release (14:14:14, NPK) fertilizer (Osmocote; Grace Sierra Horticultural Products, Milpitas, CA, USA). The plants were placed in sub-irrigated trays in a greenhouse [25 to 30 °C, 40%–90% relative humidity (RH)]. The photoperiod was 14 h, with 1800 PAR as measured at midday with a light meter. Hemp sesbania seedlings were either 10–20, 21–40, or 41–60 cm in height when treated with various concentrations of concentrations of *M. verrucaria* raw fermentation product [1.0×, 0.5× and 0.1×, where the 1.0× concentration contained the equivalent of 50.0 g mycelium (dry weight basis) L^{-1}], and the dilutions were achieved using distilled, de-ionized water. The experiment consisted of five treatments (12 plants per treatment): (1) *M. verrucaria* (1.0×) containing 0.2% (v/v) Silwet L-77™ surfactant (SW L-77; OSi Specialties, Inc., Danbury, CT, USA); (2) *M. verrucaria* at 0.5× containing SW L-77; (3) *M. verrucaria* at 0.1× containing SW L-77; (4) SW L-77 alone; and (5) distilled, de-ionized water alone. These treatments were performed in triplicate. Spray application rates delivered about 100 L ha^{-1}, and were applied with a hand held sprayer (Spray-Tool, Aervoe Industries, Gardnerville, NV, USA). After treatment, seedlings were placed in darkened dew chambers (Model I-36 DL; Percival Sci. Ind., Perry, IA, USA) in the dark at 28 °C, 100 RH for 12 h, and then moved to greenhouse benches. Plants were monitored at daily intervals for disease development for 7 days after treatment. Disease severity was based on a visual rating scale (per plant basis) [50] to estimate disease progression. A rating scale of 0 to 10 was used, with 0 being unaffected, and 2, 4, 6, 8 = 20%, 40%, 60%, and 80% leaf and stem lesion coverage/injury, respectively, and 10 = plant mortality. Percentage control and biomass reductions were determined 7 days after treatment. Plants were excised at the soil line, oven-dried for 48 h at 85 °C, weighed, and the percentage biomass reduction determined. Treatments were replicated four times, for a total of 48 individual plants per treatment. The experiment was repeated over time, and data were averaged following Bartlett's test for homogeneity of variance [51]. A randomized complete block experimental design was utilized. The mean percentage of plant mortalities and biomass reductions calculated for each treatment were subjected to arc-sin transformation. The transformed data were statistically compared using analysis of variance (ANOVA) at the 5% probability level. Values are presented as means of replicated experiments. When significant differences were detected by the *F*-test, means were separated with Fisher's protected LSD test at the 0.05 level of probability. In the disease kinetic studies, data were analyzed using standard mean errors and best-fit regression analysis.

2.3. Field Experiment Protocols

Field experiments were conducted in flooded rice field test plots at Stoneville, MS, USA, in 2006, 2007, and 2009. Pendimethalin at 1.12 kg ai ha^{-1} was applied pre-emergence for grass weed control. Rice and hemp sesbania were seeded simultaneously, with hemp sesbania seeded at a rate of ca 20 seed $meter^{-1}$ of row. Planting dates were: 18 May 2006, 21 May 2007 and 18 May 2009. Treatments were applied when the hemp sesbania plants were either, 10–20, 21–40, or 41–60 cm in height. Treatment dates for the 10–20 cm plants were 5 June 2006, 26 June 2007 and 18 July 2009. Treatment dates for the 21–40 cm plants were 4 June 2006, 25 June 2007 and 16 July 2009. Treatment dates for the 41–60 cm plants were 8 June 2006, 20 June 2007 and 20 July 2009. The experiment consisted of six treatments: (1) *M. verrucaria* (1.0×) containing SW L-77 (0.2%, v/v); (2) *M. verrucaria* at 0.5× containing SW L-77; (3) *M. verrucaria* at 0.1× containing SW L-77; (4) SW L-77 alone; (5) distilled, de-ionized water alone; and (6) acifluorfen (0.56 kg a.i. ha^{-1}). Acifluorfen is widely used for controlling hemp sesbania [12]. All spray applications were made with back-pack sprayers (Gilmour, Somerset, PA, USA) at spray volumes of ~200 L ha^{-1}. A quadrant (1.0 m^2) was randomly selected in each plot, which contained 15 to 20 hemp sesbania plants. Disease monitoring, weed control percentages and dry weight determinations were made on hemp sesbania plants within the quadrants. Weed control was determined seven days after treatment. Disease of hemp sesbania plants in the 10–20 cm growth stage was monitored at daily intervals over a 7-day period. Disease progression was based on a 0 to 10 rating scale, as described previously. Symptomatology was considered "severe" at ratings of 8–10. A randomized complete block experimental design was utilized and data were averaged over the 3 year testing period, after subjecting to Bartlett's test for homogeneity of variance [51]. Data were analyzed using analysis of variance. The percentage data (hemp sesbania injury/control and biomass reduction) were subjected to arc-sin transformation prior to analysis and treatment means and standard errors of the mean are presented. Values are presented as means of replicated experiments. When significant differences were detected by the *F*-test, means were separated with Fisher's protected LSD test at the 0.05 probability level. In the disease kinetic studies, data were analyzed using standard mean errors and best-fit regression analysis. Rice yields were recorded at the end of the season (September), and only yields from the early-season treatments (10–20 cm hemp sesbania) are presented.

3. Results and Discussion

In greenhouse experiments, application of *M. verrucaria* mycelia (1.0× rate) in 0.20% Silwet L-77 (v/v) provided 100% mortality of 10–20 cm tall hemp sesbania seedlings after 7 days, while inoculum rates of 0.5× and 0.1× controlled plants by 85% and 65%, respectively (Figure 1A). Previous studies showed that the addition of Silwet L-77 was essential to promote infection and control of various weeds by *M. verrucaria* conidial suspensions [41] and mycelial suspensions [48]. Hemp sesbania biomass (dry weights) were also reduced by 100% at the highest inoculum concentration, with 90% and 70% reductions occurring at the 0.5× and 0.1× rates, respectively (Figure 1A). No mortality or biomass reduction occurred on hemp sesbania seedlings treated with surfactant in water or with water alone

(Figure 1A–C). Efficacy of *M. verrucaria* was influenced by the stage of growth of hemp sesbania seedlings in the greenhouse and generally, weeds are more difficult to control with herbicides or bioherbicides as they become older and more mature. In the greenhouse, all *M. verrucaria*-inoculated hemp sesbania plants (regardless of size) were affected, but the levels of mortality and dry weight reduction decreased as the size of the plants increased above 20 cm (Figure 1A–C). However, by increasing inoculum concentration, significant increases in mortality and dry weight reduction were achieved (Figure 1A–C).

Figure 1. Effect of *Myrothecium verrucaria* mycelial concentration on mortality and dry weight reduction of hemp sesbania in greenhouse experiments, seven days after treatment. *M. verrucaria* mycelia were suspended in 0.20% (v/v) Silwet L-77 surfactant. **A** = hemp sesbania plants at 10–20 cm in height; **B** = hemp sesbania plants at 21–40 cm in height; and **C** = hemp sesbania plants at 41–60 cm in height. Histogram bar values for each parameter with the same letter are not significantly different at $p = 0.05$ using Fisher's least significant difference.

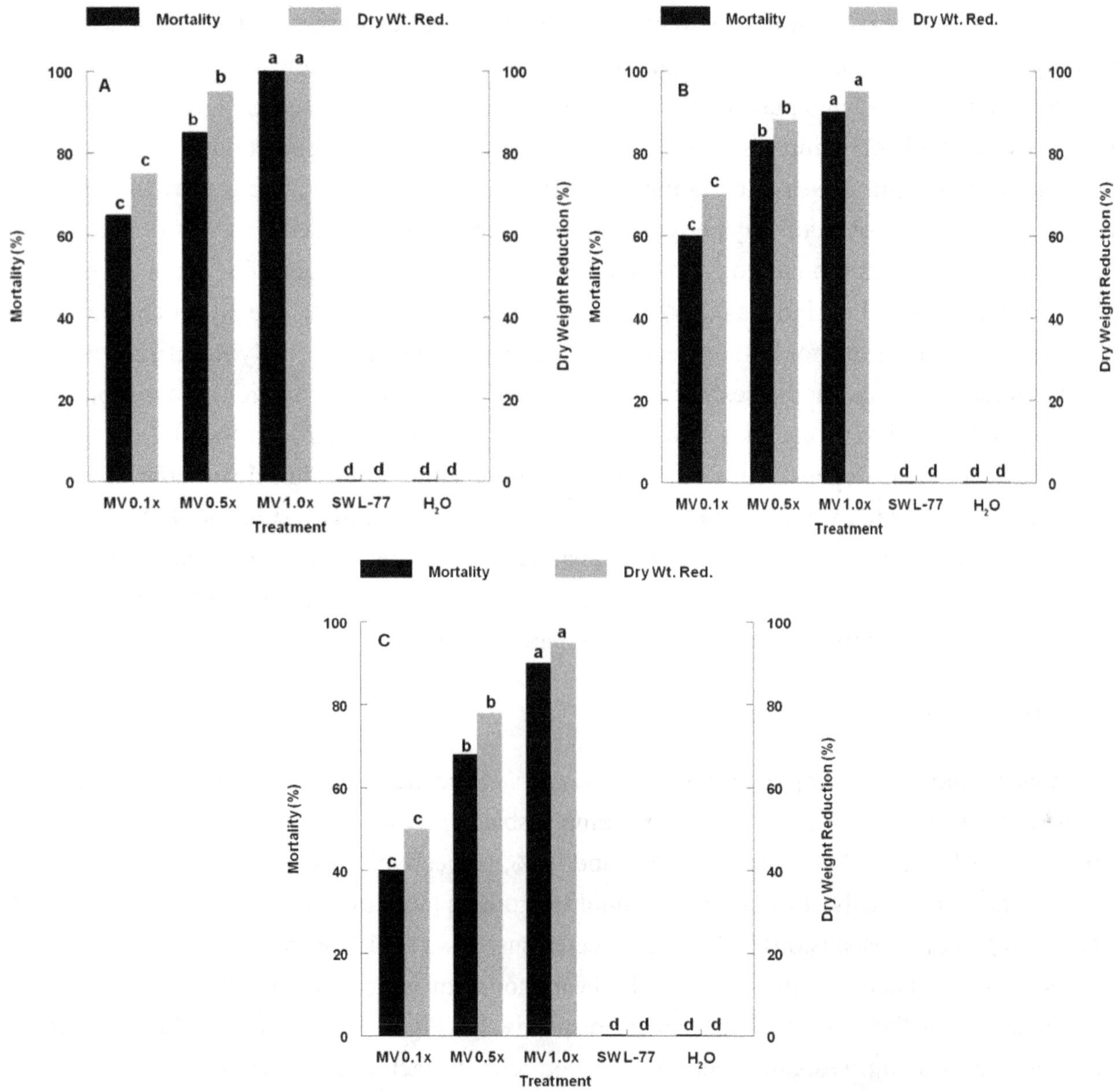

In disease kinetic studies under greenhouse conditions, disease progressed rapidly in plants treated with the highest inoculum rate (1.0×), with a 9.5 disease rating occurring after only 1 day (Figure 2). Disease progressed at a slower rate at lower inoculum concentrations, requiring 6 days to achieve disease ratings of 6.5 and 8.5, respectively, for the 0.1× and 0.5× inoculum concentrations under greenhouse conditions (Figure 2). Disease progression of infected hemp sesbania was similar to that incited by *M. verrucaria* on infected kudzu [48], and morningglories [40] in bioherbicidal studies of those weeds under greenhouse conditions.

Figure 2. Disease progression of 10–20 cm tall hemp sesbania incited by *Myrothecium verrucaria* mycelia on hemp sesbania in the greenhouse. Disease severity was based on a visual rating scale (per plant basis) [50] to estimate disease progression. A rating scale of 0 to 10 was used, as described in the Materials and Methods section. Symptomatology was considered "severe" at ratings of 8–10. The relationships for *M. verrucaria* disease progression are best described by the equations: $Y = 2.67 + 3.76X - 0.41X^2$, $R^2 = 0.72$ (solid spheres, solid line) for *M. verrucaria* applied at an inoculum rate of 1.0×; $Y = 0.36 + 2.17 - 0.15X^2$, $R^2 = 0.98$ (solid triangles, dashed line) for *M. verrucaria* applied at an inoculum rate of 0.5×; $Y = 0.65 + 1.07 - 0.03X^2$, $R^2 = 0.92$ (solid squares, dotted line) for *M. verrucaria* applied at an inoculum rate of 0.1×. Error bars = ±1 SEM.

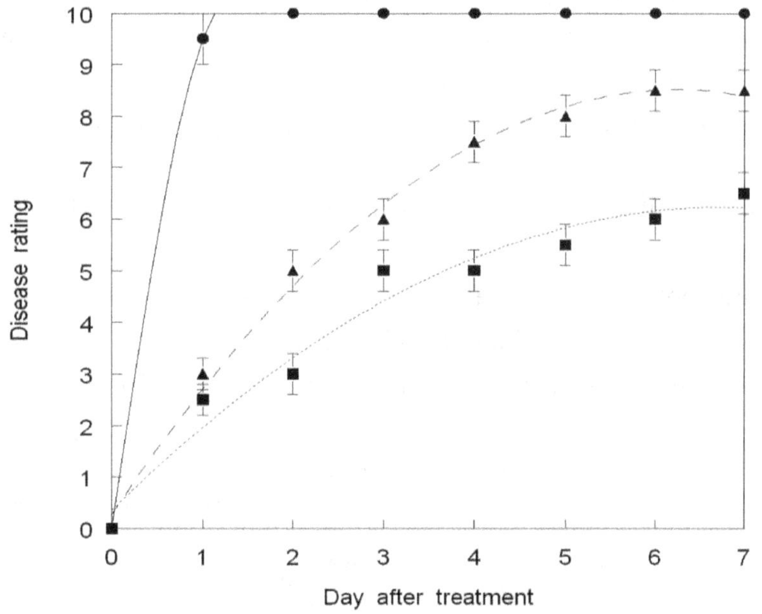

In field experiments, some pathogenesis and mortality of hemp sesbania occurred at all plant growth stages and inoculum concentrations (Figure 3A,B). Generally, *M. verrucaria* efficacy was higher on smaller plants than larger plants, regardless of inoculum concentration (Figure 3A) with concomitant increases in dry weight reductions as inoculum rates increased (Figure 3B). Lower *M. verrucaria* concentrations (0.1× and 0.5×) resulted in about 55%–70% mortality in the smallest plants tested (10–20 cm), but these fungal concentrations provided only 35% to 50% mortality in larger (41–60) plants (Figure 3A,B). However, by increasing the inoculum to 1.0×, mortality increased to 95% in the 21–40 cm plants and 90% in the 41–60 cm plants. Similar trends occurred in the dry weight reduction analyses.

Disease progression was more rapid with the high (1.0×) inoculum rate, with "severe" disease ratings occurring after only 1 day (Figure 4). The comparatively rapid rates of disease development observed in these experiments contrast greatly to disease development observed with some other bioherbicides, such as *Colletotrichum gloeosporioides* f.sp. *aeschynomenee*, or *C. truncatum*, whereas 9 days and 28 days were required to achieve equivalent levels of disease incidance on hemp sesbania in rice and soybean, respectively [38,52]. Season-long control of hemp sesbania was achieved on 10–20 cm plants treated with the high *M. verrucaria* inoculum rates (Figure 5). No visual infection of injury to rice was observed (data not shown). Rice yields were directly proportional to effective hemp sesbania control, and were not significantly different than yields from the acifluorfen-treated plots (Figure 6).

M. verrucaria application for biological control of hemp sesbania is decidedly one of the most important factors for consideration of this fungal pathogen as a bioherbicide. Application timing is also important in regard to other bioherbicidal pathogens for various weeds [26–28]. Timing plays a crucial role, both in terms of the growth of the target weed, and in the optimization of environmental conditions present during and after application. In the present studies, the rice plots were under flooded conditions when fungal applications were made, thus optimizing environmental conditions conducive for plant infection, disease development, and weed control. The present studies indicate that applications of *M. verrucaria* to smaller weeds would be most conducive for effective control. The importance of early season control of hemp sesbania in several crops, including rice, soybean, and cotton has been documented [3,4,6]. Although smaller plants are more susceptible, it may be possible to improve the efficacy of *M. verrucaria* on older or more mature plants by addition of certain adjuvants, such as unrefined corn oil or invert emulsions [34,38] that can enhance biocontrol efficacy.

Figure 3. Biological control of hemp sesbania at various growth stages in rice field plots using several *Myrothecium verrucaria* mycelial inoculum concentrations at Stoneville, MS. Acifluorfen was applied post-emergence at 0.56 kg ai ha^{-1}. **A** = hemp sesbania mortality, and **B** = hemp sesbania dry weight reduction. Histogram bar values for each parameter with the same letter are not significantly different at p = 0.05 using Fisher's least significant difference.

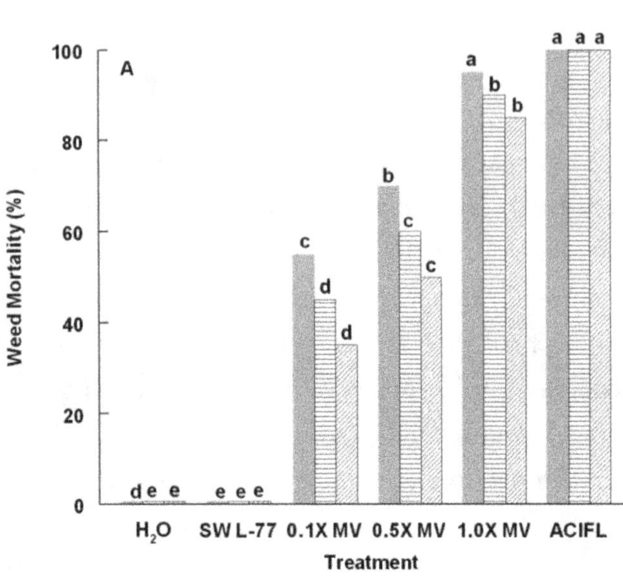

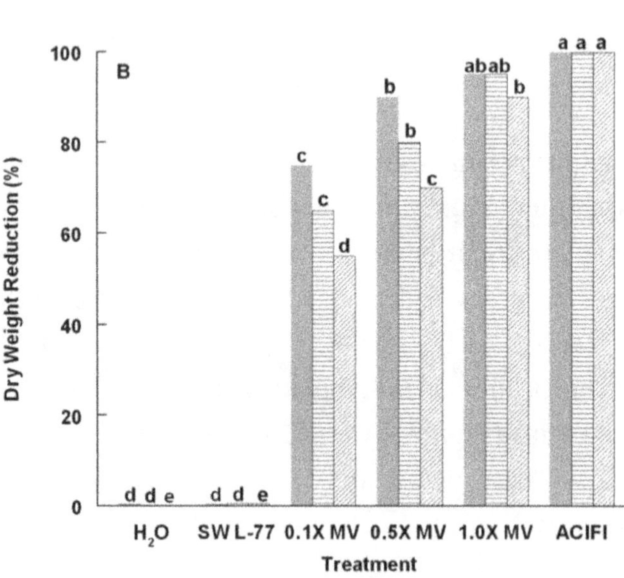

Figure 4. Disease progression on 10–20 cm tall hemp sesbania in rice field plots incited by a *Myrothecium verrucaria* mycelial formulation. Disease severity was based on a visual rating scale (per plant basis) [50] to estimate disease progression. A rating scale of 0 to 10 was used, as described in the Materials and Methods section. Symptomatology was considered "severe" at ratings of 8–10. The relationships for *M. verrucaria* disease progression are best described by the equations: $Y = 1.10 + 3.05X - 0.28X^2$, $R^2 = 0.92$ (solid spheres, solid line) for *M. verrucaria* applied at an inoculum rate of 1.0×; $Y = 0.44 + 2.16 - 0.15X^2$, $R^2 = 0.98$ (solid triangles, dashed line) for *M. verrucaria* applied at an inoculum rate of 0.5×; $Y = 0.67 + 1.06 - 0.04X^2$, $R^2 = 0.94$ (solid squares, dotted line) for *M. verrucaria* applied at an inoculum rate of 0.1×. Error bars = ±1 SEM.

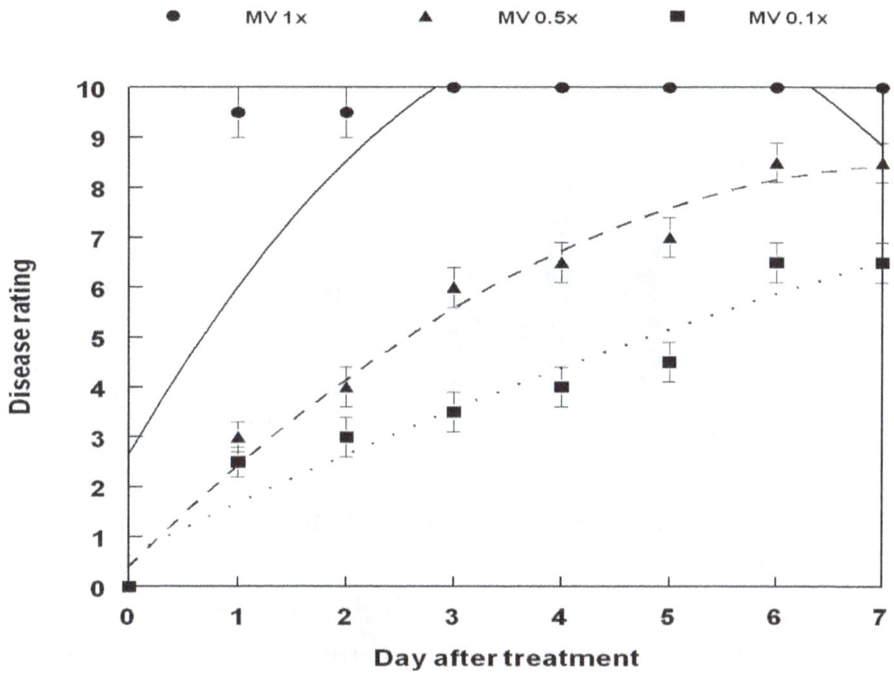

Figure 5. Photograph depicting season-long control of hemp sesbania in rice by a *Myrothecium verrucaria* mycelial formulation (**right**), and untreated (**left**).

Figure 6. The effects of biological control of hemp sesbania with *Myrothecium verrucaria* on rice yields, Stoneville, MS. Hemp sesbania mortality data was adapted from Figure 3A for direct data comparison. Yield data were taken (end of season) from plots containing hemp sesbania plants averaging 10–20 cm in height at time of treatment. Histogram bar values for each parameter with the same letter are not significantly different at $p = 0.05$ using Fisher's least significant difference.

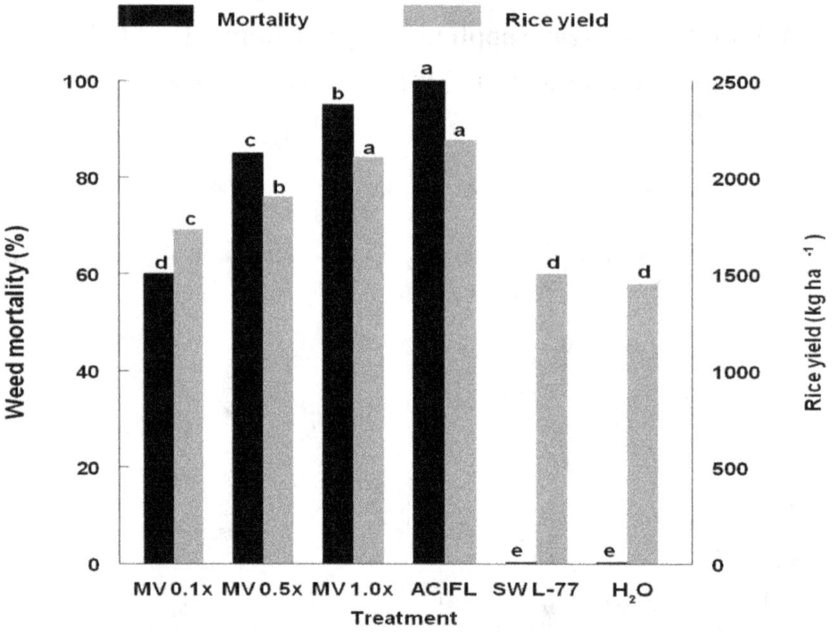

Only a few bioherbicides have been registered worldwide and about 50 have been considered to possess high potential for commercialization [28]. There are many examples of successful uses of plant pathogens for weed control using the classical and bioherbicide approaches as summarized in several reviews e.g., [25–29,31,36]. In most of these reports, the bioherbicidal pathogens exhibited high degrees of host specificity, *i.e.*, they infected and killed only one or a very few species of weeds. This trait is desirable with regard to safety issues such as avoiding infectivity to non-target plants. However in the work presented herein, we have shown that it is possible to use a relatively non-specific plant pathogen to control weeds when the crop is non-susceptible. Furthermore, we have not observed any disease symptoms or injury on susceptible crops (e.g., soybean) in adjacent plots (data not shown).

Biocontrol of weeds with pathogens also has wide interest outside the U.S. For example, researchers in academia, and public and private sectors from 14 European countries plus Israel and Egypt formed a research initiative focusing on biological control of weeds in crops [53]. This research initiative is under the European Co-operation in the Field of Scientific and Technical Research (COST) framework, and targets several major weed species [53-59].

With regard to *M. verrucaria*, many strains produce macrocyclic trichothecene mycotoxins with mammalian toxicity [43], including the isolate (IMI 361390) used in these studies which could pose a safety issue. However, tricothecenes were not detectable in sicklepod plants infected with this strain [60] and only trace amounts of trichothecenes were found in morningglory infected with another *M. verrucaria* strain (ATCC 18398) [61]. Whereas tricothecene content was relatively high in *M. verrucaria* spores, washed spores [62] and fungal mycelial formulations prepared via fermentation [48] contained little or no tricothecenes. In the present studies, no injury to rice plants or detrimental effects on rice yield were noted, and analytical tests for tricothecenes in harvested rice grains were not conducted. However, such studies should be performed in the future to assure food safety.

M. verrucaria IMI 361690 has several characteristics that make it desirable for use as a bioherbicide. Although most strains or isolates of *M. verrucaria* are weakly virulent pathogens [63–65], virulence can be influenced by manipulation of inoculum concentrations [39,41,65], and use of adjuvants, such as surfactants [39,41,48,65,66].

In recent years, the Clearfield™ system has become the predominant rice production system in the mid-south rice producing states of Arkansas, Louisiana, Mississippi, and Missouri [67]. The rice varieties utilized in the Clearfield™ system are natural mutants with tolerance to the herbicide imazethapyr (Newpath™). Although this herbicide controls many grassy and broadleaf weeds, it fails to control some weeds, especially hemp sesbania, which can result in tremendous weed infestations if other weed control measures are not utilized [68]. Furthermore, the increased use of glyphosate on glyphosate-resistant crops [69] plus the fact that hemp sesbania has a degree of inherent tolerance to this herbicide [16–18] may suggest high potential for a herbicide resistance problem in the future. Lack of effective hemp sesbania control creates a significant need for an effective weed control agent, such as *M. verrucaria*, for this troublesome weed. Future research will focus on applications of *M. verrucaria* formulations to evaluate hemp sesbania control in commercial rice fields.

4. Conclusions

M. verrucaria IMI 361690 mycelium is highly effective for biological control of hemp sesbania in rice. The mycelium can be rapidly produced on readily-available, inexpensive agricultural products, such as soy flour and corn meal. Weeds even larger than 60 cm in height were effectively controlled by using inoculum concentrations of 50 g L^{-1} dry equivalent rates. No visual infection or injury to rice was observed, and levels of weed control and rice yields from plots where weeds were effectively controlled were similar to those that occurred in acifluorfen-treated plots. It is also noteworthy that this research study is one of the few demonstrating an increase in rice yield achieved with a bioherbicide. Rapid infection and mortality caused by *M. verrucaria* suggests the role of possible phytotoxin(s). More research is required to elucidate the intricacies of the infection processes.

These results suggest that a mycelial formulation of *M. verrucaria* has potential as a bioherbicide for controlling hemp sesbania in rice.

Acknowledgments

The authors thank Terry Newton and Eric Smith for valuable technical assistance.

Conflicts of Interest

The authors declare no conflict of interest.

References

1. Woon, C.K. Effect of two row spacings and hemp sesbania competition on sunflower. *J. Agron. Crop Sci.* **1987**, *159*, 15–20.
2. *Poisonous Plants of the Southern United States*; Bulletin No. 818; Agricultural Extension Service, University of Tennessee: Knoxville, TN, USA, 1980; p. 12.
3. Dowler, C.C. Weed survey—Southern states. *Proc. South. Weed Sci. Soc.* **1992**, *45*, 392–407.
4. Lorenzi, H.J.; Jeffery, L.S. *Weeds of the United States and Their Control*; Van Nostrand Reinhold: New York, NY, USA, 1987.
5. Lovelace, M.L.; Oliver, L.R. Effects of interference and tillage on hemp sesbania and pitted morningglory emergence and seed production. *Proc. South. Weed Sci. Soc.* **2000**, *53*, 202.
6. King, C.A.; Purcell, L.C. Interference between hemp sesbania (*Sesbania exaltata*) and soybean (*Glycine max*) in response to irrigation and nitrogen. *Weed Sci.* **1997**, *45*, 91–97.
7. Norsworthy, J.K.; Oliver, L.R. Hemp sesbania interference in drill-seeded glyphosate-resistant soybean. *Weed Sci.* **2000**, *50*, 34–41.
8. Everest, J.W.; Powe, T.A., Jr.; Freeman, J.D. *Sesbania*. In *Poisonous Plants of the Southeastern United States*; Alabama Cooperative Extension System [ACES] Publications: Auburn, AL, USA, 2010. Available online: http://www.aces.edu/pubs/docs/A/ANR-0975/ (accessed on 17 August 2013).
9. Bar, A.R.; Baggie, I.; Sanginga, N. The use of Sesbania (*Sesbania rostrata*) and urea in lowland rice production in Sierra Leone. *Agrofor. Syst.* **2000**, *48*, 111–118.
10. Herrera, W.T.; Garrity, D.P.; Vejpas, C. Management of *Sesbania rostrata* green manure crops grown prior to rain-fed lowland rice on sandy soils. *Field Crops Res.* **1997**, *49*, 259–268.
11. Shaw, D.R.; Arnold, J.C. Weed control from herbicide combinations with glyphosate. *Weed Technol.* **2002**, *16*, 1–6.
12. Vidrine, P.R.; Reynolds, D.B.; Griffin, J.L. Postemergence hemp sesbania (*Sesbania exaltata*) control in soybean (*Glycine max*). *Weed Technol.* **1992**, *6*, 374–377.
13. Norris, J.L.; Shaw, D.R.; Snipes, C.E. Weed control from herbicide combinations with three formulations of glyphosate. *Weed Technol.* **2001**, *15*, 552–558.
14. Pellerin, K.J.; Webster, E.P.; Zhang, W.; Blouin, D.C. Herbicide mixtures in water-seeded imidazolinone-resistant rice (*Oryza sativa*). *Weed Technol.* **2003**, *17*, 836–841.
15. Pellerin, K.J.; Webster, E.P.; Zhang, W.; Blouin, D.C. Potential use of imazethapyr mixtures in drill-seeded imidazolinone-resistant rice. *Weed Technol.* **2004**, *18*, 1037–1042.
16. Jordan, D.L.; York, A.C.; Griffin, J.L.; Clay, P.A.; Vidrine, P.R.; Reynolds, D.B. Influence of application variables of efficacy on glyphosate. *Weed Technol.* **1997**, *11*, 354–362.
17. Lich, J.M.; Renner, K.A.; Penner, D. Interaction of glyphosate with postemergence soybean (*Glycine max*) herbicides. *Weed Sci.* **1997**, *45*, 12–21.
18. Oliver, L.R.; Taylor, S.E.; Gander, J.R. Influence of application timing and rate of glyphosate on weed control in soybean. *Proc. South. Weed Sci. Soc.* **1996**, *49*, 57.
19. LaMastus, F.E.; Shaw, D.R.; Smith, M.C. Influence of application timing and rate on weed control in Roundup-Ready soybean. *Proc. South. Weed Sci. Soc.* **1998**, *51*, 8.

20. Miller, D.K.; Milligan, J.L.; Wilson, C.F. Evaluation of reduced rate preemergence herbicides in Roundup-Ready soybean. *Proc. South. Weed Sci. Soc.* **1998**, *51*, 271–272.
21. Vidrine, P.R.; Griffin, J.L.; Jordan, D.L.; Miller, D.K. Postemergence weed control in soybeans utilizing glyphosate and chlorimuron. *Proc. South. Weed Sci. Soc.* **1997**, *50*, 175.
22. Ellis, J.M.; Griffin, J.L. Benefits of soil-applied herbicides in glyphosate-resistant soybean (*Glycine max*). *Weed Technol.* **2002**, *16*, 541–547.
23. Common and chemical names of herbicides approved by the Weed Science Society of America. *Weed Sci.* **2010**, *58*, 511–518.
24. Hallett, S.G. Where are the bioherbicides? *Weed Sci.* **2005**, *53*, 404–415.
25. Rosskopf, E.N.; Charudattan, R.; Kadir, J.B. Use of Plant Pathogens in Weed Control. In *Handbook of Biological Control*; Bellows, T.S., Fisher, T.W., Eds.; Academic Press: New York, NY, USA, 1999; pp. 891–918.
26. Charudattan, R. Biological control of weeds by means of plant pathogens: Significance for integrated weed management in modern agro-ecology. *BioControl* **2001**, *46*, 229–260.
27. Charudattan, R. Ecological, practical, and political inputs into selection of weed targets: What makes a good biological control target? *Biol. Control.* **2005**, *35*, 183–196.
28. Weaver, M.A.; Lyn, M.E.; Boyette, C.D.; Hoagland, R.E. Bioherbicides for Weed Control. In *Non-Chemical Weed Management*; Updhyaya, M.K., Blackshaw, R.E, Eds.; CABI, International: Cambridge, MA, USA, 2007; pp. 93–110.
29. Hoagland, R.E. *Microbes and Microbial Products as Herbicides*; American Chemical Society: Washington, DC, USA, 1990.
30. Boyette, C.D.; Abbas, H.K. Host range alteration of the bioherbicidal fungus *Alternaria crassa* with fruit pectin and plant filtrates. *Weed Sci.* **1994**, *42*, 487–491.
31. Boyette, C.D. Host range and virulence of *Colletotrichum truncatum*, a potential mycoherbicide for hemp sesbania (*Sesbania exaltata*). *Plant Dis.* **1991**, *75*, 62–64.
32. Boyette, C.D.; Quimby, P.C., Jr.; Bryson, C.T.; Egley, G.H.; Fulgham, F.E. Biological control of hemp sesbania (*Sesbania exaltata*) under field conditions with *Colletotrichum truncatum* formulated in an invert emulsion. *Weed Sci.* **1993**, *41*, 497–500.
33. Abbas, H.K.; Boyette, C.D. Solid substrate formulation of the mycoherbicide *Colletotrichum truncatum* for hemp sesbania (*Sesbania exaltata*) control. *Biocontrol Sci. Technol.* **2000**, *10*, 297–304.
34. Boyette, C.D.; Hoagland, R.E.; Weaver, M.A. Biocontrol efficacy of *Colletotrichum truncatum* for hemp sesbania (*Sesbania exaltata*) control is enhanced with unrefined corn oil and surfactants. *Weed Biol. Manag.* **2007**, *7*, 70–76.
35. Cartwright, K.; Boyette, D.; Roberts, M. Lockdown: Collego bioherbicide gets a second act. *Phytopathology* **2010**, *100*, S162.
36. Templeton, G.E.; Smith, R.J., Jr.; TeBeest, D.O. Perspectives on mycoherbicides two decades after discovery of the Collego pathogen. In Proceedings of the VII International Symposium on the Biological Control of Weeds, Rome, Italy, 6–11 March 1988; pp. 553–558.
37. Boyette, C.D.; Bowling, A.J.; Vaughn, K.C.; Hoagland, R.E.; Stetina, K.C. Induction of infection of *Sesbania exaltata* by *Colletotrichum gloeosporioides* f. sp. *aeschynomene* formulated in an invert emulsion. *World J. Microbiol. Biotechnol.* **2010**, *26*, 951–956.

38. Boyette, C.D.; Gealy, D.; Hoagland, R.E.; Vaughn, K.C.; Bowling, A.J. Hemp sesbania (*Sesbania exaltata*) control in rice (*Oryza sativa*) with the bioherbicidal fungus *Colletotrichum gloeosporioides* f. sp. *aeschynomene* formulated in an invert emulsion. *Biocontrol Sci. Technol.* **2011**, *21*, 1399–1407.
39. Boyette, C.D.; Walker, H.L.; Abbas, H.K. Biological control of kudzu (*Pueraria lobata*) with an isolate of *Myrothecium verrucaria*. *Biocontrol Sci. Technol.* **2002**, *12*, 75–82.
40. Hoagland, R.E.; McCallister, T.S.; Boyette, C.D.; Weaver, M.A.; Beecham, R.V. Effects of *Myrothecium verrucaria* on morning-glory (*Ipomoea*) species. *Allelopath. J.* **2011**, *27*, 151–162.
41. Walker, H.L.; Tilley, A.M. Evaluation of an isolate of *Myrothecium verrucaria* from sicklepod (*Senna obtusifolia*) as a potential mycoherbicide agent. *Biol. Control* **1997**, *10*, 104–112.
42. Anderson, K.I.; Hallett, S.G. Bioherbicidal spectrum and activity of *Myrothecium verrucaria*. *Weed Sci.* **2004**, *22*, 623–627.
43. Abbas, H.K.; Johnson, B.J.; Shier, W.T.; Tak, H.; Jarvis, B.B.; Boyette, C.D. Phytotoxicity and mammalian cytotoxicity of macrocyclic trichothecenes from *Myrothecium verrucaria*. *Phytochemistry* **2002**, *59*, 309–313.
44. Conway, K.E. Evaluation of *Cercospora rodmanii* as a biological control of water hyacinth. *Phytopathology* **1976**, *66*, 914–917.
45. Boyette, C.D.; Weidemann, G.J.; TeBeest, D.O.; Quimby, P.C., Jr. Biological control of jimsonweed (*Datura stramonium*) with *Alternaria crassa*. *Weed Sci.* **1991**, *39*, 673–667.
46. Ghorboni, R.; Seel, W.; Litterick, A.; Leifert, C. Evaluation of *Alternaria alternata* for biological control of *Amaranthus retroflexus*. *Weed Sci.* **2000**, *48*, 474–480.
47. Elzein, A.; Kroschel, J. *Fusarium oxysporum* Foxy 2 shows potential to control both *Striga heronthica* and *S. asiatica*. *Weed Res.* **2004**, *44*, 433–438.
48. Boyette, C.D.; Weaver, M.A.; Hoagland, R.E.; Stetina, K.C. Submerged culture of a mycelial formulation of a bioherbicidal strain of *Myrothecium verrucaria* with mitigated mycotoxin production. *World J. Microbiol. Biotechnol.* **2008**, *24*, 2721–2726.
49. Tuite, J. *Plant Pathological Methods: Fungi and Bacteria*; Burgess Publication Co.: Minneapolis, MN, USA, 1969.
50. Horsfall, J.G.; Barratt, R.W. An improved grading system for measuring diseases. *Phytopathology* **1945**, *35*, 655.
51. Steele, R.G.D.; Torrey, J.H.; Dickeys, D.A. Multiple Comparisons. In *Principles and Procedures of Statistics—A Biometrical Approach*; McGraw Hill: New York, NY, USA, 1997; p. 365.
52. Boyette, C.D.; Hoagland, R.E. Biological Control of Hemp sesbania (*Sesbania exaltata*) and sicklepod (*Senna obtusifolia*) in soybean with anthracnose pathogen mixtures. *Weed Technol.* **2010**, *24*, 551–556.
53. Müller-Schärer, H. Finding solutions for biological control of weeds in European crop systems. *BioControl* **2001**, *46*, 125–126.
54. Amsellem, Z.; Barghouthi, S.; Cohen, B.; Goldwasser, Y.; Gressel, J.; Hornok, L.; Kerenyi, A.; Kleifeld, Y.; Klein, O.; Kroschel, J.; *et al.* Recent advances in the biocontrol of *Orobanche* (broomrape) species. *BioControl* **2001**, *46*, 211–228.

55. Bürki, H.M.; Lawrie, J.; Greaves, M.P.; Down, V.M.; Jüttersonke, B.; Cagán, L.; Vráblová, V.; Ghorbani, R.; Hassan, E.A.; Schroeder, D. Biocontrol of *Amaranthus* spp. in Europe: State of the art. *BioControl* **2001**, *46*, 197–210.
56. Défago, G.; Ammon, H.U.L.; Cagán, B.; Draeger, B.; Greaves, M.P.; Guntli, D.; Hoeke, D.; Klimes, L.; Lawrie, J.; Moënne-Loccoz, Y.; *et al.* Towards the biocontrol of bindweeds with a mycoherbicide. *BioControl* **2001**, *46*, 157–173.
57. Frantzen, J.; Paul, N.D.; Müller-Schärer, H. The system management approach of biological weed control: Some theoretical considerations and aspects of application. *BioControl* **2001**, *46*, 139–155.
58. Netland, J.; Dutton, L.C.; Greaves, M.P.; Baldwin, M.; Vurro, M.; Evidente, J.; Einhorn, G.; Scheepens, P.C.; French, L.W. Biological control of *Chenopodium album* L. in Europe. *BioControl* **2001**, *46*, 175–196.
59. Scheepens, P.C.; Müller-Schärer, H.; Kempenaar, C. Opportunities for biological weed control in Europe. *BioControl* **2001**, *46*, 127–138.
60. Abbas, H.K.; Tak, H.; Boyette, C.D.; Shier, W.T.; B.B. Jarvis. Macrocyclic trichothecenes are undetectable in kudzu (*Pueraria montana*) plants treated with a high-producing isolate of *Myrothecium verrucaria*. *Phytochemistry* **2001**, *58*, 269–276.
61. Milhollon, R.W.; Berner, D.K.; Paxson, L.K.; Jarvis, B.B.; Bean, G.W. *Myrothecium verrucaria* for control of annual morningglories in sugarcane. *Weed Technol.* **2003**, *17*, 276–283.
62. Weaver, M.A.; Boyette, C.D.; Hoagland, R.E. Bioherbicidal activity from washed spores of *Myrothecium verrucaria*. *World J. Microbiol. Biotechnol.* **2012**, *28*, 1941–1946.
63. Domsch, K.H.; Gams, W.; Anderson, T.H. *Myrothecium.* In *Compendium of Soil Fungi*; Academic Press: New York, NY, USA, 1980; pp. 481–487.
64. Nguyen, T.H.; Mathur, S.B.; Neergaard, P. Seed-borne species of *Myrothecium* and their pathogenic potential. *Trans. Br. Mycol. Soc.* **1973**, *61*, 347–354.
65. Yang, S.; Jong, S. C. Host range determination of *Myrothecium verrucaria* isolated from leafy spurge. *Plant Dis.* **1995**, *79*, 994–997.
66. Weaver, M.A.; Jin, X.; Hoagland, R.E.; Boyette, C.D. Improved bioherbicidal efficacy by *Myrothecium verrucaria* via spray adjuvants or herbicide mixtures. *Biol. Control* **2009**, *50*, 150–156.
67. Shivrain, V.K.; Burgos, N.R.; Moldenhauer, K.A.K.; McNew, R.W.; Baldwin, T.L. Characterization of spontaneous crosses between Clearfield rice (*Oryza sativa*) and red rice (*Oryza sativa*). *Weed Technol.* **2006**, *20*, 576–584.
68. Scott, R.C.; Meins, K.B.; Smith, K.L. Tank-mix partners with Newpath herbicide for hemp sesbania control in a Clearfield rice-production system. *Ark. Agric. Res. Ser.* **2005**, *54*, 225–229.
69. Powles, S.B. Evolved glyphosate-resistant weeds around the world: Lessons to be learnt. *Pest Manag. Sci.* **2008**, *64*, 360–365.

3

Benefits of Transgenic Insect Resistance in *Brassica* Hybrids under Selection

Cynthia L. Sagers [1,†,*], Jason P. Londo [2,†], Nonnie Bautista [3], Edward Henry Lee [4], Lidia S. Watrud [4] and George King [5]

[1] Department of Biological Sciences, University of Arkansas, Fayetteville, AR 72701 USA
[2] United States Department of Agriculture, Agricultural Research Service, Geneva, NY 14456 USA; E-Mail: jason.londo@ars.usda.gov
[3] Plant Biology Division, Institute of Biological Sciences, University of the Philippines Los Baños, Laguna 4031, Philippines; E-Mail: nsbautista@yahoo.com
[4] US Environmental Protection Agency, Western Ecology Division, National Health and Environmental Effects Research Lab, Corvallis, OR 97333, USA; E-Mails: lee.ehenry@epa.gov (E.H.L.); watrud.lidia@epa.gov (L.S.W.)
[5] CSS-Dynamac, Corvallis, OR 97333, USA; E-mail: king.george@epa.gov

† These authors contributed equally to this work.

* Author to whom correspondence should be addressed; E-Mail: csagers@uark.edu

Academic Editor: Peter Langridge

Abstract: Field trials of transgenic crops may result in unintentional transgene flow to compatible crop, native, and weedy species. Hybridization outside crop fields may create novel forms with potential negative outcomes for wild and weedy plant populations. We report here the outcome of large outdoor mesocosm studies with canola (*Brassica napus*), transgenic canola, a sexually compatible weed *B. rapa*, and their hybrids. *Brassica rapa* was hybridized with canola and canola carrying a transgene for herbivore resistance (*Bt Cry1Ac*) and grown in outdoor mesocosms under varying conditions of competition and insect herbivory. Treatment effects differed significantly among genotypes. Hybrids were larger than all other genotypes, and produced more seeds than the *B. rapa* parent. Under conditions of heavy herbivory, plants carrying the transgenic resistance were larger and produced more seeds than non-transgenic plants. Pollen derived gene flow from transgenic canola to *B. rapa* varied between years (5%–22%) and was not significantly impacted by herbivory. These results confirm that canola-weed hybrids benefit from transgenic resistance and are aggressive competitors with congeneric crops and ruderals. Because some crop and crop-weed hybrids may be competitively superior, escapees may alter the composition and ecological functions of plant communities near transgenic crop fields.

Keywords: *Brassica*; *Bt Cry1Ac*; feral species; herbivory; *Plutella xylostella*; risk assessment; weed evolution

1. Introduction

Hybridization is an essential tool in the traditional development of new crop varieties. Hybrids frequently express novel traits or combination of traits, which allows selection of new varieties more suited to the local environment. Similarly, hybridization in nature followed by natural selection produces genotypes more closely adapted to local conditions [1]. As such, hybridization is now understood as an important mechanism of plant evolution and diversification [2]. In agricultural systems, a number of important weeds have evolved from hybrids of wild and domestic species [3,4]. The study of crop-wild, and crop-weed hybrids has gained renewed interest following the introduction of biotech crops. Since then, non-transgenic crops have been found to carry genes engineered for beneficial traits, which arose through hybridization among cultivars. Examples include cotton, creeping bentgrass, canola, alfalfa, corn, soybean, and sugar beet (reviewed in [5]). On occasion, transgenic crop-crop or crop-wild hybrids have been found following small-scale field trials of genetically modified cultivars [6]. Despite the importance of hybridization for the evolution of new weed species [4], the ecologies of crop-weed and crop-wild hybrids have received limited study outside of agricultural fields. This project was undertaken to investigate the ecologies of hybrids of domesticated and feral *Brassica* species, including hybrids carrying the pesticide gene *Bt Cry1Ac*, in a simulated natural environment subjected to competition and herbivory.

A chief concern of crop-weed hybridization is the uncertain ecologies of transgenic hybrids in the native landscape [7]. The frequency of transgenic forms will depend upon continued migration of alleles from cultivated plants, the benefits of the transgene under selection, and the costs of the transgene when no selective pressure is present [8]. Costs may be due to pleiotropic effects of the transgene, an increase in gene dosage, or mutational changes induced by plant transformation [9]. If costs are high and beneficial selection rare, fitness of the hybrid will be reduced relative to non-transgenic form and its frequency should decrease. If the converse is true, high benefits and low costs, the transgenic form is expected to increase in frequency. Moreover, the costs of transgene expression by commercial varieties are expected to be small as a result of strong artificial selection for production efficiency in commercial cultivars [8].

In ecological and evolutionary studies, traits are deemed beneficial when they promote the survival and reproduction of their bearers. For example, among the plant-incorporated proteins (PIPs), the family of *Bt* transgenes indirectly increases survival and reproduction by limiting damage by herbivores. Reduced herbivory often results in substantial increases in plant biomass and seed output (see [10] for review). Changes in viability and fertility could have impacts on the population biology of natural populations; should resistant plants be more likely to survive, grow larger and produce more flowers. It follows that the frequency of transgenic resistance in the population would increase under selection [11]. Therefore, plants carrying a PIP for herbivore resistance, by their larger size and greater flower production, could sire more offspring than non-*Bt* forms.

We report herein our investigations of the effects of a genetically engineered herbivore resistance on crop and wild *Brassica* species and their hybrids under selection by herbivores and competitors. We grew transgenic, insect-resistant canola, non-transgenic canola, and *B. rapa* L. (Brassicaceae) with their hybrids in competition with congeners, or with common ruderal species in of the western U.S. In addition, we challenged these populations with diamondback moth (*Plutella xylostella*) (Lepidoptera: Plutellidae), a *Bt*-susceptible herbivore that is specialized on brassicaceous species. Our aims were to test the following predictions:

(1) the transgene will be beneficial in the presence of herbivores;
(2) the *Bt Cry1Ac* transgene is costly in the absence of herbivores, and
(3) selection by herbivores for insecticidal transgenes will increase the rate of transgene flow.

2. Results and Discussion

2.1. MANOVA Effects by Year and Univariate ANOVA Effects by Year and Genotype

Since the levels of herbivores applied were different in each year, it is necessary to analyze the data by Year. In the MANOVA by Year, Genotype had significant effects on aboveground biomass (ABM) and seed production (ESN) in both years of the study (Table 1). The main effects of Herbivory and Competition on ABM and ESN were significant only in the second year (Table 1). Interaction effects were significant for Genotype X Herbivory for ABM in 2007, ESN in 2006, and ESN in 2007. Finally, the interaction effects of Genotype X Competition was significant for ESN in 2006 and 2007 at the 0.10 level of significance (Table 1). Because genotype so clearly impacted the other effects of Herbivory and Competition, univariate ANOVA was performed by Year and by genotype (Table 2). In this analysis, Herbivory significantly impacted the different traits, though only in the year with higher herbivore density; 2007. Herbivory impacted the ABM measures of *B. rapa* and the non-GM F_1 in 2007, and the ESN measures of the GM F_1, *B. rapa,* non-GM F_1 and non-GM Westar genotypes. Competition effects were more variable by Year and significantly impacted genotypes in no discernable pattern (Table 2). The Herbivore X Competition interaction was not significant for any genotypes in either year. Details of the analyses and their impact on different measures of community performance are discussed below.

Table 1. MANOVA by Year for aboveground biomass (ABM) and estimated seed number (ESN) for experimental genotypes (G) growing under herbivory (H) and Competition (C) treatments. The effects of Herbivory, Competition and their interaction were analyzed at the level of the mesocosm as between-subject factors using the whole-plot error term for a split-plot design. The effects of Competition, Genotype and their interactions with Herbivory treatment were tested as within-subject factors at the tub level using the split-plot error term.

2006		**ABM**		**ESN**	
Between-subject factors	*df*	*F*	*p*	*F*	*p*
Herbivory (H)	1	0.94	0.370	2.76	0.148
Competition (C)	2	2.49	0.125	0.47	0.638
C X H	2	1.24	0.324	0.88	0.444
Within-subject factors					
Genotype (G)	4	475.19	<0.0001	130.40	<0.0001
G X H	4	0.68	0.625	3.62	0.067
G X C	8	0.72	0.673	2.76	0.047
G X C X H	8	0.35	0.931	1.37	0.291
2007		**ABM**		**ESN**	
Between-subject factors	*df*	*F*	*p*	*F*	*p*
Herbivory (H)	1	5.89	0.051	10.56	0.018
Competition (C)	2	3.57	0.061	2.65	0.115
C X H	2	0.10	0.908	0.65	0.540
Within-subject factors					
Genotype (G)	4	407.42	<0.0001	191.96	<0.0001
G X H	4	3.60	0.051	9.62	0.004
G X C	8	1.03	0.452	2.15	0.092
G X C X H	8	0.35	0.936	0.39	0.907

Table 2. Univariate ANOVA by Genotype, by Year, for ABM (g) and ESN (seeds) for experimental genotypes (G) growing under herbivory (H) and Competition (C) treatments with interactions (HxC). Significant effects shown in bold at a *p*-value of 0.1.

		GM Westar		**GM F1**		***B. rapa***		**Non-GM F1**		**Westar**	
		2006	**2007**	**2006**	**2007**	**2006**	**2007**	**2006**	**2007**	**2006**	**2007**
ABM	Herbivory (H)	0.537	0.720	0.391	0.863	0.727	**0.012**	0.927	**0.016**	0.613	0.587
	Competition (C)	0.718	0.867	0.278	**0.030**	0.711	0.235	**0.094**	0.194	0.532	0.397
	H X C	0.423	0.917	0.284	0.926	0.686	0.736	0.806	0.798	0.647	0.702
ESN	Herbivory (H)	0.191	0.248	0.108	**0.099**	0.307	**0.000**	0.970	**0.054**	0.813	**0.001**
	Competition (C)	0.861	0.850	0.344	**0.005**	0.229	0.615	**0.041**	0.228	0.835	0.143
	H X C	0.630	0.859	0.439	0.322	0.774	0.942	0.838	0.988	0.618	0.426

2.1.1. Heterosis

Genotype had a significant effect on ABM and ESN in both years (Table 1). F_1 hybrids were the largest plants in the study and *B. rapa* plants were the smallest (Figure 1). Both GM, and non-GM F_1 hybrid genotypes were significantly larger than the average *B. rapa* parent, and significantly larger than their respective crop parent in 2007 (Figure 1). Contrast *p*-values are shown in Table 3. These results are largely consistent with previous studies reporting vegetative heterosis of F_1 hybrids between *B. rapa* and canola [8,12]. Regarding seed production, GM and non-GM F_1 hybrids produced significantly fewer seeds than the crop genotypes (GM Westar and Westar) demonstrating the negative fecundity effects of forming a triploid hybrid [13], but significantly more than *B. rapa* in both years ($p < 0.001$ for all contrasts) (Figure 2). This large reduction in seed production (relative to *B. napus*) has been observed in field examination of herbicide resistant hybrids produced from crop *B. napus* X wild *B. rapa* plants in Canada [13]. It is important to note that advanced backcrosses of these two species regain fecundity within a few generations.

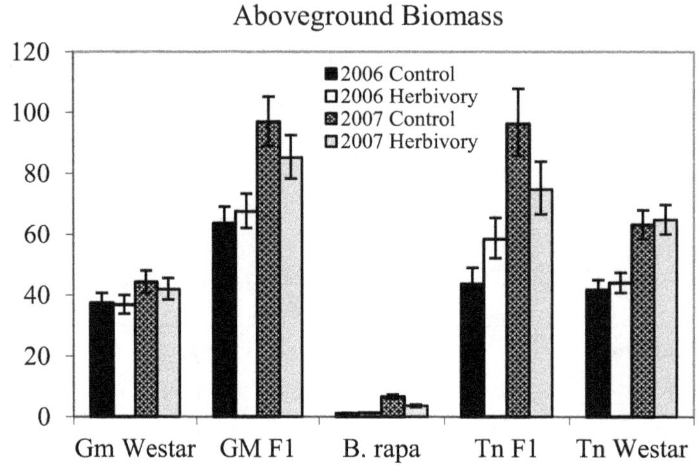

Figure 1. ABM of experimental genotypes growing under control and experimental herbivory treatments in 2006 (low herbivore densities) and 2007 (high herbivore densities). Bars represent mean (±1 SE). Closed bars = control; open bars = herbivory.

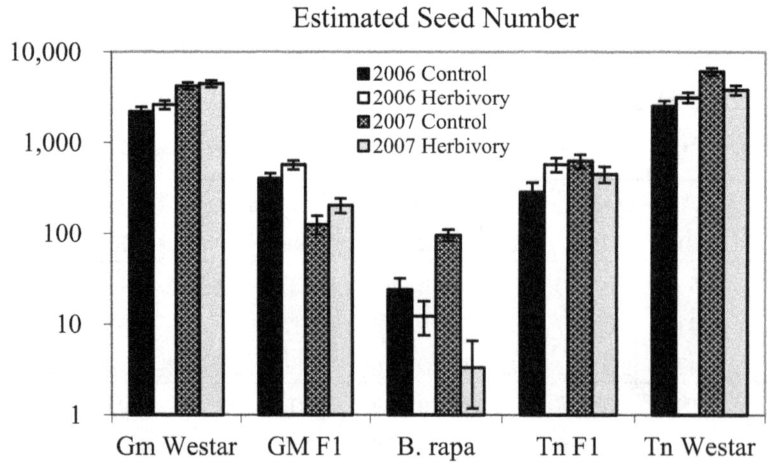

Figure 2. ESN of experimental genotypes under control and experimental herbivory treatments in 2006 (low herbivore densities) and 2007 (high herbivore densities). Bars represent mean (±1 SE). Closed bars = control; open bars = herbivory; *Y*-axis displayed in log scale.

Table 3. *Post hoc* contrasts between genotypes for ABM (g) and ESN (seeds) in each year for experimental genotypes growing under herbivory and control treatments. Values in bold indicate a significant Bonferroni corrected *p*-value of 0.01.

Contrast	ABM 2006		ABM 2007		ESN 2006		ESN 2007	
	Control	Herbivory	Control	Herbivory	Control	Herbivory	Control	Herbivory
GM Westar *vs.* GM F1	**0.0039**	**0.0049**	**0.0001**	**0.0005**	**0.0003**	**0.0001**	**0.0001**	**0.0001**
GM Westar *vs.* non-GM F1	0.4526	**0.0049**	**0.0007**	0.0136	**0.0011**	**0.0001**	**0.0001**	**0.0001**
GM Westar *vs.* Westar	0.372	0.1771	**0.0034**	**0.01**	0.7	0.1849	**0.0011**	0.3396
GM Westar *vs. B. rapa*	**0.0001**	**0.0001**	**0.0001**	**0.0001**	**0.0002**	**0.0001**	**0.0001**	**0.0001**
GM F1 *vs.* non-GM F1	0.0755	0.4247	0.9272	0.1862	0.3987	0.9542	**0.0054**	0.0212
GM F1 *vs.* Westar	**0.0069**	0.0232	**0.0003**	**0.0004**	**0.0003**	**0.0001**	**0.0001**	**0.0001**
GM F1 *vs. B. rapa*	**0.0001**	**0.0001**	**0.0001**	**0.0001**	**0.0012**	**0.0001**	0.0219	**0.0001**
Non-GM F1 *vs.* Westar	0.8273	0.1249	**0.0081**	0.0715	**0.0013**	**0.0003**	**0.0001**	**0.0001**
Non-GM F1 *vs. B. rapa*	**0.0001**	**0.0001**	**0.0001**	**0.0001**	**0.0002**	**0.0001**	**0.0015**	**0.0001**
Westar *vs. B. rapa*	**0.0001**	**0.0001**	**0.0001**	**0.0001**	**0.0002**	**0.0001**	**0.0001**	**0.0001**

2.1.2. Herbivory

The MANOVA by Year indicated that Herbivory significantly reduced ABM and ESN of the mesocosms in 2007, while Genotype X Herbivory terms had significant effects on ABM in 2007 and ESN in both years (Table 1). The significant main and interaction effects seen in 2007, relative to 2006 suggest that the difference in herbivore densities between 2006 (low densities) and 2007 (high densities) may have contributed to the changes observed. Given the dramatic visual observations of herbivory damage in the mesocosms in 2007, we believe herbivore density was largely responsible for the significant effects observed in that year although we are unable to rule out additional effects due to year.

The MANOVA analysis by Genotype shows how resistant and non-resistant genotypes responded to Herbivory in each year. Neither the GM Westar commercial line nor the GM F_1 hybrid line showed an effect of Herbivory on ABM measures in either year. Similarly, ESN measures were not significantly changed in 2006 for these genotypes, though a significant effect on ESN (increased seed number) in the GM F_1 was observed in 2007. These results demonstrate a general resistance of these genotypes to the applied stress and in the case of the GM F_1, increased fecundity. In contrast, non-GM lines (*B. rapa*, non-GM F_1, Westar) showed a significant effect of Herbivory for both response variables in 2007 with a no effect in the first year. These results mirror those observed for this transgene in field based experiments of Sunflower (*Helianthus annuus*) where transgenic plants were more resistant to lepidopteran herbivores [14].

Fertility and biomass of susceptible crop genotypes (*B. rapa*, non-GM F_1 hybrid, Westar) tolerated herbivory in 2006, but showed reduced growth and productivity in 2007 (Table 2). In contrast, the resistant, transgenic crop genotypes (GM Westar, GM F_1 hybrid) tolerated herbivory and showed no reduction of growth or fertility in the presence of diamondback moth in either year of the study.

Brassica rapa responded similarly to the herbivory treatment in both years of the study. Seed counts for *B. rapa* in herbivore-treated relative to control mesocosms were reduced by 29% in 2006, and 65% in 2007 (Figure 2). Furthermore, *B. rapa* biomass was significantly less in the herbivore treatment in 2007 ($p = 0.012$), but not in 2006. Overall, the effects of herbivory on *B. rapa* were more substantial than for any other genotype in the study, including the other vulnerable genotypes (non-GM F_1 and Westar).

2.1.3. Costs of *Bt Cry1Ac* Gene

In 2006, the genetically modified (GM) Westar canola growing in the herbivore-free mesocosms produced was smaller and produced fewer seeds than the non-GM Westar, though not significantly so ($p = 0.372$; $p = 0.700$). However, ABM and ESN measures in 2007 was significantly lower for the GM Westar in this contrast ($p = 0.003$; $p = 0.001$) *versus* the non-GM crop. As control mesocosms are not confounded by year or herbivory effects, we interpret this to indicate that constitutive expression of the inserted transgene, or the physical effects of transformation (e.g., site of insertion) of the Westar line may have a detrimental effect on the GM Westar in the environmental and competitive conditions of our mesocosm greenhouses. As a result, GM plants in this study might be expected to be smaller and produce fewer seeds than non-GM plants, all else equal. Costs of genetic engineering have not previously been reported for these lines [15] or others [8,13] in the *Brassica* system, nor in Sunflower [14]. They were observed here perhaps because of the stresses imposed by the mesocosm environments [16]. Similar observations of fitness costs are not uncommon and have been observed in other studies, for example, studies on herbicide resistance traits (reviewed by [17]). In these studies, fitness costs become measurable when variable stress is included as a factor in experimental designs. Detection of costs may have important implications for risk assessment and these results therefore merit further study.

The above results indicate that the costs and benefits of insecticidal transgenes vary in space and time. As demonstrated with the *B. napus* genotypes in the second year of the study, but not the first, the costs of genetic modification were overcome by benefits of reduced herbivory. The result was a net benefit of the transgene when herbivores were present. As a result, increases in biomass and fecundity associated with the transgenic phenotype may translate to fitness benefits of crop canola and its hybrids growing outside of cultivation. Therefore, environmental variation is expected to be an important factor determining whether a transgene or any other beneficial domesticated allele will pose a risk [8,18].

2.2. Competition

The main effect of Competition in the overall MANOVA was significant for ABM and ESN (Table 1) in 2007, but not in 2006. In general, all plants tended to be largest under control conditions and smallest in competition with ruderal weed species (data not shown). However, when analyzed independently, genotypes differed in response to competition: parental genotypes (*B. rapa*, GM Westar and Westar) showed no significant effect of Competition for either response variable. In contrast, the effects of competition treatments were significant for hybrids. ABM and ESN differed significantly among competition treatments for the GM F_1 (ABM $p = 0.030$; ESN $p = 0.005$) in 2007 and non-GM F_1 genotypes in 2006 (ABM $p = 0.094$; ESN $p = 0.041$).

Further, *post hoc* analysis showed that both F_1 genotypes were smallest (ABM) when grown with ruderals relative to control competition, and competition with congeners (GM F_1 in 2007: $p = 0.034$, $p = 0.013$; non-GM F_1 in 2006: $p = 0.044$, $p = 0.086$, respectively). In addition, competition had an effect on ESN for the non-GM F_1 genotype in 2006 ($p = 0.041$). Non-GM F_1 lines produced significantly fewer seeds in ruderal relative to control treatments in 2006 ($p = 0.033$). The effects on the non-GM F_1 were not significant after Bonferonni adjustment nor were the effects on the GM F_1 ESN but the trend is worth noting given the restricted sample sizes used here. As there was no significant Herbivory X Competition interaction, effects are averaged over herbivore treatment.

Canola and F_1 hybrids are potentially aggressive competitors of *B. rapa* in the wild. Crops and crop-weed hybrids are physically large, with hybrids up to 50× the size of the *B. rapa* parent and larger even than the crop parent. However, we found little effect of the competition treatments on *B. rapa* or it was similarly affected by all treatments. Design limitations precluded additional competition treatments, but future studies should include a treatment of *B. rapa* in the absence of competition from crop and crop-weed hybrids.

2.3. Transgene Flow under Selection

All the *B. rapa* plants that survived to the end of the study and that produced seeds produced at least one transgenic F_1 hybrid seed. Gene flow rates, estimated as the proportion of *GFP+* progeny produced by each individual of *B. rapa*, ranged from 0%–100%. A significant increase in mean transgene flow rate was observed in the second year of the experiment (from 5.1% to 22.3%) ($\chi^2 = 3.87$, $df = 1$, $p = 0.049$). The cause of this increase is uncertain, but in neither year of the study did we detect a significant effect of herbivory on transgene flow rate.

High rates of outcrossing in *B. rapa* may be related to its breeding biology. *Brassica rapa* is self-sterile and obligately-outcrossing. Because *B. rapa* must outcross, the frequency of hybrids in a population will be a function of the availability of congeneric pollen in the pollen cloud [19]. The larger the canola population, the greater the potential hybridization rate and fewer the proportion of pure *B. rapa* seed in each successive generation [19]. The implications for this result are compelling, but additional studies of free-living populations are necessary to confirm these largely theoretical predictions.

3. Materials and Methods

3.1. Experimental Design

Fitness associated traits of transgenic, non-transgenic plants and their hybrids under selective pressure of an herbivore were evaluated in experiments carried out at the US EPA facilities in Corvallis, OR, USA. Experiments were performed during the springs of 2006 and 2007 in eight free-standing, outdoor mesocosms (detailed in [19]) using a split-plot design to test the effects of herbivory and intra-genotype competition on the vegetative and reproductive responses of five *Brassica* genotypes including *Brassica napus* L. (canola) and *Brassica rapa*. Each mesocosm was approximately

3.1 m diameter × 3.3 m height, with three large tubs each containing approximately 1 m^3 of sandy loam soil. Four of the eight mesocosms were randomly assigned to the herbivory treatment (*i.e.*, presence of herbivore) and the other four mesocosms were assigned to the control (*i.e.*, absence of herbivore). *Plutella xylostella* L. (Diamondback moth) larvae were added to the treated mesocosms to simulate herbivory under field conditions. Measures of Diamondback moth (DBM) densities in the field range from 1–100 larvae per plant [20]. DBM densities in in 2006 were approximately 100 larvae per tub; approximately 6 per plant to simulate low selective pressure. In 2007, densities were approximately 2000 per tub; approximately 100 larvae per plant to simulate high selective pressure. To promote pollination and prevent pollen limitation, approximately 1000 *Musca domestica* L. (housefly) larvae were introduced to each mesocosm each year of the study as soon as the first flowers were evident (following [21]). Three individuals of each of five plant genotypes were planted in each mesocosm tub in a grid pattern to simulate a plant community. Within a mesocosm, each tub was randomly assigned to one of three competition treatments: no additional plant competitors ("control"), competition with additional *B. rapa* plants, or competition with ruderal species. To examine intra-generic competition effects, nine individuals of *B. rapa* were planted in and around the original planting grid. In the third treatment, three individuals of each of three ruderal species were planted within and around the experimental plants. Ruderal species were chosen to represent broadly distributed weeds of the western U.S. common on disturbed soils such as fallow fields and road verges: *Achillea millefolium* L. (common yarrow), *Panicum capillare* L. (witchgrass) and *Lapsana communis* L. (nipplewort). Each mesocosm was enclosed with 8 mil PVC film, exposed to ambient light, and regulated by evaporative coolers to approximate ambient temperature. Mesocosms were equipped with pollen filters and insect netting to prevent escape of *Brassica* pollen and seeds and to prevent movement of pollinators between units.

Seedlings were started in greenhouses, marked individually and transplanted at the four-leaf stage. Plants were grown under ambient light from mid-April to mid-July in 2006, and from mid-March to mid-July in 2007. When plants began to senesce, irrigation was stopped and plants were allowed to desiccate. Aboveground biomass was harvested two weeks later and dried in a forced air oven at 60 °C for 10 days. Seed numbers were estimated from the mass of 100 seed counts. All *B. rapa* seeds were screened for *GFP* expression. Transgenic hybrids were identified in qualitative assays for *GFP* fluorescence using 10× microscopy under long-wave ultraviolet light [22]. Gene flow rates (% transgenic seeds per *B. rapa* plant) and the total numbers of transgenic seeds produced per mesocosm were recorded.

3.2. Plant Materials

The five *Brassica* genotypes used in these experiments represent three parental lines: (1) *B. rapa*; (2) non-transgenic *B. napus* (canola) variety Westar (hereafter, "Westar"); and (3) transgenic Westar (hereafter, "GM Westar"); and two hybrids: (4) *B. rapa* X Westar (hereafter, "non-GM F_1"); and (5) an F1 hybrid-*B. rapa* X GM Westar (hereafter, "GM F_1"). *B. rapa* is a naturalized weed introduced from Asia, nearly cosmopolitan in the New World and frequently found in disturbed habitats (http://plants.usda.gov/java/). The species is considered a noxious weed in parts of the U.S. and an agronomic pest in Canada [23]. *Brassica rapa* L. is considered self-sterile, although evidence of minimal selfing has been reported [18]. *B. rapa* seeds used in these experiments were collected from a persistent population on the campus of Oregon State University, Corvallis, OR, USA. *B. rapa* spontaneously hybridizes with *Brassica napus* L. (canola) in the field [22] at rates ranging from 0%–55% in field experiments and greenhouse studies [24–26]. F_1 generation hybrids with canola are triploid and vegetatively vigorous, but often express reduced fertility relative to the weedy or native parent species [13,15]. However, one study suggests that fertility is quickly reestablished in early backcross generations [8].

Brassica napus (canola) is an oilseed crop grown extensively in regions of the northern U.S. and southern Canada. Canola is an allotetraploid hybrid of *B. oleracea* L. and *B. rapa*, is predominantly self-fertile, but is sexually compatible with all cole crops and with a number of species listed by the USDA as noxious weeds (*B. rapa*, *Sinapis arvensis* L., *Erucastrum gallicum* (Willd.) O.E. Schulz) [24–26]. Canola is considered an annual, but secondary seed dormancy has now been confirmed [27]. Mature plants can overwinter as rosettes [17] and volunteers are routinely reported years after a canola harvest [21]. A substantial portion (1%–30%) of each crop is lost during harvest [27,28] and mature crops are frequently left unharvested because of unfavorable harvest conditions [29]. Overwintering and seed dormancy increase the phenological overlap with sexually compatible relatives thereby improving the likelihood of crop-weed hybridization [23].

The canola variety "Westar" was genetically modified with markers useful in the tracking of pollen movement and transgene flow [15,18]. Three transgenes were under the control of the CaMV 35S promoter in separate cassettes on a single T-DNA vector: *mgfp5-er* and a synthetic truncated *Bt cry1Ac* [12]. The pSAM12 plasmid (described in [30]) used to transform *B. napus* contained *GFP*, *Bt*, and kanamycin resistance cassettes in the T-DNA, enabling all three traits to be inserted in a single, genetically linked locus [12]. The *GFP* and *Bt* transgenes were shown to be genetically linked and functional through the BC4 generation [31]. This experimental transgenic Westar line was generated in the lab of Dr. Charles Neal Stewart, Jr., and is used with his permission. The GM Westar line selected for these experiments was shown through a series of selfed generations to be homozygous for all transgenes. Hybrid lines were developed by manual pollination under greenhouse conditions following Bautista *et al.* [32]. Eighteen *B. rapa* plants were hand-crossed on sequential days with pollen from GM Westar and Westar to generate F_1 hybrid generations. The hybrid status of a subset of individuals was verified with flow cytometry [33] (data not shown).

3.3. Statistical Analysis

Measures of vegetative biomass (ABM) and estimated seed number (ESN) were analyzed using multivariate analysis of variance (MANOVA) by Year with Genotype as the within-subject factor and Herbivory and Competition as the between-subject factors. It was not logistically possible to test different levels of herbivores in the same year due to limitations on the number of available mesocosm testing greenhouses. Because herbivore densities were different in 2006 and 2007, herbivory effects cannot be directly compared between years and were analyzed separately. While measurements were taken from individual plants, the main effect of Herbivory was analyzed at the mesocosm level using the whole-plot error term for a split-plot design. The main effect of Competition and its interaction with Herbivory were tested at the tub level using the split-plot error term. The MANOVA tests for

Genotype and its interactions with Herbivory and Competition were also performed at the tub level using the split-plot error term. When there were significant interactions, further analysis (ANOVA) was performed as on data subsets [34]. Since ABM and ESN measures differed by several orders of magnitude between genotypes, data were transformed prior to statistical analysis using the natural log transformation (ABM) and square root transformations (ESN). Values were back-transformed for reporting the results. Because the number of mesocosms limits the replication of the experiment, a p-value > 0.10 was deemed sufficient to detect significant effects. A one-sided F-test was used when testing for the effects of Herbivory. For *post hoc* contrasts between genotypes, a Bonferroni modified p-value was used (p = 0.1/10, contrasts = 0.01). The MANOVA and ANOVA analyses were generated using SAS/STAT software (PROC GLM) with planned contrast statements to examine differences between individual genotypes by Year. Wilcoxon rank sum test (exact) was used to test for differences between years and treatments in the frequency of transgene flow to *B. rapa*. We used mesocosm averages with SAS/STAT software (PROC NPAR1WAY), Version 9.13 of the SAS System for Windows (SAS 9.13, Cary, NC, USA).

4. Conclusions

This study examined the ecology of transgenic and non-transgenic *Brassica napus* crops and sexually compatible *Brassica rapa* weeds in an open pollination system competing with natives and congeners, and under selection by insect herbivores. The results suggest that canola, regardless of transgene phenotype, which has escaped from cultivation, has the potential to thrive in more naturalized landscapes. Moreover, crop-wild hybrids are able to compete effectively with both crop and wild congeners. These outcomes are more evident in the presence of a biotic stressor, the diamondback moth, differences in plant community composition and competition, and are impacted by the phenotype of herbivore resistance. Hybrid forms, whether they carried the transgene or not, had much greater vegetative and reproductive fitness traits compared with the wild/weedy *B. rapa*, demonstrating the potential for extirpation of *B. rapa* in populations where hybrid forms occur.

A recent review underscores the finding that few transgenes have found their way to native species or weeds [35]. Instead, transgene flow is reported more frequently among conspecific crops species where barriers to hybridization are limited. We suggest, however, that the long-term effects of agricultural species on native plant communities will require far more intense scrutiny as it is likely that rare events occur [36] and remain undetected for some time. The value of mesocosms studies such as these lays in the ability to safely simulate the conditions that could possibly promote the spread of a beneficial transgene and observing the initial shorter term responses at the population and community levels.

Acknowledgments

This research was funded in part by the USEPA Office of Research and Development, USEPA GRO, NSF Undergraduate Research Fellowship, and the USDA. Dynamac Corporation provided technical support. Seeds were generously provided by Charles Neil Stewart, Jr. and Mike Bollman. Christina Hauther and Kristina Walker assisted with data collection. Much of this work was completed while Cynthia Sagers, Nonnie Bautista and Jason Londo were fellows of the National Research Council at the USEPA. Paul Arriola and three anonymous reviewers provided useful comments on the manuscript. This research has been funded by USEPA and the USDA (USDA CREES NRI 35615-19216). It has been subjected to review by the National Health and Environmental Effects Research Laboratory's Western Ecology Division and approved for publication. Approval does not signify that the contents reflect the views of the Agency, nor does mention of trade names or commercial products constitute endorsement or recommendation for use.

Author Contributions

Cynthia L. Sagers and Jason P. Londo conceived of the experiment. Cynthia L. Sagers drafted the manuscript. Jason P. Londo and Edward Henry Lee analyzed the data. Jason P. Londo designed the tables and figures. Nonnie Bautista developed the plant materials used in the experiment. Edward Henry Lee advised on the experimental design. George King and Lidia S. Watrud designed the planting arrays and managed the pre-submission review.

Conflict of Interest

The authors declare no conflict of interest.

References

1. Rieseberg, L.H.; Raymond, O.; Rosenthal, D.M.; Lai, Z.; Livingstone, K.; Nakazato, T.; Durphy, J.L.; Schwarzbach, E.E.; Donovan, L.A.; Lexer, C.; *et al.* Major ecological transitions in annual sunflowers facilitated by hybridization. *Science* **2003**, *301*, 1211–1216.
2. Grant, V. *Plant Speciation*, 2nd ed.; Columbia University Press: New York, NY, USA, 1981; p. 435.
3. Harlan, J.R. The possible role of weed races in the evolution of cultivated plants. *Euphytica* **1965**, *14*, 173–176.
4. Harlan, F.R.; de Wet, M.J. Sympatric evolution in sorghum. *Genetics* **1974**, *78*, 473–474.
5. Mallory-Smith, C.; Zapiola, M. Gene flow from glyphosate-resistant crops. *Pest. Manag. Sci.* **2008**, *64*, 428–440.
6. Watrud, L.S.; Lee, E.H.; Fairbrother, A.; Burdick, C.; Reichman, J.R.; Bollman, M.; Storm, M.; King, G.; van de Water, K.P. Evidence for landscape-level, pollen-mediated gene flow from genetically modified creeping bentgrass with CP4 EPSPS as a marker. *Proc. Natl. Acad. Sci. USA* **2004**, *101*, 14533–14538.
7. Snow, A.A.; Andow, D.A.; Gepts, P.; Hallerman, E.M.; Power, A.; Tiedje, J.M.; Wolfenbrger, L.L. Genetically engineered organisms and the environment: Current status and recommendations. *Ecol. Appl.* **2005**, *15*, 377–404.
8. Snow, A.A.; Andersen, B.; Jørgensen, R.B. Costs of transgenic herbicide resistance introgressed from *Brassica napus* into weedy *Brassica rapa*. *Mol. Ecol.* **1999**, *8*, 605–615.
9. Bergelson, J.; Purrington, C.B.; Palm, C.J.; Lopez-Gutierrez, J.C. Costs of resistance: A test using transgenic *Arabidopsis thaliana*. *Proc. R. Soc. Lond. B Biol.* **1996**, *263*, 1659–1663.
10. Strauss, S.Y.; Rudgers, J.A.; Lau, J.A.; Irwin, R.E. Direct and ecological costs of resistance to herbivory. *Trends Ecol. Evol.* **2002**, *17*, 278–285.

11. Lenski, R.E. Coevolution of bacteria and phage—Are there endless cycles of bacterial defenses and phage counterdefenses. *J. Theor. Biol.* **1984**, *108*, 319–325.
12. Wei, W.; Darmency, H. Gene flow hampered by low seed size of hybrids between oilseed rape and five wild relatives. *Seed Sci. Res.* **2008**, *18*, 115–123.
13. Warwick, S.I.; Legere, A.; Simard, M.-J.; James, T. Do escaped transgenes persist in nature? The case of an herbicide resistance transgene in a weedy *Brassica rapa* population. *Mol. Ecol.* **2008**, *17*, 1387–1395.
14. Snow, A.A.; Pilson, D.; Rieseberg, L.H.; Paulsen, M.J.; Pleskac, N.; Reagon, M.R.; Wolf, D.E.; Selbo, S.M. A *Bt* Transgene Reduces Herbivory and Enhances Fecundity in Wild Sunflowers. *Ecol. Appl.* **2003**, *13*, 279–286.
15. Halfhill, M.D.; Richards, H.A.; Mabon, S.A.; Stewart, C.N., Jr. Expression of *GFP* and *Bt* transgenes in *Brassica napus* and hybridization with *Brassica rapa*. *Theor. Appl. Genet.* **2001**, *103*, 659–667.
16. Mercer, K.L.; Andow, D.A.; Wyse, D.L.; Shaw, R.G. Stress and domestication traits increase the relative fitness of crop-wild hybrids in sunflower. *Ecol. Lett.* **2007**, *10*, 383–393.
17. Vila-Aiub, M.M.; Neve, P.; Powles, S.B. Fitness costs associated with evolved herbicide resistance alleles in plants. *New Phytol.* **2009**, *184*, 751–767.
18. Halfhill, M.D.; Sutherland, J.P.; Moon, H.S.; Poppy, G.M.; Warwick, S.I.; Weissinger, A.K.; Ruffy, T.W.; Raymer, P.L.; Stewart, C.N., Jr. Growth, productivity, and competitiveness of introgressed weedy *Brassica rapa* hybrids selected for the presence of *Bt. cry1Ac* and *GFP* transgenes. *Mol. Ecol.* **2005**, *14*, 3177–3189.
19. Waschmann, R.S.; Watrud, L.S.; Reece, L.R.; Shiroyama, T. Sunlit mesocosms designed for pollen confinement and risk assessment of transgenic crops. *Aerobiologia* **2010**, *26*, 311–325.
20. Bigger, D.S.; Fox, L.R. High-density populations of diamondback moth have broader host-plant diets. *Oecologia* **1997**, *112*, 179–186.
21. Halfhill, M.; Millwood, R.; Weissinger, A.; Warwick, S.; Stewart, C.N., Jr. Additive transgene expression and genetic introgression in multiple green-fluorescent protein transgenic crop × weed hybrid generations. *Theor. Appl. Genet.* **2003**, *107*, 1533–1540.
22. Jørgensen, R.B.; Andersen, B. Spontaneous hybridization between oilseed rape (*Brassica napus*) and weedy *Brassica campestris*: A risk of growing genetically modified oilseed rape. *Am. J. Bot.* **1994**, *81*, 1169–1175.
23. Simard. M.-J.; Légère, A.; Warwick, S. Transgenic *Brassica napus* fields and *Brassica rapa* weeds in Quebec: Sympatry and weed-crop *in situ* hybridization. *Can. J. Bot.* **2006**, *84*, 1842–1851.
24. Rackow, G.; Woods, D.L. Outcrossing in rape and mustard under Saskatchewan prairie conditions. *Can. J. Plant Sci.* **1987**, *678*, 147–151.
25. Warwick, S.I.; Simard, M.-J.; Légère, A.; Beckie, H.J.; Braun, L.; Zhu, B.; Mason, P.; Séguin-Swartz, G.; Stewart, C.N., Jr. Hybridization between transgenic *Brassica napus* L. and its wild relatives: *B. rapa* L., *Raphanus raphanistrum* L., *Sinapis arvensis* L., and *Erucastrum gallicum* (Willd.) O.E. Schulz. *Theor. Appl. Genet.* **2003**, *107*, 528–539.

26. FitzJohn, R.G.; Armstrong, T.T.; Newstrom-Lloyd, E; Wilton, A.D.; Cochrane, M. Hybridisation within *Brassica* and allied genera: Evaluation of potential for transgene escape. *Euphytica* **2007**, *58*, 209–230.
27. Gulden, R.H.; Shirtliffe, S.J.; Thomas, A.G. Harvest losses of canola (*Brassica napus*) cause large seedbank inputs. *Weed Sci.* **2003**, *51*, 83–86.
28. López-Granados, F.; Lutman, P. Effect of environmental conditions on the dormancy and germination of volunteer oilseed rape seed (*Brassica napus*). *Weed Sci.* **1998**, *46*, 419–423.
29. National Agricultural Statistics Service. Available online: http://www.nass.usda.gov/ (accessed on 8 October 2014).
30. Harper, B.K.; Mabon, S.A.; Leffel, S.M.; Halfhill, D.; Richards, H.A.; Moyer, K.A.; Stewart, C.N., Jr. Green fluorescent protein in transgenic plants indicates the presence and expression of a second gene. *Nat. Biotechnol.* **1999**, *17*, 1125–1129.
31. Zhu, B.; Lawrence, J.R.; Warwick, S.I.; Mason, P.; Braun, L.; Halfhill, M.D.; Stewart, C.N., Jr. Stable *Bacillus thuringiensis* (*Bt*) toxin content in interspecific F1 and backcross populations of wild *Brassica rapa* after *Bt.* gene transfer. *Mol. Ecol.* **2004**, *13*, 237–241.
32. Bautista, N.S.; Sagers, C.L.; Lee, E.H.; Watrud, L.S. Flowering times in genetically modified *Brassica* hybrids in the absence of selection. *Can. J. Plant Sci.* **2010**, *90*, 185–187.
33. Londo, J.P.; Bautista, N.S.; Sagers, C.L.; Lee, E.H.; Watrud, L.S. Glyphosate drift promotes changes in fitness and transgene gene flow in canola (*Brassica napus*) and hybrids. *Ann. Bot.* **2010**, *106*, 957–965.
34. Snedecor, G.; Cochran, W. *Statistical Methods*, 7th ed.; Iowa State University Press: Ames, IA, USA, 1980; p. 507.
35. Kwit, C.; Moon, H.S.; Warwick, S.I.; Stewart, C.N., Jr. Transgene introgression crop relatives: Molecular evidence and mitigation strategies. *Trends Biotechnol.* **2011**, *29*, 284–293.
36. De Wet, J.M.; Harlan, J.R. Weeds and Domesticates: Evolution in the man-made habitat. *Econ. Bot.* **1975**, *29*, 99–108.

No Effect Level of Co-Composted Biochar on Plant Growth and Soil Properties in a Greenhouse Experiment

Hardy Schulz [1,*], Gerald Dunst [2] and Bruno Glaser [1]

[1] Soil Biogeochemistry, Martin-Luther-University Halle-Wittenberg, Von-Seckendorff-Platz 3, Halle 06120, Germany; E-Mail: bruno.glaser@landw.uni-halle.de

[2] Sonnenerde, Oberwarterstraße 100, Riedlingsdorf A-7422, Austria; E-Mail: g.dunst@sonnenerde.at

* Author to whom correspondence should be addressed; E-Mail: hardy.schulz@gmail.com

---

Abstract: It is claimed that the addition of biochar to soil improves C sequestration, soil fertility and plant growth, especially when combined with organic fertilizers such as compost. However, little is known about agricultural effects of small amounts of composted biochar. This greenhouse study was carried out to examine effects of co-composted biochar on oat (*Avena sativa* L.) yield in both sandy and loamy soil. The aim of this study was to test whether biochar effects can be observed at very low biochar concentrations. To test a variety of application amounts below 3 Mg biochar ha^{-1}, we co-composted five different biochar concentrations (0, 3, 5, 10 kg Mg^{-1} compost). The biochar-containing compost was applied at five application rates (10, 50, 100, 150, 250 Mg ha^{-1} 20 cm^{-1}). Effects of compost addition on plant growth, Total Organic Carbon, N_{tot}, pH and soluble nutrients outweighed the effects of the minimal biochar amounts in the composted substrates so that a no effect level of biochar of at least 3 Mg ha^{-1} could be estimated.

Keywords: biochar; compost; no-effect-level; greenhouse; C management

---

1. Introduction

Many studies on biochar effects in different soil substrates have been scientifically examined during the last decade, the majority thereof proving positive effects on plant growth and soil properties [1–3].

In a recent meta-analysis study, *Jeffery et al.* [4] reviewed 177 treatments from 16 individual studies and found only one with negative impacts on plant growth but several studies showing no biochar effect on plant growth.

Usually biochars are low in nutrients, depending on feedstock and pyrolysis temperature [5,6]. This limited supply of nutrients implies additional fertilization if biochar is applied for agricultural purposes. Recent studies suggested adding biochar to compost [7] or even better co-composting biochar [8,9] as a preferable alternative to input intensive or finite (phosphorus) fertilizer. Another study claims that biochar increases the nutrient retention of the existing nutrients in compost due to the increase of biochar surface oxidation when biochar is applied into the fresh compost mixture. In other words: abiotic and biotic processes during composting lead to the formation of oxygen-containing functional groups and therewith to an increase of nutrient holding capacity [10].

Research already opposed maximum biochar application amounts, as shown by Schulz and Glaser [9] who applied biochar amounts of up to 90 Mg ha^{-1} in the form of co-composted biochars, which induced increased plant growth, and the more biochar added to the soil, the more carbon storage potential there was. However, from a farmer's perspective minimal biochar amounts are desirable due to economic reasons. The economic cost of biochar is in a range of \$200–\$2,000 per Mg (worldwide, data from online market research). In addition, companies being able to supply more than 1 Mg per day are still rare in Europe [11].

Due to our knowledge, little is known on threshold amounts of biochar for positive agronomic effects. Only one other study is published with similarly small biochar application amounts, still this is not comparable to our setup as they calculated per hectare amounts but applied the biochar in relatively small bands only surrounding the sown seeds (approximately one Mg ha^{-1}[12]).

Our study was designed by combining the knowledge of synergistic effects that composting has on biochar with the need to find no effect level (NOEL) for biochar amendments. Therefore, we investigated the effects of both (i) biochar addition rate and (ii) co-composted biochar application amount on oat (*Avena sativa* L.) yield. We hypothesized that (1) co-composted biochar amended soil increases the TOC (with positive effects on soil water status); (2) retains more nutrients in the available form and (3) results in higher crop yields.

2. Materials and Methods

2.1. Soil Substrates

For our study we used a sandy and a loamy substrate which had not been used for agricultural purposes prior to the experiment. The substrates were collected at Kiesgrube ZAPF, Weidenberg, Germany and Ökologisch Botanischer Garten, University of Bayreuth, Germany, respectively. Selected basic properties of soil substrates are given in Table 1. The very poor sandy substrate (which was washed sand-mix originally intended for concrete mixes) was representative of nutrient-poor infertile soil, while the loamy substrate represented soils with sufficient nutrient supply common in Central Europe. Strongly contrasting contents of organic material and clay size particles of the two substrates were supposed to induce different responses comparable to natural soil types.

Table 1. Chemical composition of the two soil substrates and the biochar composts are shown. "CO" is compost without biochar. The number following "BC-" denotes the approximate fraction of biochar in the composted product as "kg biochar per Mg". "n.a." means not analyzed. "BET" is BET surface area, "±se" means plus minus standard error ($n = 5$).

	Al	Ca	K	Mg	Na	P	Biochar	TOC	N	C/N	Ash	NO3	NH4	BET ± se
	[g kg^{-1}]	[g kg^{-1}]	[g kg^{-1}]	[g kg^{-1}]	[g kg^{-1}]	[g kg^{-1}]	[g kg^{-1}]	[g kg^{-1}]	[g kg^{-1}]			[g kg^{-1}]	[g kg^{-1}]	[m^2 g^{-1}]
Sand	0.068	0.118	0.008	0.025	0.007	0.008	0	0.96	n.a.	n.a.	n.a.	n.a.	n.a.	n.a.
Loam	0.683	2.511	0.202	0.333	0.030	0.091	0	16.09	n.a.	n.a.	n.a.	n.a.	n.a.	n.a.
CO	11.0	3390	302	103	45.4	24.7	0	112.83	9.46	13.41	78.40	0.25	0.06	2.3 ± 0.3
BC-03	7.9	3610	312	103	46.7	22.0	3	120.52	9.85	13.63	77.80	0.32	0.04	11.6 ± 0.8
BC-05	8.0	3510	292	100	44.4	25.3	5	117.36	9.69	13.48	77.40	0.34	0.03	12.7 ± 0.1
BC-10	7.8	3590	325	107	51.0	23.9	10	122.11	9.43	14.31	76.20	0.36	0.06	12.9 ± 0.7

2.2. Biochar Composts

The biochar was an activated carbon from a commercial producer (carbopal®, Donau Carbon GmbH, Frankfurt, HE, Germany, ash content <6%, bulk density ~0.6 g/cm^3, surface area ~900 m^2/g, specific surface 1200 m^2/g, bulk density ~375 kg/m^3). Compost input material consisted of 50% sewage sludge (25% dry matter), 35% chopped wood (60% dm) and 15% rest soil or woody debris (leftovers from composting). After piling 20 Mg compost raw material to six meter wide and three meter high piles for two weeks, the piles were diverted into three meter wide and 1.5 m high mounds and mixed twice a week. After the biochar was added to respective piles in amounts of 3.5 and 10 kg biochar per Mg compost (BC-03, BC-05 and BC-10, respectively) and composted together for two weeks (mixed once a week) before the final phase of composting was induced by piling six meter wide and three meter high mounds (mixed every third week). Properties of individual biochar-amended composts are given in Table 1.

2.3. Greenhouse Experiment

The study was set up in a greenhouse at an average temperature of around 22 °C, with 200 mL of water irrigation every other day, and constant light conditions (400 W sodium discharge lamp, 8 h per day) for the whole duration of the experiment. For the experiment, we used commercial plastic pots with a total volume of 1000 cm^3 and a diameter of 13 cm, with a surface area of 133 cm^2. The perforated bottoms were covered with fine gauze, hindering the loss of particulate matter but allowing leaching of water. One kilogram of dry matter of the substrate was placed in the pots. The biochar compost types were applied in five application rates (equivalent to 10, 50, 100, 150, 250 Mg ha^{-1} 20 cm^{-1} in five replicates); hence, the respective biochar component application rates were between 0.03 and 2.5 Mg ha^{-1} (Figure 1). Soil samples were taken at time zero, after mixing and before sowing. All pots were arranged in a randomized block design and 10 oat (*Avena sativa* L.) seeds were sown in each pot, similar to common oat sowing in the field at 500–700 seeds per square meter. The survival rate was noted at harvest time and plants were cut just above the ground leading to the biomass data. Seeds were separated manually afterwards and weighed separately.

2.4. Soil and Plant Analyses

Three months after sowing, the plants' heights were recorded and we harvested above-ground biomass. Plant biomass was dried at 65 °C and then weighted. Results were scaled up to Mg ha^{-1} using the pot surface area. Composted biochars and soil samples were analyzed using the Mehlich-III-extraction method [13]. To do so, 2.5 g of soil was passed through a 2 mm sieve into 125 mL Erlenmeyer flasks, and 30 mL of Mehlich-III-extractant (0.2 M CH_3COOH, 0.25 M NH_4NO_3, 0.015 M NH_4F, 0.013 M HNO_3 and 0.001 M EDTA.) added. The suspension was shaken for 5 min on a rotating shaker with 120 rpm. After filtrating through No. 42 Whatman filter paper, filtrates were analyzed by ICP–OES (BayCEER, University of Bayreuth). Total organic carbon (TOC) and total nitrogen (N) were measured by dry combustion with a VARIOMAX CNS elemental analyzer (Elementar, Hanau, Germany).

Figure 1. Individual amounts of applied compost and biochar (CO = pure compost, BC-03 = compost with 3 kg Mg^{-1} *w/w* biochar, BC-05 = compost with 5 kg Mg^{-1} *w/w* biochar, BC-10 = compost with 10 kg Mg^{-1} *w/w* biochar) at 5 application amounts (10, 50, 100, 150, 250 Mg ha^{-1}) calculated as per hectare amounts (in Mg ha^{-1}).

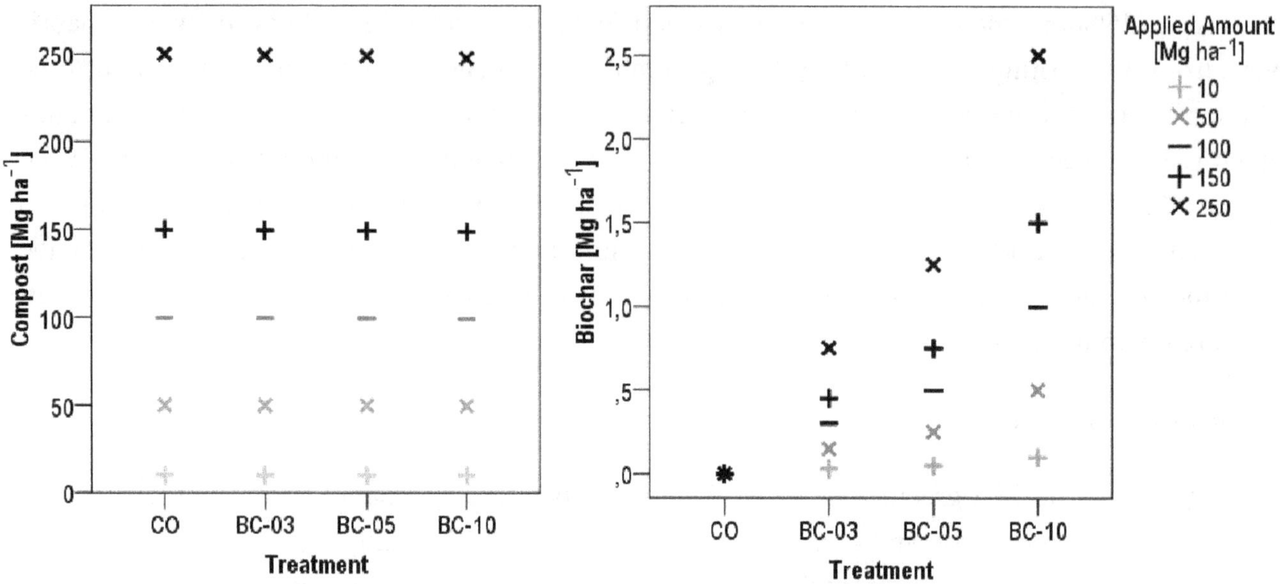

2.5. Statistical Analysis

Data were analyzed using simple linear regressions (SLR) with the equivalent per hectare amounts of composted biochars or composts to analyze biochar and compost effects separately; regression coefficients are indicated if significant (justification for this procedure is found in setup description, Figure 1). Asterisks *, **, *** indicate $p < 0.05$, $p < 0.01$, $p < 0.001$, respectively; not significant data is indicated by "n.s." in the tables. The values behind "±"symbols in the text represent one standard error of the mean ($n = 5$). All analyses were performed with SPSS Statistics 17 (IBM).

3. Results

3.1. Plant Growth

3.1.1. Oat Grain Yield

The oat grain yield ranged between 0.00 and 0.14 Mg ha^{-1} on sandy substrate (control: 0.02 ± 0.00 Mg ha^{-1}) and between 0.04 and 0.19 on loamy substrate (control: 0.06 ± 0.00 Mg ha^{-1}; Figure 2). Compost significantly increased grain yield (sandy: $p < 0.001$; loamy substrate: $p = 0.001$; Table 2), while no effect of biochar on oat yield could be proven ($p > 0.05$ at all applied amounts and on both substrates; Table 2).

Figure 2. Grain biomass (**top**) and plant biomass (**bottom**) of oat (Avena sativa L.) in Mg ha^{-1} on sandy (**left**) and loamy substrate (**right**) depicted for five treatments (CO = pure compost, BC-03 = compost with 3 kg Mg^{-1} *w/w* biochar, BC-05 = compost with 5 kg Mg^{-1} *w/w* biochar, BC-10 = compost with 10 kg Mg^{-1} *w/w* biochar, in five application amounts (10, 50, 100, 150, 250 Mg ha^{-1}) *versus* control (CTRL = no amendment) ($n = 5$).

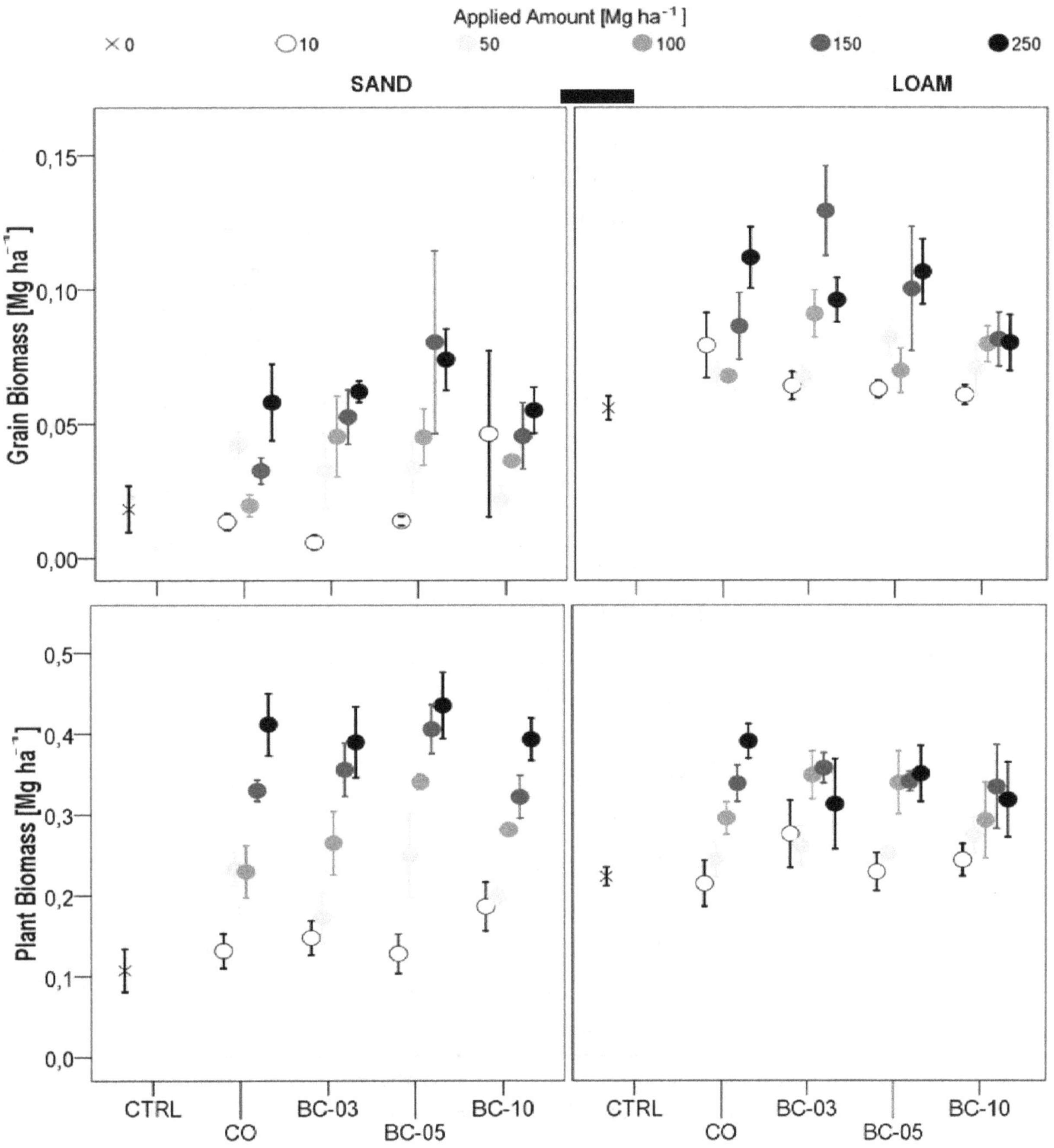

Table 2. Linear regression of plant and soil data calculated with per hectare amounts of the applied biochar composts. "CO" stands for regressions with the compost amounts and the variables, "BC" for the biochar amount and the variables. If "BC" had significant influence on the variables, the respective application amount is indicated by the superscript number.

Substrate	Variable	Regression (CO)		Regression (BC)	
Sand	Seed yield	$1.85 + 0.02 \times CO$	***	n.s	
	Biomass	$16.48 + 0.10 \times CO$	***	n.s.	
	Plant height	$73.00 + 0.11 \times CO$	***	$101.31 - 8.11 \times BC^{150}$	*
	TOC	$2.71 + 0.01 \times CO$	***	$0.97 + 33.50 \times BC^{10}$	**
	TN	$0.22 + 0.00 \times CO$	***	n.s.	
	pH	$8.54 + 0.00 \times CO$	***	n.s.	
	P	$0.056 + 0.001 \times CO$	***	n.s.	
	K	$0.009 + 0.000 \times CO$	***	$0.018 - 0.130 \times BC^{10}$ $0.011 + 0.011 \times BC^{50}$ $0.030 + 0.008 \times BC^{250}$	* * *
	Mg	$0.043 + 0.000 \times CO$	***	$0.079 + 0.015 \times BC^{250}$	*
	Ca	$0.574 + 0.008 \times CO$	***	n.s.	
	Na	$0.011 + 0.000 \times CO$	***	n.s.	
	Al	$0.105 - 0.000 \times CO$	n.s.	n.s.	
Loam	Seed yield	$6.66 + 0.01 \times CO$	**	n.s	
	Biomass	$26.18 + 0.04 \times CO$	***	$41.04 - 4.69 \times BC^{250}$	*
	Plant height	$87.49 + 0.02 \times CO$	n.s.	n.s.	
	TOC	$18.99 + 0.03 \times CO$	***	$25.19 - 9.79 \times BC^{100}$	*
	TN	$1.66 + 0.00 \times CO$	***	n.s.	
	pH	$7.21 + 0.00 \times CO$	***	n.s.	
	P	$0.131 + 0.001 \times CO$	***	n.s.	
	K	$0.193 + 0.000 \times CO$	***	$0.240 - 0.662 \times BC^{50}$	**
	Mg	$0.345 + 0.000 \times CO$	***	n.s.	
	Ca	$2.942 + 0.009 \times CO$	***	$3.105 + 0.897 \times BC^{50}$	*
	Na	$0.04 + 0.000 \times CO$	***	$0.043 - 0.174 \times BC^{10}$	**
	Al	$0.668 - 0.000 \times CO$	***	n.s.	

Significant differences are marked with asterisks: *, **, *** indicate $p < 0.05$, $p < 0.01$, $p < 0.001$, respectively; n.s. indicates "not significant". Seed yield = separated seeds, Biomass = complete above ground biomass.

3.1.2. Plant Biomass

Total above-ground biomass yield ranged between 0.02–0.54 Mg ha^{-1} on sandy substrate (control: 0.11 ± 0.01 Mg ha^{-1}) and between 0.10–0.48 Mg ha^{-1} on loamy substrate (control: 0.22 ± 0.01 Mg ha^{-1}; Figure 2). Compost application significantly increased oat biomass both on sandy ($p < 0.001$) and loamy substrates (Table 2). Biochar showed no significant effect on plant biomass on sandy substrate, while on loamy substrate biomass yield was significantly lower at the highest applications amounts (250 Mg ha^{-1}; $p = 0.04$) but no clear tendency was detected looking at increasing biochar amounts (Table 2).

3.1.3. Plant Height

Plant height increased on both substrate types with nearly all amendments resulting in heights between 31.0–119.0 cm on sandy substrate (control: 62.6 ± 3.1 cm) and between 50.0–122.0 cm on loamy substrate (control: 78.6 ± 3.5 cm; Figure 3). Raising the total amounts of compost significantly increased plant heights only on sandy substrate ($p < 0.001$), on loamy substrate the effect was only visible as a tendency ($p = 0.15$; Table 2). Biochar showed only significantly negative effect on plant heights in one application amount on sandy substrate (150 Mg ha^{-1}; $p = 0.04$), leading to the conclusion there was no trend or tendency of biochar influencing plant heights.

Figure 3. Plant height of oat (*Avena sativa* L.) in cm on sandy (**top**) and loamy substrate (**bottom**) depicted for five treatments (CO = pure compost, BC-03 = compost with 3 kg Mg^{-1} *w/w* biochar, BC-05 = compost with 5 kg Mg^{-1} *w/w* biochar, BC-10 = compost with 10 kg Mg^{-1} *w/w* biochar) in five application amounts (10, 50, 100, 150, 250 Mg ha^{-1}) *versus* control (CTRL = no amendment) ($n = 5$).

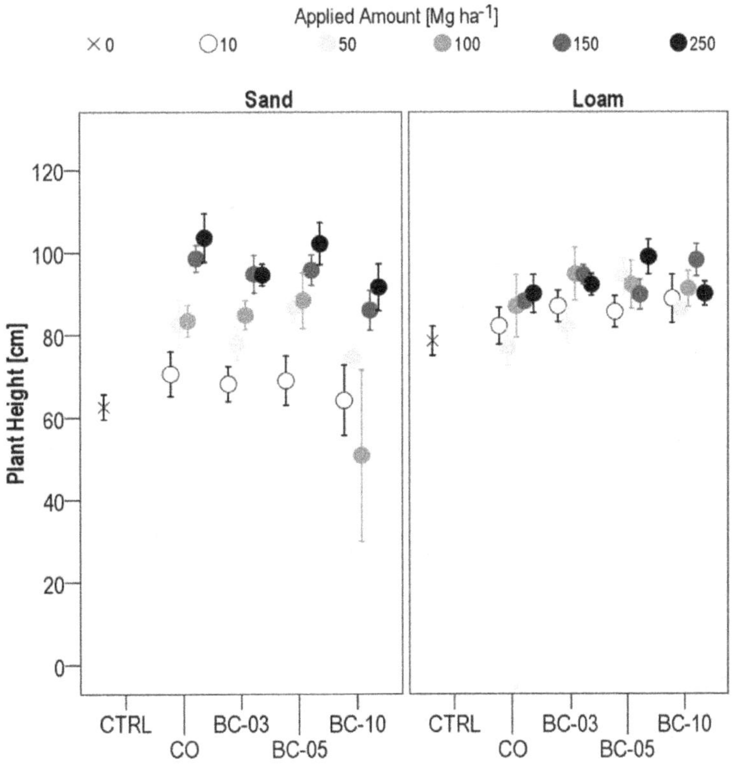

3.2. Changes in Soil Properties

3.2.1. Total Organic Carbon (TOC)

The TOC contents of sandy substrate ranged between 0.2 and 8.9 g kg^{-1} (control: 1.1 ± 0.3 g kg^{-1}) and between 4.0 and 31.1 g kg^{-1} on loamy substrate (control: 18.1 ± 0.4 g kg^{-1}; Figure 4). Compost amendments significantly increased TOC contents on both sandy and on loamy substrates ($p < 0.001$), while no significant biochar effect could be observed (Table 2).

3.2.2. Total Nitrogen (N_{tot})

N_{tot} ranged from 0.00–0.71 g kg^{-1} on sandy substrate (control: 0.0 ± 0.0 g kg^{-1}) and from 0.20–2.49 g kg^{-1} on loamy substrate (control: 1.52 ± 0.27 g kg^{-1}; Figure 5). Significant influence on N_{tot} content was proven for compost on sandy and loamy substrate equally ($p < 0.001$; Table 2). Differences between the compost and the respective biochar compost applications were marginal and not significant; hence the applied low amounts of biochar did not influence N_{tot}. (Figure 5).

Figure 4. Total organic carbon (TOC) in g kg^{-1} on sandy (**top**) and loamy substrate (**bottom**) depicted for five treatments (CO = pure compost, BC-03 = compost with 3 kg Mg^{-1} *w/w* biochar, BC-05 = compost with 5 kg Mg^{-1} *w/w* biochar, BC-10 = compost with 10 kg Mg^{-1} *w/w* biochar) in five application amounts (10, 50, 100, 150, 250 Mg ha^{-1}) *versus* control (CTRL = no amendment) (n = 5).

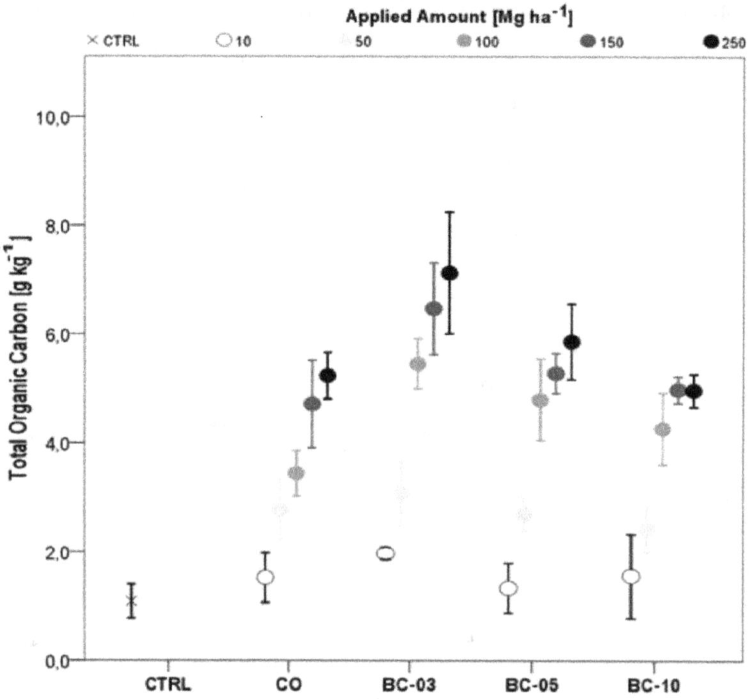

Figure 4. *Cont.*

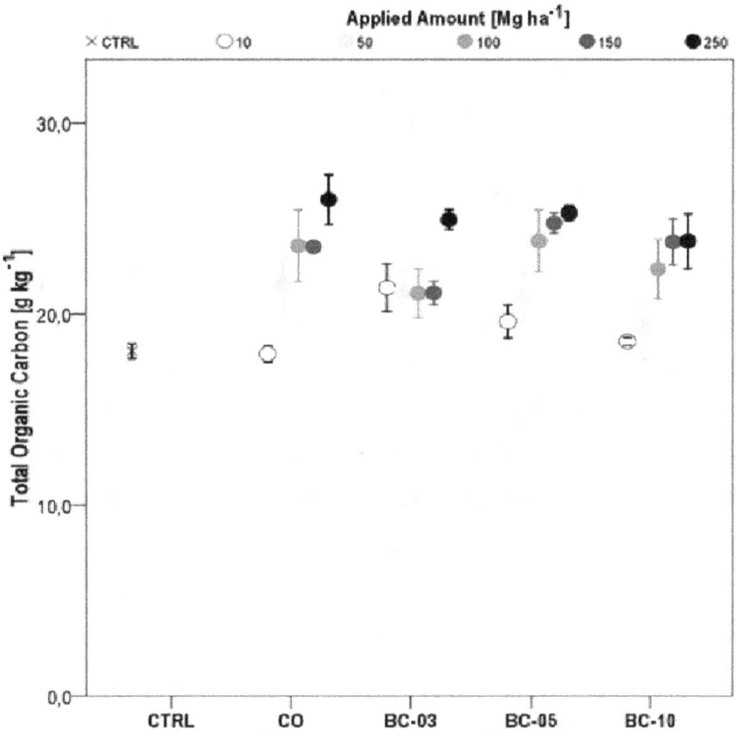

Figure 5. Total nitrogen (N_{tot}) in g kg^{-1} on sandy (**a**) and loamy substrate (**b**) depicted for 5 treatments (CO = pure compost, BC-03 = compost with 3 kg Mg^{-1} *w/w* biochar, BC-05 = compost with 5 kg Mg^{-1} *w/w* biochar, BC-10 = compost with 10 kg Mg^{-1} *w/w* biochar) in five application amounts (10, 50, 100, 150, 250 Mg ha^{-1}) *versus* control (CTRL = no amendment) (n = 5).

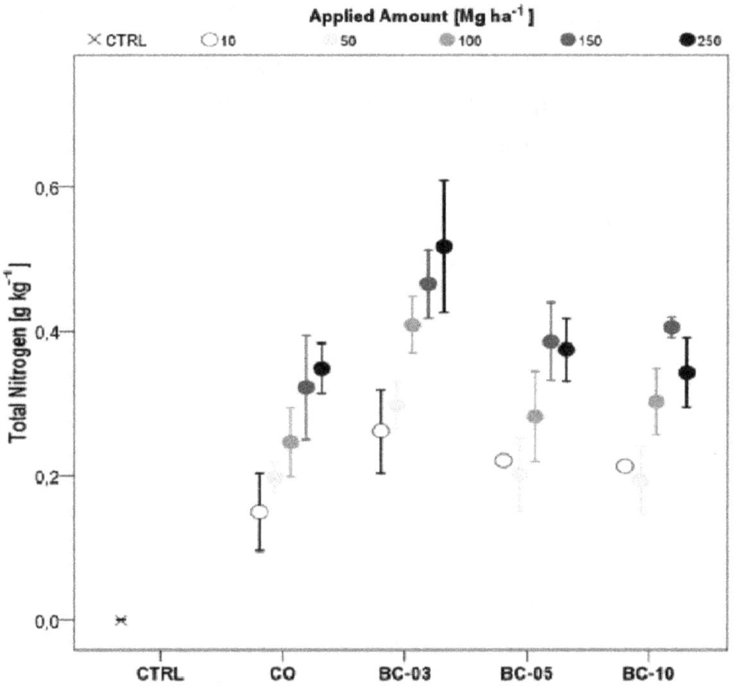

3.2.3. Soil Reaction (pH)

PH values ranged from 6.95–8.80 (sandy substrate control: 8.53 ± 0.04; mean: 8.33 ± 0.03) and 6.76–8.14 (loamy substrate control: 6.88 ± 0.04; mean: 7.31 ± 0.02). Alkalinity (rising pH) was significantly influenced to a similar degree in both substrates by compost ($p < 0.001$).

3.2.4. Plant-Available Nutrients and Aluminum

Compost amendment enriched both substrate types significantly with phosphorus ($p < 0.001$ on both substrates) boosting the phosphorus (P) content by factors of 2.3–30.1 compared to sandy control with factors of 1.2–3.4 compared to loamy control; however, there was no biochar effect. The contents of available potassium (K) were elevated by factors of 1.4–3.0 on sand which was very significant in relation to compost additions; biochar amendments were proven to elevate K contents significantly at 50 and 250 Mg ha^{-1} application amounts while they showed a negative impact at 10 Mg ha^{-1} which brings us to the conclusion that there is no clear effect of biochar on K status in sandy substrate. Potassium load was increased only by factors 1.0–1.2 on loamy substrate, where compost contents significantly increased K at all application amounts and biochar amounts at 50 Mg ha^{-1} significantly decreased K with no other statistically significant influences in biochar. Plant-available calcium (Ca), magnesium (Mg) and sodium (Na) contents were elevated with the highest statistical significance by the compost content of our amendments on both substrates ($p < 0.001$, respectively); on sandy substrate biochar showed one exceptional significant response and elevated Mg contents at one particular application level (Figure 6, Table 2) while biochar increased Ca and decreased Na content significantly at one particular application level in each case on loamy substrate (Figure 7, Table 2). Contents of available Aluminum (Al) decreased the more compost was added to our two substrates ($p < 0.001$ respectively); biochar did not show an effect that was statistically discernible on both substrates. Calcium content rose significantly after all applications especially on sandy substrate, leading to 17.7 times higher Ca contents at the highest application amounts, whereas on loamy substrate the factor was 2.1 at the same rate. This definitely had a positive influence on the Al-Ca-ratio, neutralizing the Aluminum. Ratios of Al to Ca were not critical to plant growth at any treatment level whatsoever.

Figure 6. Plant-available nutrients and Aluminum (in cmolc kg^{-1} soil) on sandy substrate depicted for five treatments (CO = pure compost, BC-03 = compost with 3 kg Mg^{-1} *w/w* biochar, BC-05 = compost with 5 kg Mg^{-1} *w/w* biochar, BC-10 = compost with 10 kg Mg^{-1} *w/w* biochar) in five application amounts (10, 50, 100, 150, 250 Mg ha^{-1}) *versus* control (CTRL = pure sandy substrate) (n = 5).

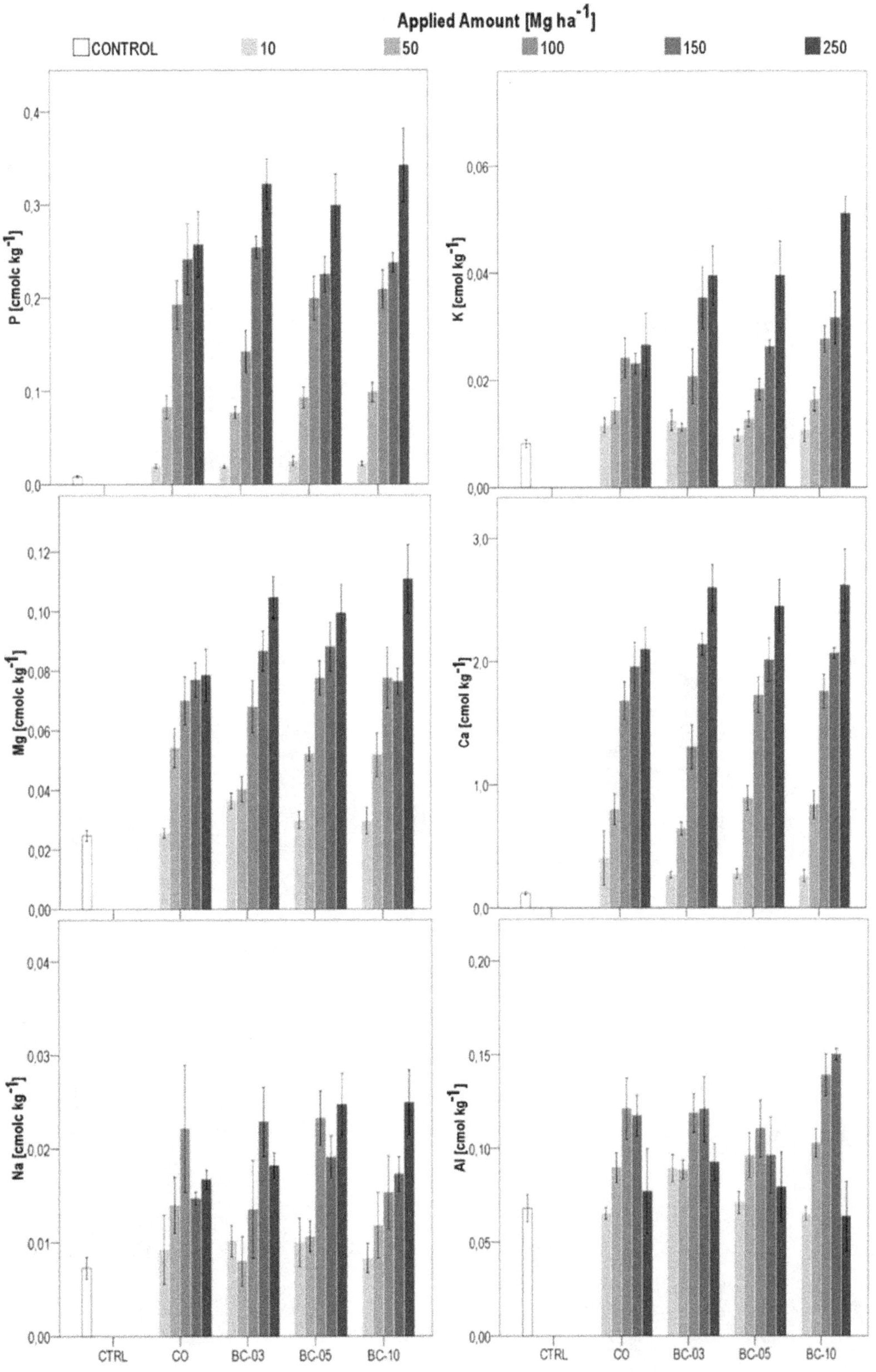

Figure 7. Plant-available nutrients and Aluminum (in cmolc kg^{-1} soil) on loamy substrate depicted for five treatments (CO = pure compost, BC-03 = compost with 3 kg Mg^{-1} *w/w* biochar, BC-05 = compost with 5 kg Mg^{-1} *w/w* biochar, BC-10 = compost with 10 kg Mg^{-1} *w/w* biochar) in five application amounts (10, 50, 100, 150, 250 Mg ha^{-1}) *versus* control (CTRL = pure loamy substrate) (n = 5).

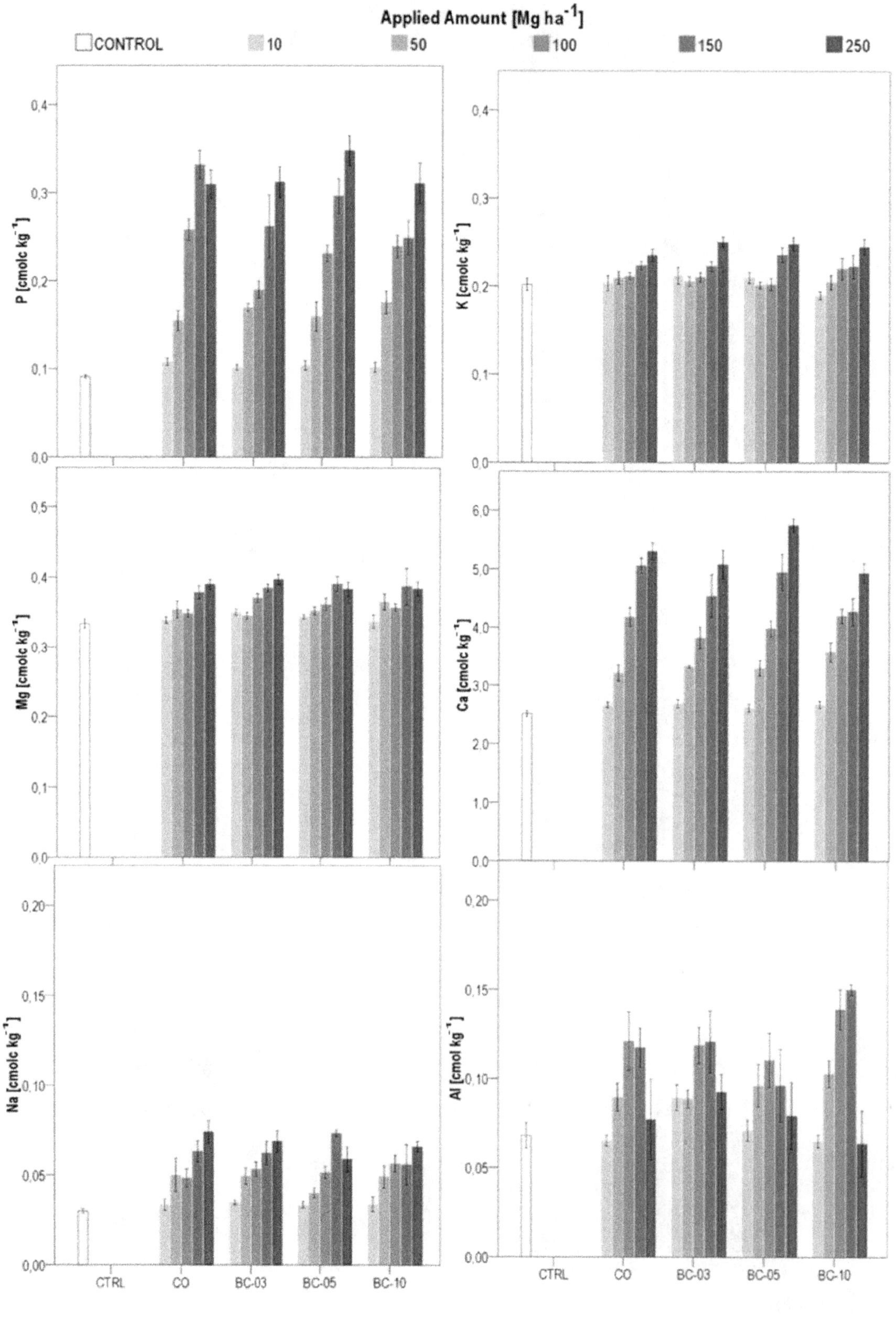

4. Discussion

Plant growth significantly increased with increasing compost amendment in both soil substrates (Figures 1–3). However, we could not prove any biochar effect on plant growth in our study which is in contrast to most other reported biochar research [2–4,7]. This is probably due to the extremely low amounts of biochar of 0.03–2.5 Mg ha^{-1} used in the different compost application amounts. The biochar effect is masked by compost. Additionally, a special type of biochar was used (activated carbon) which is known to be valuable for element sorption but perhaps this is not the case in a plant-available form. Another reason why we could not detect a significant influence by the biochar could be the limited duration of our trial. Several authors discussed reactions of biochar in soils over time increasing its impact through surface oxidation and bio-activation with soil microbes and fungi growing on the biochar [14–16].

Plant growth results of the different biochar composts showed increases in much larger magnitudes on sandy substrate than on loamy substrate, which was suggested by [17] who wrote that soil fertility of poorer soils would improve more in reaction to organic amendments. The different reactions of the two soil substrates could be also proven in a further greenhouse study by Schulz and Glaser [7] by using similar soil substrates and gaining similar results comparing the soil substrates' differing responses. In the study mentioned, we found the alterations of TOC, N_{tot}, soil reaction and plant-available nutrients appearing in much bigger orders on sandy substrate following compost and composted biochar applications. This difference in the effects could be related to the low baseline of the pure sand regarding initial nutrient status, clay minerals and organic components. It could also be connected to the initially high soil reaction of the sandy substrate (pH around 8 in sandy substrate, contrasting a pH around 7 in loamy substrate).

It is difficult to relate the results of our minimal biochar additions to the frequently published proofs that biochar applications to soil increase agricultural productivity (e.g. [1,3,4,18–20] due to the higher biochar application amounts used in these studies and because their biochar effects were not masked with the compost effects. Steiner *et al.* [21] reported cumulative yield increases of rice and sorghum on a Brazilian Amazon Oxisol of approximately 75% after four growing seasons over two years, when 11 Mg ha^{-1} biochar was applied at the beginning of the experiment. In a degraded Kenyan Oxisol, Kimetu *et al.* [22] found a doubling of cumulative maize yield after three repeated biochar applications of 7 Mg ha^{-1} over two years corresponding to a total of 21 Mg ha^{-1}.

If biochar was applied in higher amounts than in our study, soil nutrient availability has repeatedly been increased in highly weathered tropical soils comparable (Lehmann *et al.* [23] with ~560 Mg ha^{-1}; Lehmann *et al.* [18] with 67.6–135.2 Mg ha^{-1}; Steiner *et al.* [21] 2008 with 11 Mg ha^{-1}). Similar amounts as in our study were tested in the trial from Iswaran *et al.* [24] where they showed increased biomass production in a poor sandy soil after adding small amounts of charcoal of 0.5 Mg ha^{-1} together with sufficient artificial fertilization. The positive effect of charcoal was attributed to its positive effect on Rhizobium abundance by poisoning Rhizobium antagonists with charcoal inherent phenolic substances. As we did not apply legumes and, furthermore, did not experience other negative effects of biochar induced poisoning of soil biota, we cannot relate the data from Iswaran *et al.* [24] to our results.

In many studies, biochar incorporation has been shown to induce soil alkalization which can increase soil nitrification [18,25–30], moreover also the high sorption capacity caused by aromaticity of the biochar could have an influence on nutrient cycling[1]—none of these effects could be achieved by our small application amounts in relation to amounts of compost added and the initial alkaline substrates. Neither did the increased porosity (indicated by the BET surfaces of the co-composted biochars, Table 1) significantly influence the sorption capacity as suggested by the marginal and non-linear differences in our nutrient data.

The compost addition positively and significantly influenced plant growth and soil properties as expected after long-term experience in compost applications [30,31]. Compost improved oat yield significantly stronger on sandy substrate than on loamy substrate, which could be attributed to the very low content of nutrients and organic matter in the pure sandy substrate where any low amendment would alter the conditions for plant growth [7]. Nitrogen loads of our compost products were designed for optimum nitrogen supply from the first year on, because—unlike natural/agricultural conditions—we did not need to consider water protection guidelines (adding 100–2500 kg N ha^{-1} at one time, as we did, would be far above the European guidelines). The same total application amounts of composted biochars (BC-03, -05, -10) and the pure composts (CO) improved the soils to a similar degree; there are no statistical differences regarding plant biomass or seed yield, nutrient loads, organic matter or soil reaction between the treatments containing biochar and those that lack of it. Clearly, we owe the effects our amendments had on all measured parameters to the compost shares of our amendments. We attribute this absent biochar effect to the low amounts of added biochar (<3 Mg ha^{-1}). It can be stated that investments for biochar amendments below €2,000 per hectare are irrelevant for improving plant growth and soil quality at given actual costs for biochar of around €300–800 per Mg biochar. Farmers' costs could be lowered if the biochar is produced locally and from farmyard waste or in a projected future when biochar would be accounted for actual carbon offset. Around €27.600 per hectare would be necessary to invest for the biochar application amounts which showed the biggest effect on grain yield (*Avena sativa* L.) in the study from Schulz and Glaser [9]. There, the strongest effect on grain yield (*Avena sativa* L.) was measured after applications of composted biochar comprising of 92 Mg biochar ha^{-1} and 107 Mg compost ha^{-1} (leading to a 300% higher yield on sandy substrate compared to the pure compost) leaving us with impossible investments for farmers. The meta-analysis study of Jeffery *et al.* [4] marked the best results at application amounts of 100 Mg biochar ha^{-1}, which requires investments of money no farmer would spend easily. One feasible option might be the application of 1 Mg every year until a certain stock is established, or as discussed in Blackwell *et al.* [12] in form of bandings and thereby closer to the plants growing space. Agronomic considerations including increased crop productivity, reduced fertilizer and pesticide use need to be made at the farm scale.

5. Conclusions

We proved that low level biochar applications had no immediate effects on plant growth and soil fertility both in sandy and loamy soils. Our data suggests that co-composted biochar application could only be a better way to enhance plant yields and soil parameters if applied in doses higher than

2.5 Mg ha^{-1} or applied differently, e.g. as suggested by Blackwell [32], or loaded with nutrients (biochar activation). We found no negative effects of the applied activated carbon.

Due to the proclaimed longevity of the biochar in soils, all commercial "Terra Preta" producers should be obliged to thoroughly test their products and to provide convincing results of the claimed benefits, e.g. by providing scientific results with proper experimental setup and statistical design.

Acknowledgments

The authors acknowledge the German Ministry for Education and Research (BMBF) for financial support within the coordinated project "Climate protection: CO_2 sequestration by use of biomass in a PYREG reactor with steam engine" (01LY0809F). We are indebted to Jie Liu for the lab work, Daniel Fischer for compost analyses and to Ananda Erben and Georg Lemmer for help at the greenhouse.

Conflicts of Interest

The authors declare no conflict of interest.

References

1. Glaser, B.; Lehmann, J.; Zech, W. Ameliorating physical and chemical properties of highly weathered soils in the tropics with charcoal—A review. *Biol. Fertil. Soils* **2002**, *35*, 219–230.
2. Sohi, S.P.; Krull, E.; Lopez-Capel, E.; Bol, R. A review of biochar and its use and function in soil. *Adv. Agron.* **2010**, *105*, 47–82.
3. Waters, D.; Zwieten, L.; Singh, B.; Downie, A.; Cowie, A.; Lehmann, J. Biochar in Soil for Climate Change Mitigation and Adaptation. In *Soil Health and Climate Change*; Singh, B.P., Cowie, A.L., Chan, K.Y., Eds.; Springer: Berlin/Heidelberg, Germany, 2011; pp. 345–368.
4. Jeffery, S.; Verheijen, F.G.A.; van der Velde, M.; Bastos, A.C. A quantitative review of the effects of biochar application to soils on crop productivity using meta-analysis. *Agric. Ecosyst. Environ.* **2011**, *144*, 175–187.
5. Chan, K.Y.; Xu, Z. Biochar: Nutrient Properties and Their Enhancement. In *Biochar for Environmental Management: Science and Technology*; Lehmann, J., Joseph, S., Eds.; Earthscan: London, UK, 2009; pp. 67–84.
6. Singh, B.; Singh, B.P.; Cowie, A.L. Characterisation and evaluation of biochars for their application as a soil amendment. *Soil Res.* **2010**, *48*, 516–525.
7. Schulz, H.; Glaser, B. Effects of biochar compared to organic and inorganic fertilizers on soil quality and plant growth in a greenhouse experiment. *J. Plant Nutr. Soil Sci.* **2012**, *175*, 410–422.
8. Fischer, D.; Glaser, B. Synergisms between Compost and Biochar for Sustainable Soil Amelioration. In *Management of Organic Waste*; Kumar, S., Bharti, A., Eds.; Intech: Shanghai, China, 2012.
9. Schulz, H.; Dunst, G.; Glaser, B. Positive effects of composted biochar on plant growth and soil fertility. *Agron. Sustain. Dev.* **2013**, *33*, 817–827.

10. Wiedner, K.; Baumgartl, M.-L.; Favilli, F.; Criscuoli, I.; Walther, S.; Fischer, D.; Miglietta, F.; Glaser, B. Surface Oxidation of Modern and Fossil Biochars. In Proceedings of the Eurosoil, Bari, Italy, 2–6 July 2012.
11. Wiedner, K.; Naisse, C.; Rumpel, C.; Pozzi, A.; Wieczorek, P.; Glaser, B. Chemical modification of biomass residues during hydrothermal carbonization—What makes the difference, temperature or feedstock? *Org. Geochem.* **2013**, *54*, 91–100.
12. Blackwell, P.; Krull, E.; Butler, G.; Herbert, A.; Solaiman, Z. Effect of banded biochar on dryland wheat production and fertiliser use in south-western Australia: An agronomic and economic perspective. *Soil Res.* **2010**, *48*, 531–545.
13. Mehlich, A. Mehlich 3 soil test extractant: A modification of Mehlich 2 extractant. *Commun. Soil Sci. Plant Anal.* **1984**, *15*, 1409–1416.
14. Ding, W.-C.; Zeng, X.-L.; Wang, Y.-F.; Du, Y.; Zhu, Q.-X. Characteristics and performances of biofilm carrier prepared from agro-based biochar. *China Environ. Sci.* **2011**, *31*, 451–1455.
15. Nguyen, B.T.; Lehmann, J.; Hockaday, W.C.; Joseph, S.; Masiello, C.A. Temperature sensitivity of black carbon decomposition and oxidation. *Env. Sci. Tec.* **2010**, *44*, 3324–3331.
16. Cheng, C.-H.; Lehmann, J.; Engelhard, M.H. Natural oxidation of black carbon in soils: Changes in molecular form and surface charge along a climosequence. *Geochimica et Cosmochimica Acta* **2005**, *72*, 1598–1610.
17. Glaser, B.; Birk, J.J. State of the scientific knowledge on properties and genesis of Anthropogenic Dark Earths in Central Amazonia (terra preta de Índio). *Geochim. Cosmochim. Acta* **2012**, *82*, 39–51.
18. Lehmann, J.; Pereira da Silva, J.; Steiner, C.; Nehls, T.; Zech, W.; Glaser, B. Nutrient availability and leaching in an archaeological Anthrosol and a Ferralsol of the Central Amazon basin: Fertilizer, manure and charcoal amendments. *Plant Soil* **2003**, *249*, 343–357.
19. Marris, E. Putting the carbon back: Black is the new green. *Nature* **2006**, *442*, 624–626.
20. Blackwell, P.; Riethmuller, G.; Collins, M. Biochar Application to Soil. In *Biochar for Environmental Management: Science and Technology*; Lehmann, J., Joseph, S., Eds.; Earthscan: London, UK, 2009; 67–84.
21. Steiner, C.; Teixeira, W.; Lehmann, J.; Nehls, T.; de Macêdo, J.; Blum, W.; Zech, W. Long term effects of manure, charcoal and mineral fertilization on crop production and fertility on a highly weathered Central Amazonian upland soil. *Plant Soil* **2007**, *291*, 275–290.
22. Kimetu, J.M.; Lehmann, J.; Ngoze ,S.O.; Mugendi, D.N.; Kinyangi, J.M.; Riha, S.; Verchot, L.; Recha, J.W.; Pell, A.N. Reversibility of soil productivity decline with organic matter of differing quality along a degradation gradient. *Ecosystems* **2008**, *11*, 726–739.
23. Lehmann, J.; da Silva, J.P., Jr.; Rondon, M.; Cravo, M.S.; Greenwood, J.; Nehls, T.; Steiner, C. Slash-and-char—A Feasible Alternative for Soil Fertility Management in the Central Amazon. In Proceedings of the 17th World Congress of Soil Science, Bangkok, Thailand, 14–21 August 2002.
24. Iswaran, V.; Jauhri, K.S.; Sen, A. Effect of charcoal, coal and peat on the yield of moong, soybean and pea. *Soil Biol. Biochem.* **1980**, *12*, 191–192.
25. Yamato, M.; Okimori, Y.; Wibowo, I.; Anshori, S.; Ogawa, M. Effects of the application of charred bark of Acacia mangium on the yield of maize, cowpea and peanut, and soil chemical properties in South Sumatra, Indonesia. *J. Soil Sci. Plant Nutr.* **2006**, *52*, 489–495.

26. DeLuca, T.H.; MacKenzie, M.D.; Gundale, M.J. Biochar Effects on Soil Nutrient Transformations. In *Biochar for Environmental Management: Science and Technology*; Lehmann, J., Joseph, S., Eds.; Earthscan: London, UK, 2009; pp. 251–270.
27. Topoliantz, S.; Ponge, J.; Ballof, S. Manioc peel and charcoal: A potential organic amendment for sustainable soil fertility in the tropics. *Biol. Fertil. Soils* **2005**, *41*, 15–21.
28. Oguntunde, P.G.; Fosu, M.; Ajayi, A.E.; Giesen, N. Effects of charcoal production on maize yield, chemical properties and texture of soil. *Biol Fertil Soils* **2004**, *39*, 295–299.
29. Hua, L.; Wu, W.X.; Liu, Y.X.; McBride, M.; Chen, Y.X. Reduction of nitrogen loss and Cu and Zn mobility during sludge composting with bamboo charcoal amendment. *Environ. Sci. Pollut. R.* **2009**, *16*, 1–9.
30. Amlinger, F.; Peyr, S.; Geszti, J.; Dreher, P.; Karlheinz, W.; Nortcliff, S. Beneficial Effects of Compost Application on Fertility and Productivity of Soils. In *Federal Ministry for Agricultural and Forestry, Environment and Water Management*; Lebensministerium: Vienna, Austria, 2007.
31. Diacono, M.; Montemurro, F. Long-term effects of organic amendments on soil fertility. *J. Sustain. Agric.* **2011**, *2*, 761–786.
32. Blackwell, P.; Shea, S.; Storer, P.; Solaiman, Z.; Kerkmans, M.; Stanley, I. Improving Wheat Production with Deep Banded Oil Mallee Charcoal in Western Australia. In Proceedings of the First Asia Pacific Biochar Conference, Terrigal, Australia, 29 April–2 May 2007.

5

A Review of Nutrient Management Studies Involving Finger Millet in the Semi-Arid Tropics of Asia and Africa

Malinda S. Thilakarathna and Manish N. Raizada *

Department of Plant Agriculture, University of Guelph, 50 Stone Road East, Guelph, ON N1G 2W1, Canada; E-Mail: mthilaka@uoguelph.ca

* Author to whom correspondence should be addressed; E-Mail: raizada@uoguelph.ca

Academic Editor: Yantai Gan

Abstract: Finger millet (*Eleusine coracana* (L.) Gaertn) is a staple food crop grown by subsistence farmers in the semi-arid tropics of South Asia and Africa. It remains highly valued by traditional farmers as it is nutritious, drought tolerant, short duration, and requires low inputs. Its continued propagation may help vulnerable farmers mitigate climate change. Unfortunately, the land area cultivated with this crop has decreased, displaced by maize and rice. Reversing this trend will involve achieving higher yields, including through improvements in crop nutrition. The objective of this paper is to comprehensively review the literature concerning yield responses of finger millet to inorganic fertilizers (macronutrients and micronutrients), farmyard manure (FYM), green manures, organic by-products, and biofertilizers. The review also describes the impact of these inputs on soils, as well as the impact of diverse cropping systems and finger millet varieties, on nutrient responses. The review critically evaluates the benefits and challenges associated with integrated nutrient management, appreciating that most finger millet farmers are economically poor and primarily use farmyard manure. We conclude by identifying research gaps related to nutrient management in finger millet, and provide recommendations to increase the yield and sustainability of this crop as a guide for subsistence farmers.

Keywords: finger millet; nutrient management; nitrogen; phosphorus; potassium; micronutrients; farmyard manure; biofertilizers; organic fertilizer; integrated nutrient management

1. Introduction

Finger millet (*Eleusine coracana* (L.) Gaertn) is a major food crop of the semi-arid tropics of Asia and Africa and has been an indispensable component of dryland farming systems [1–3]. Its name is derived from the seedhead, which has the shape of human fingers (Figure 1). Locally, the crop is called ragi (India); koddo (Nepal); dagussa, tokuso, barankiya (Ehiopia); wimbi, mugimbi (Kenya); bulo (Uganda); kambale, lupoko, mawale, majolothi, amale, bule (Zambia); rapoko, zviyo, njera, rukweza, mazhovole, uphoko, poho (Zimbabwe); mwimbi, mbege (Tanzania); and kurakkan (Sri Lanka) [4]. The crop was domesticated in the highlands of Ethiopia and Uganda 5000 years ago, but reached India 3000 years ago [4,5]. Today, the crop is ranked fourth globally in importance among the millets, after sorghum, pearl millet, and foxtail millet [6]. It is cultivated in more than 25 countries, mainly in Africa (Ethiopia, Eritrea, Mozambique, Zimbabwe, Namibia, Senegal, Niger, Nigeria, and Madagascar) and Asia (India, Nepal, Malaysia, China, Japan, Iran, Afghanistan, and Sri Lanka) [7,8]. In India, finger millet is primarily grown in the states of Karnataka, Andhra Pradesh, Odisha, and Tamil Nadu [8]. In Eastern Africa, the major producers are Uganda, Ethiopia, and Kenya [9]. Finger millet production data for the last five years in major finger millet producing countries (India, Nepal, and Ethiopia) is listed in Table 1.

Figure 1. Illustrations of finger millet cultivation. Typical finger millet seed heads at a young stage (**A**) and at maturity (**B**) in farmers' fields in Nepal. The seed heads resemble the fingers of a human hand. (**C**) Finger millet growing in a terraced field on a smallholder farm in Nepal. (**D**) Mixed-cropping of finger millet with soybean in a terraced field of a smallholder farmer in Nepal. Picture sources: LI-BIRD photo bank.

Table 1. Production data for finger millet in selected major finger millet producing nations (2009/2010–2013/2014).

Country	Year	Area under cultivation ('000 ha)	Production ('000 tones)	Average yield (kg ha^{-1})	References
India	2009/2010	1268	1889	1489	[10]
	2010/2011	1286	2194	1705	[10]
	2011/2012	1176	1929	1641	[10]
	2012/2013	1179	1785	1514	[10]
	2013/2014	1138	1688	1483	[11]
Nepal	2009/2010	268.5	299.5	1116	[12]
	2010/2011	269.8	302.7	1122	[12]
	2011/2012	278.0	315.1	1133	[12]
	2012/2013	274.4	305.6	1114	[12]
	2013/2014	271.2	304.1	1121	[12]
Ethiopia	2009/2010	369.0	524.2	1421	[13]
	2010/2011	408.1	634.8	1556	[13]
	2011/2012	432.6	651.8	1507	[14]
	2012/2013	431.5	742.3	1720	[15]
	2013/2014	454.7	849.0	1867	[16]

Nutritionally, finger millet is primarily consumed as a porridge in Africa, but in South Asia as bread, soup, roti (flat bread), and to make beer [4]. Interestingly, new food products made from finger millets are also becoming popular among younger people, including noodles, pasta, vermicelli, sweet products, snacks, and different bakery products [17,18]. In some nutritional components, finger millet is a superior crop compared to some major cereal crops especially polished rice [17]. Among the other millets, finger millet has a high amount of calcium (0.38%), fiber (18%), phenolic compounds (0.3%–3%), and sulphur containing amino acids [17,19–21]. Finger millet also has high amounts of tryptophan, cysteine, methionine, and total aromatic amino acids compared to the other cereals, and thus is an important crop in poor nations to alleviate malnutrition [4]. As a result, unlike many crops grown by subsistence farmers, finger millet remains highly valued in traditional production systems, especially for its nutrient benefits to pregnant women and children for whom it is used as a weaning food [4,18]. As finger millet seeds can be stored for more than five years due to low vulnerability to insect damage [4,20], it provides food security for poor farmers. Although finger millet plays a very important role especially in the diet of rural peoples, it has become a less important cereal crop due to high demand for rice and maize cultivation [17] and lack of adequate male labour [1]. Many of the management practices are conducted by women including land preparation, seeding/transplanting, harvesting, and threshing (Figure 2). Therefore, improvements in productivity of finger millet will benefit the food production systems of Asian and African nations while enhancing local nutrition.

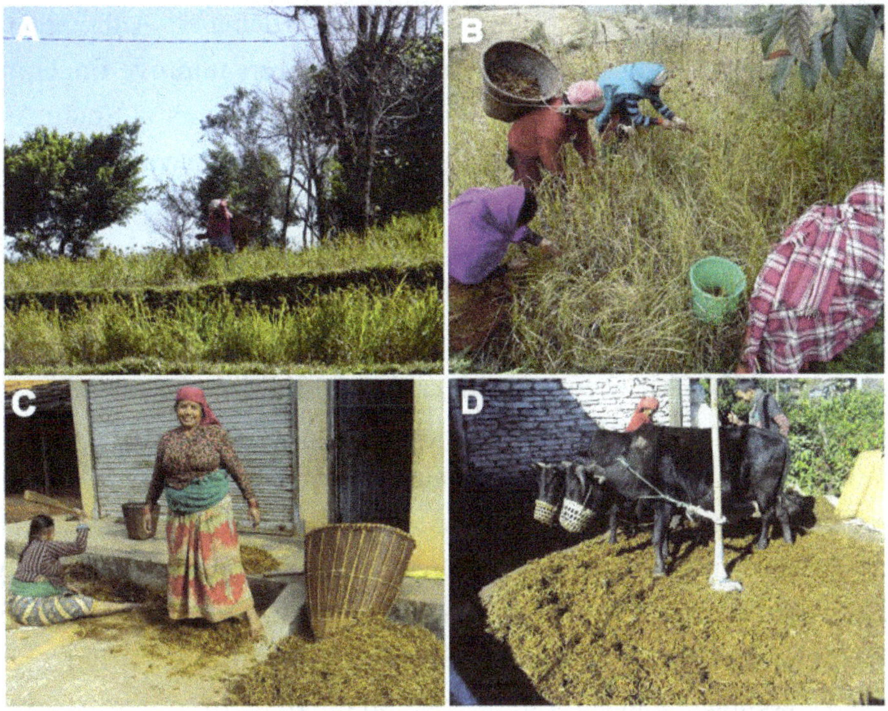

Figure 2. Manual labour associated with finger millet production. (**A**, **B**) Harvesting of finger millet by women in a terraced field of a smallholder farmer in Nepal; (**C**) Hand threshing of finger millet with a wooden pole by women in Nepal; (**D**) Threshing of finger millet with oxen in Nepal. Picture sources: LI-BIRD photo bank.

The striking feature of finger millet is its ability to adjust to different agro-climatic conditions [22]. Once adequate moisture is available (minimum water requirement is 400 mm) and the temperature is above 15 °C, finger millet can be grown throughout the year [22]. It is well adapted to higher elevations and is grown in the Himalayas up to an altitude of 2400 m [4]. Finger millet is drought tolerant [8,22], disease resistant [1], effective in suppressing weed growth [23], and able to grow on marginal lands with poor soil fertility. Finger millet varieties are primarily grouped into two types based on crop duration: early maturity (90–100 days) and late maturity (110–120 days) [22]. It can be established either by broadcasting the seeds or transplanting, where the yield is higher when transplanted in rows compared to broadcasting [22,24].

Though finger millet is valued by traditional farmers as a low fertilizer input crop [4], under these conditions, it suffers from low yields [20,22]. Most of the soils in the semi-arid tropics, where finger millet is grown, are deficient in major and micronutrients, mainly due to continuous cropping, low use of mineral fertilizer, poor recycling of crop residues, and low rates of organic matter application which can limit yield potential [25]. Therefore, it is important to optimize nutrient management practices and other related factors affecting finger millet cultivation in order to attain better yields under the comparatively marginal local growing conditions. Unfortunately, compared to the major cereal crops, the recommendations available for nutrient management in finger millet are scarce, limiting the ability of agricultural extension officers to assist subsistence farmers.

The land available for agriculture is declining especially in the semi-arid tropics mainly due to increases in population. On the other hand, the food productivity of staple food crops in these regions (e.g., finger millet) has to be increased in order to meet food demand. Although many research findings suggest that increased application rates of inorganic fertilizers improve finger millet yield and productivity, it is not a practical option for many poor finger millet farmers in South Asia and Africa, as they cannot afford inorganic fertilizer. Therefore, integrated nutrient management (INM) may be a sustainable option for finger millet farmers in these regions. The main objectives of INM are improvements in plant performance and resource use efficiency while minimizing negative environmental impacts [26,27]. These can be achieved through use of all possible sources of nutrients to meet crop demand, matching soil nutrient availability with crop demand (spatially and temporally), and minimizing nitrogen losses [26,28]. The major advantages of INM are increases in yield, water use efficiency, grain quality, economic return, and sustainability [26].

The objective of this paper is to review the literature concerning nutrient management of finger millet in the semi-arid tropic regions of Asia and Africa, including the use of inorganic fertilizers (macronutrients and micronutrients), farmyard manure, green manures, organic by-products, and biofertilizers. The review further discusses the benefits and concerns of INM as well as different cropping systems, and reviews the limited data that exists on varietal effects. The review concludes with identifying gaps and recommendations for improving productivity of this crop.

2. Current Literature Concerning Inorganic and Organic Nutrient Management

As illustrated below, most of the major inorganic fertilizer studies related to finger millet have focused on testing the effects of N, P, and NPK together (Table 2). Most of the studies related to micronutrients have focused on zinc (Zn) and boron (B). In addition, significant attention has been paid to measuring the effects of different organic fertilizers on finger millet growth and yield, including the use of FYM, green manures, and bio-fertilizers (Table 2).

Table 2. Effect of different fertilizer management practices on finger millet growth, yield, and soil nutrition.

Area/Country	Soil properties	Cropping system	Nutrient treatments	Key results	References
			NPK + FYM		
Bangalore, Southern India	Alfisol soil Soil organic carbon (SOC) = 0.30%–0.45% Available soil N = 163–204 kg N ha^{-1}	Finger millet	Control 100% NPK (50:50:25 kg N, P_2O_5, K_2O ha^{-1}) FYM at 10 t ha^{-1} FYM at 10 t ha^{-1} + 50% NPK FYM at 10 t ha^{-1} + 100% NPK Control Maize residues (MR) at 5 t ha^{-1} MR at 5 t ha^{-1} + 50% NPK MR at 5 t ha^{-1} + 100% NPK	Higher grain yield was found with FYM (10 t ha^{-1}) + 100% NPK (3167 kg ha^{-1}) and MR (5 t ha^{-1}) + 100% NPK (2518 kg ha^{-1}) compared to the recommended NPK fertilizer (1826 and 1965 kg ha^{-1} respectively). Similar trend was found for soil NPK.	[29]
Bangalore, Southern India	Alfisol soil SOC = 0.46% Available soil N = 302 kg N ha^{-1}	Finger millet monocropping under rainfed conditions	Control (no NPK fertilizer or organic amendments) FYM 10 t ha^{-1} + 50% NPK FYM 10 t ha^{-1} + 100% NPK FYM 10 t ha^{-1} Recommended dose of NPK (50:50:25 kg N, P_2O_5, K_2O ha^{-1})	Higher grain yield (mean grain yield of 3281kg ha^{-1}) and sustainable yield index were achieved with integrated nutrient management (INM) (FYM at 10 t ha^{-1} + 100% NPK) than recommended NPK. Application of FYM improved soil C stock (35% of soil C buildup after 27 years).	[30]
Tamil Nadu, Southern India	Fine Montmorillonitic, isohyperthermic SOC = 4% (control) and 0.62% (NPK + FYM)	Finger millet-maize rotation	50% NPK (90:45:17.5 kg N, P_2O_5, K_2O ha^{-1}) 100% NPK 150% NPK 100% NPK + hand weeding 100% NP 100% N 100% NPK + FYM at 10 t ha^{-1} 100% NPK (-S) Control	Application of 100% NPK + FYM increased soil organic C level (6.2 *vs.* 5.3 g kg^{-1} soil), CEC (34.4 *vs.* 28.9 cmol (p+) kg^{-1} soil), available soil N (197 *vs.* 165 kg ha^{-1}), P (26.2 *vs.* 16.8 kg ha^{-1}), K (650 *vs.* 596 kg ha^{-1}), and micronutrients (Cu, Mn, Fe, Zn) compared to NPK fertilizer alone. 100% N and control treatments resulted in lowest soil organic carbon level.	[31,32]

Table 2. *Cont.*

Area/Country	Soil properties	Cropping system	Nutrient treatments	Key results	References
Karnataka, India	Alfisol, sandy loam soil SOC = 0.42% Available soil N = 302 kg N ha^{-1}	Finger millet-groundnut rotation	No fertilizer FYM (10 t ha^{-1}) 100 % NPK (50:21.8:20.7 kg N, P, K ha^{-1}) FYM + 50% NPK FYM + 100% NPK	Higher grain yield was achieved with NPK + FYM treatment (3957 kg ha^{-1}) compared to the recommended NPK (2578 kg ha^{-1}). SOC increased by 41% with INM after 13 years of crop rotation.	[33]
Bangalore, India	Red sandy loam SOC = 0.34% Available soil N = 172 kg N ha^{-1}	Finger millet-groundnut	NPK (100:50:50 kg N, P_2O_5, K_2O ha^{-1}) NPK + FYM (7.5 t ha^{-1})	Application of FYM + NPK improved finger millet yield, soil NPK content, and microbial biomass. INM maintained neutral pH, where the recommended NPK treatment caused acidic conditions.	[34]
Karnataka, India	Red sandy loam Available soil N = 329 kg N ha^{-1}	Finger millet	Recommended NPK (50:40:25 kg N, P_2O_5, K_2O ha^{-1}) through fertilizer Farmer practice (20 N, 21 P_2O_5 kg ha^{-1}) 50% N through FYM + 50% NPK through fertilizer Recommended N + P and K through FYM	Application of 50% N through FYM + 50% NPK produced slightly greater yield (30.3 quintiles ha^{-1}) than recommended NPK through fertilizer (28.7 quintiles ha^{-1}). In comparison to recommended NPK, the above treatment had slightly high/similar plant height (77 *vs.* 75 cm), straw yield (36.2 *vs.* 35.0 quintiles ha^{-1}), and benefit/cost ratio (3.2 *vs.* 3.0).	[35]
Bangalore, India	Fine, mixed isothermic Kandic Paleustalfs	Finger millet	50% NPK 100% NPK 150% NPK 100% NPK + HW (not defined) 100% NPK + Lime 100% NP 100% N 100% NPK + FYM 100% NPK (S-free) 100% NPK + FYM+ Lime Control	In comparison to 100% NPK, INM treatments (100% NPK + FYM + lime and 100% NPK + FYM) showed increased root biomass (10.7 *vs.* 9.2 quintiles ha^{-1}), root N (0.53–0.54 *vs.* 0.51%), K (1.04–1.05 *vs.* 0.9%), Ca (0.54–0.58 *vs.* 0.48%), Mg (0.31–0.34 *vs.* 0.26%), and micronutrient content (Fe, Cu, Zn, Mn).	[36]

Table 2. *Cont.*

Area/Country	Soil properties	Cropping system	Nutrient treatments	Key results	References
Bangalore, India	Red sandy loam soil SOC = 0.34% Available soil N = 172 kg N ha^{-1}	Finger millet under irrigation	NPK (100:50:50 kg N, P_2O_5, K_2O ha^{-1}) NPK + FYM (7.5 t ha^{-1})	Application of 100% NPK + FYM increased millet yield (3086 kg ha^{-1}) by 9.5% compared to NPK alone (2946 kg ha^{-1}). INM also increased the number of tillers per hill, ear length, ear weight, 1000 grain weight, threshing percent, and number of fingers per ear head.	[37]
Pakhribas and Dordor Gaun, Nepal	Sandy loam and silt loam Organic matter and N% = 1.33, 0.08% (Pakhribas) and 0.82, 0.08% (Dordor Gaun)	Maize-finger millet cropping system	No inputs Farmer practices for fertilizer-T1 Farmer practices for FYM-T2 50% (T1 + T2) 50% T1 50% T2 25% (T1 + T2)	Highest millet yield was associated with the FYM applied to previous maize crop than inorganic fertilizer treatment. FYM application reduced the rate of C and N losses.	[38]
Pakhribas and Dordor Gaun, Nepal	Sandy loam and silt loam SOC and total N at top 25 cm soil = 1.32, 0.08% (Pakhribas) and 0.82, 0.08% (Dordor Gaun)	Maize-finger millet	No fertilizer NPK fertilizer (90:30:30 kg ha^{-1}) FYM alone (90 kg N ha^{-1}) NPK + FYM (different ratios)	A trend of high yield was found in finger millet plots, which were previously manured, compared to inorganic fertilizer applied plots. Recovery of fertilizer applied to maize by subsequent finger millet crop was very low (3%).	[39]
Dhankuta, Eastern Nepal	Dystochrept (sandy clay loam texture) Organic matter = 1.9%	Maize/millet rotation	NPK FYM (0, 15, 25 t ha^{-1}) NPK + FYM Lime	Millet yield increased following maize treated with FYM (by 705 kg ha^{-1}) or FYM + inorganic fertilizer (by 631 kg ha^{-1}), compared to inorganic fertilizer alone.	[40]

Table 2. *Cont.*

Area/Country	Soil properties	Cropping system	Nutrient treatments	Key results	References
			Micronutrients		
Karnataka, India	Alfisols and Inceptisols SOC = 0.37%	Finger millet under rainfed	Farmers' inputs (FI) + NP (60:130 kg N, P_2O_5 ha^{-1}) FI + S, B, Zn (30:0.5:10 kg S, B, Zn ha^{-1}) FI + NP + S, B, Zn (60:130:30:0.5:10 kg N, P_2O_5, S, B, Zn ha^{-1}).	In comparison to FI, combined application of FI + NP + S, B, Zn fertilizers enhanced grain yield (3350 *vs.* 2150 kg ha^{-1}), straw yield (6650 *vs.* 4630 kg ha^{-1}), and uptake of N (31 *vs.* 20 kg ha^{-1}), P (7.5 *vs.* 5.2 kg ha^{-1}), K (17 *vs.* 11 kg ha^{-1}), S (2.9 *vs.* 1.8 kg ha^{-1}), and Zn (49 *vs.* 36 kg ha^{-1}).	[25]
Karnataka, India	Vertisol and Alfisol (sandy, loam, clay) SOC = <0.5% Available soil N = <280 kg N ha^{-1}	Finger millet under rainfed conditions	Farmer practice (N + P) Farmer practice + Zn, B, S (10:0.5:30 kg Zn:B:S ha^{-1})	In comparison to farmers' practice, farmer practice + Zn, B, S increased finger millet grain yield (3354 *vs.* 2142 kg ha^{-1}), stover biomass (6654 *vs.* 4630 kg ha^{-1}), total biomass (10008 *vs.* 6772 kg ha^{-1}), and plant uptake of Zn (322 *vs.* 193 g ha^{-1}), B (21 *vs.* 17 g ha^{-1}), and S (16 *vs.* 10 kg ha^{-1}).	[41]
			NPK + FYM + Bio-fertilizers/Green manures		
Wakawali, India	Terraced upland	Finger millet	Recommended fertilizer (RF) (80:40:00 kg N, P_2O_5, K_2O ha^{-1}) FYM at 5 t ha^{-1} + RF FYM at 5 t ha^{-1} + 75% RF + bio-fertilizers (*Azospirillum* + PSB) FYM at 10 t ha^{-1} + bio-fertilizers FYM at 15 t ha^{-1} + bio-fertilizers	Higher grain yield obtained with FYM at 5 t ha^{-1} + 75% NPK + bio-fertilizers	[42]
Karnataka, India	Red sandy loam soil Available soil N = 59 kg N ha^{-1}	Finger millet under rainfed conditions	Recommended NPK (50:37.5:25 kg NPK ha^{-1}) RF + *Azotobacter* (1 kg ha^{-1} root dipping) RF + gypsum (500 kg ha^{-1}) RF + *Azotobacter* + $ZnSO_4$ (10 kg ha^{-1}) RF + *Azotobacter* + gypsum RF + $ZnSO_4$ + gypsum RF + *Azotobacter* + $ZnSO_4$ + gypsum	In comparison to RF treatment, INM (RF + $ZnSO_4$ + gypsum + *Azotobacter*) had higher number of tillers (8 *vs.* 4 plant^{-1}), ear head (8 *vs.* 3 plant^{-1}), number of fingers (43 *vs.* 25 plant^{-1}), yield (61 *vs.* 49 quintiles ha^{-1}), and benefit:cost ratio (2.5 *vs.* 2.2).	[43]

Table 2. *Cont.*

Area/Country	Soil properties	Cropping system	Nutrient treatments	Key results	References
Odisha, India	Red lateritic sandy loam	Finger millet	Farmers' input (FI) (FYM at 2 t ha^{-1} + 17:12:0 kg N, P_2O_5, K_2O) RF (40:20:20 kg N, P_2O_5, K_2O ha^{-1}) FI + 2.5 kg ha^{-1} each of PSB and *Azotobacter* FI + *Gliricidia* at 5 t ha^{-1} 50% RF + 2.5 t ha^{-1} FYM + 2.5 kg ha^{-1} each of PSB and *Azotobacter* 50% RF + 2.5 t ha^{-1} *Gliricidia* + 2.5 kg ha^{-1} each of PSB and *Azotobacter*	In comparison to FI, INM (50% RF + 2.5 t ha^{-1} *Gliricidia* + 2.5 kg ha^{-1} each of PSB and *Azotobacter*) treatment increased shoot and root growth (10.9, 3.5 *vs.* 9.7, 2.8 g $plant^{-1}$), yield parameters, grain yield (4.0 *vs.* 3.5 t ha^{-1}), benefit:cost ratio (2.4 *vs.* 2.1), soil moisture, SOC (0.46 *vs.* 0.41%), soil available N (278 *vs.* 240 kg ha^{-1}), available P (14.7 *vs.* 12.1 kg ha^{-1}), and available K (307 *vs.* 279 kg ha^{-1}).	[8]
Bangalore, Southern India	Red sandy clay loam SOC = 0.36% Available soil N = 175 kg N ha^{-1}	Finger millet-pigeon pea rotation	100% N through urea 50% N through urea + 25% N through FYM + 25% N through *Gliricidia* 50% N through FYM + 50% N through *Gliricidia*	INM had higher grain yield (2666 kg ha^{-1}) with a 29% increase compared to 100% N supply through urea (2067 kg ha^{-1}). In comparison to sole organic N treatment, INM had higher tiller number (5.8 *vs.* 4.9 $plant^{-1}$), grain yield (2666 *vs.* 1665 kg ha^{-1}), and benefit:cost ratio (3.23 *vs.* 1.71).	[44]
Bangalore, Southern India	Sandy and gravel soil SOC = 0.44% Available soil N = 307 kg N ha^{-1}	Finger millet	Recommended FYM + 100% NPK Recommended FYM + 100% N through Pongamia cake Recommended FYM + 100% N through Mahua cake Recommended FYM + 100% N through Neem cake	Recommended FYM + Neem cake equivalent to 100% N increased finger millet yield (2454 *vs.* 2175 kg ha^{-1}), soil available N (391 *vs.* 315 kg ha^{-1}), available P (50 *vs.* 30 kg P_2O_5 ha^{-1}), and available K (391 *vs.* 260 kg K_2O ha^{-1}) compared to the 100% NPK + recommended FYM.	[45]
Karnataka, India	Sandy loam SOC = 0.57% Available soil N = 205 kg N ha^{-1}	Finger millet following potato	100% NPK 100% N as FYM 100% NPK + FYM (10 t ha^{-1}) 25%–50% N as composted weeds	In comparison to 100% inorganic fertilizer, 100% N as FYM had higher finger millet grain yield (4.77 *vs.* 4.13 t ha^{-1}), soil N (133 *vs.* 107 kg ha^{-1}), soil P (27 *vs.* 21 kg ha^{-1}), and soil K (174 *vs.* 142 kg ha^{-1}).	[46]

Table 2. *Cont.*

Area/Country	Soil properties	Cropping system	Nutrient treatments	Key results	References
Bangalore, India	Red sandy loam Soil organic matter = 0.48% Available soil N = 268 kg N ha^{-1}	Horse gram in the previous season	FYM Biogas slurry Poultry manure City waste compost Agrimagic Green manure 100% RF (50:40:25 NPK kg ha^{-1}) 100% RF + FYM 100% NK	100% RF + FYM at 7.5 t ha^{-1} had higher grain yield (3660 kg ha^{-1}) compared to other treatments (1400–3200 kg ha^{-1}). Finger millet supplied with poultry manure produced higher grain yield (2970 kg ha^{-1}) than FYM or green manure treatments (2200–2300 kg ha^{-1}).	[47]
Tamil Nadu, India	Not available	Rice-Finger millet	Recommended fertilizer + FYM (12.5 t ha^{-1}) RF + composted coirpith (12.5 t ha^{-1})	Finger millet yield was higher under FYM + RF (2816 kg ha^{-1}) and composted coirpith + RF (2739 kg ha^{-1}).	[48]
Tamil Nadu, Southern India	Not available	Finger millet	Control (no organics) FYM at 12.5 t ha^{-1} Pressmud at 12.5 t ha^{-1} Composted coirpith at 12.5 t ha^{-1} Gypsum at 500 kg ha^{-1}	Application of pressmud and composted coirpith significantly improved finger millet yield (3316, 3385 *vs.* 2593 kg ha^{-1}), soil N (135, 136 *vs.* 120 kg ha^{-1}), soil P (6.18, 10.3 *vs.* 5.7 kg ha^{-1}), soil K (134, 138 *vs.* 116 kg N ha^{-1}) macro and micronutrients (Zn, Cu, Mg, Fe), soil organic carbon (0.45, 0.46 *vs.* 0.25%), pH (8.33, 8.27 *vs.* 8.45%), EC (0.25 *vs.* 0.41 dSm^{-1}), and microbial population (bacteria, fungi, actinomycete) compared to the control.	[49]

Abbreviations: FYM: farmyard manure; MR: maize residues; FI: farmers' inputs; RF: recommended fertilizer; PSB: phosphorus solubilizing bacteria; SOC: soil organic carbon; INM: integrated nutrient management.

2.1. Nitrogen (N)

Finger millet responds well to N application [6,22,50], since many of the soils in the semi-arid regions of Asia are deficient in N [25]. Studies concerning N management in finger millet are mainly focused on the amount of N applied, timing of application, and varietal responses to N. Rao *et al.* [51] reported increases in yield and grain protein content in finger millet due to N fertilizer application rates of up to 40 kg N ha^{-1} in Andhra Pradesh, India. The authors claimed that the economic optimum rate of N fertilizer for finger millet was 43.5 kg ha^{-1} under rainfed conditions. Hegde and Gowda [22] reported that finger millet grain yield was 23.1 kg per kg N at 20 kg N ha^{-1}, while the yield benefit declined to 19.9 kg per kg N at 60 kg N ha^{-1}. These results suggest that application of the correct dose of N fertilizer is important in order to maximize the profits of poor finger millet farmers. It is also important to note that the application of inorganic N fertilizer can delay flowering and physiological maturity by 1–2 weeks [24], which can affect the final yield. The latter study also found that application of inorganic N alone (22.5–45 kg N ha^{-1}) did not increase the grain yield compared to the no fertilizer application under conditions of seed broadcasting and row planting. Therefore, the authors claimed that N application alone is not economical in finger millet cultivation. Based on a long-term field experiment with finger millet, Hemalatha and Chellamuthu [32] found that continuous application of inorganic N fertilizer alone reduced the soil organic carbon level due to low dry matter production and reduced return of crop residues to the field (Table 2).

In addition to the amount of N supplied, the timing of N application is also important for finger millet. The importance of applying N starts with seed germination, a challenge for small seed crops like finger millet especially under nutrient deficient conditions. The application of inorganic N fertilizer at the time of planting stimulates better crop emergence especially in N deficient soil [20]. Hegde and Gowda [22] also claimed that incorporation of N fertilizer during seeding enhanced finger millet yield by 30% compared to broadcasted fertilizer. Synchronizing N supply with crop N demand is essential to maximize yield and N use efficiency. Hegde and Gowda [22] reported that application of N on sandy loam soils at 50 kg ha^{-1} produced a finger millet grain yield of 2430 kg when applied at planting, whereas the yield increased to 2650 kg ha^{-1} when the application time was split (at planting and 25–30 days after planting). Therefore, split application of N fertilizer enhances finger millet yield production and possibly reduces N losses as well.

2.2. Phosphorus (P)

Although P is one of the major macronutrients required by finger millet, limited research has been conducted to evaluate the significance of P on finger millet growth and yield. Nevertheless P is one of the highly limited nutrients in farmers' fields in semi-arid regions of Asia [25]. Based on multi location field experiments conducted in Eastern Uganda, Tenywa *et al.* [24] found that application of P fertilizer (20–40 kg P_2O_5 ha^{-1}) increased the growth and yield of finger millet compared to the no fertilizer control under row planting conditions. However, Hedge and Gowda [22] reported a reduction in finger millet grain yields from 16.3 to 14.7 kg per kg P_2O_5 when the P application rate was increased from 30 to 60 kg ha^{-1} P_2O_5. Similar to inorganic N, this result suggests that application of excess P does not improve yield, but rather that application of balanced fertilizer is crucial.

Organic practices have been shown to be important for P nutrition in finger millet. Based on a long-term field study at Tamil Nadu, India, Hemalatha and Chellamuthu [31,32] found that continuous application of 100% NPK + FYM increased P availability (Table 2), which agrees with previous findings by Govindappa *et al.* [47]. This could be due possibly to the solubilisation of P by organic acids released during organic matter decomposition. An earlier study by Subramanian and Kumaraswami [52] also reported that application of NPK, along with FYM increased the uptake of P by finger millet. This highlights the importance of applying FYM along with inorganic fertilizer to improve P availability for finger millet. With respect to lessons learned from crop rotations, in Eastern Uganda, Ebanyat *et al.* [53] found that application of P to legume crops generally enhanced the yield of the subsequent finger millet crop; six different legumes were tested in parallel including cowpea, pigeonpea, and groundnut (peanut). Addition of P increased the amount of N fixed by legume crops, resulting in a positive effect on yield of the subsequent finger millet crop. In addition to the benefit provided by N, residual P may have also had a positive effect on finger millet yield. Generally the yield response of finger millet to the addition of P to the previous legume crop was higher in fields with low soil fertility. The authors also found that P supplied to the previous legume crops increased the N use efficiency in finger millet, but the results were not consistent across the different legume species or soil fertility types tested. Although the application of P to legumes increased the yield of the subsequent finger millet crop, it may not be profitable due to the extra cost associated with P fertilizer [53].

2.3. Nitrogen, Phosphorus, and Potassium (K)

NPK has been shown to be important for early establishment of finger millet. Based on a well-planned, three-year study conducted in farmers' fields in Eastern Zimbabwe, Rurinda *et al.* [20] found that finger millet emergence was low without inorganic NP fertilizer or with manure (<15%) compared to fertilization with either NP fertilizer or manure + fertilizer (>70%). The data suggests that application of manure alone may not be beneficial to finger millet, perhaps because the nutrients are not readily available to the seedling. This result highlights the importance of supplying starter NPK mineral fertilizer for better finger millet establishment. The authors also found that agronomic N use efficiency (kg grain yield produced per kg N applied) decreases at high NPK rates, thus identification of the optimum fertilizer requirement is very important in order to maximize crop productivity. Hegde and Gowda [22] reported that the required application rate of NPK fertilizer depends on whether the conditions are wet or dry, with a higher rate of fertilizer required under irrigated conditions (100, 50, 50 kg N, P_2O_5, K_2O) compared to dryland conditions (50, 37.5, 25 kg N, P_2O_5, K_2O) in order to achieve their respective yield potentials (*i.e.*, there is more biomass produced under wet conditions). Similarly Sankar *et al.* [29] found that the benefit of applying inorganic fertilizer increased under high moisture availability compared to low moisture availability (<500 mm).

Application of the major macronutrients to finger millet alone does not necessarily provide better yields, rather the application of balanced nutrients is important as already noted above. Using a soil management study in Eastern Uganda, Tenywa *et al.* [24] found that application of N or P alone did not increase finger millet growth and yield compared to non-fertilized plants. However, they found that application of N + P and manure + P produced better growth and yield in finger millet compared to N or manure alone, highlighting the importance of balanced nutrient management. Based on a two year field trial, Bhoite and Nimbalkar [54] reported that application of 60 kg N ha^{-1} and 20 kg P_2O_5 produced the best finger millet yield in Maharashtra, India. Based on a 25 year long term experiment conducted under rainfed conditions on alfisols in Bangalore (Southern India), it was observed that application of $N:P_2O_5:K_2O$ at 50:50:25 kg ha^{-1} increased finger millet yield and soil fertility status compared tonon-fertilized plants [29]. The authors further observed that application of FYM and maize residues along with NPK enhanced the yield and soil fertility status. Long term application of 100% NPK along with FYM (10 t ha^{-1}) also increased the available soil nutrients (N, P, K, Ca, Mg) [32].

2.4. Micronutrients

Soil micronutrients are commonly deficient in South Asia [25,41] and Sub Saharan Africa [55]. Most of the micronutrient studies related to finger millet have concentrated on zinc (Zn) and boron (B). Based on soil tests with 1617 farmers in the semi-arid tropics of India, Srinivasarao *et al.* [41] found that Zn and B deficiency ranged from 2%–100% and 0%–100% respectively in farmers' fields, depending on the geographic region. The authors considered the following minimum levels to be critical for available Zn and B in farmers' fields, respectively: 0.75 mg Zn kg^{-1} soil (DTPA extractable), 0.58 mg B kg^{-1} soil (hot water extractable). Similarly, based on surface soil testing (802 soil samples), Rao *et al.* [25] also found that farmers' fields were deficient in Zn (34%–88% of fields tested) and B (53%–96%) in the semi-arid regions of Karnataka, India. Srinivasarao *et al.* [41] found that application of Zn, B, and sulfur (S) along with N and P, enhanced finger millet grain yield (56%), stover biomass (44%), total biomass (48%), and plant uptake of Zn (66%) and B (22%) compared to the addition of N and P alone (Table 2). Based on a three-year experiment, Rao *et al.* [25] found that application of B and Zn along with farmer inputs (farmers chose their fertilizer types and application rates), N, P, and S fertilizer increased grain yield, straw productivity, and nutrient uptake (N, P, S, B, and Zn) of finger millet compared to the farmers' traditional inputs (Table 2). Similarly Maury and Verma [56] claimed that application of Zn along with NPK fertilizer favored the uptake of NPK, but reduced the Zn content, but further methodological details were not available to explain this result. Ramachandrappa *et al.* [57] reported that soil application of $ZnSO_4$ (12.5 kg ha^{-1}) and borax (10 kg ha^{-1}) along with the recommended NPK increased finger millet yield in B and Mo deficient soils.

Other than micronutrient fertilizers, FYM may be a good source of essential micronutrients for millet growth. Based on a long-term field experiment in Tamil Nadu, India, it was found that continuous application of 100% NPK + FYM (10 t ha^{-1}) increased some of the available micronutrients (Fe, Zn, Mn, and Cu) in a finger millet-maize cropping system [32] (Table 2).

2.5. Farmyard Manure (FYM) + Inorganic Fertilizer

2.5.1. Yield Response of FYM + NPK

A significant number of studies have been conducted to evaluate the effect of farmyard manure along with recommended inorganic fertilizer on finger millet growth and yield (Table 2). Some of these studies have been noted above. Generally application of FYM along with recommended NPK fertilizer enhances finger millet yield and soil fertility [25,29,30,34,37,42,58]. Kumara *et al.* [37] found that application of FYM along with the recommended NPK fertilizer increased yield parameters of finger millet under an irrigated system in Bangalore, India (ear length, 1000 grain weight, number of fingers per ear head, ear weight per plant, and grain weight per plant). The same trend was observed by Govindappa *et al.* [47], where application of FYM (7.5 t ha^{-1}) along with the recommended NPK increased dry matter production, grain weight, grain yield, and straw yield of finger millet (Table 2). The authors also found that poultry manure was a better source of manure than FYM or green manure for finger millet to achieve better growth and yield. Based on a very long-term comprehensive study (1984–2008), Sankar *et al.* [29] found that application of FYM and maize residues increased millet yield as well as sustainability in rainfed semiarid tropical alfisols (Table 2). Furthermore, based on a long-term field experiment, Pushpa *et al.* [36] found that application of FYM along with 100% NPK + lime increased root growth and root nutrient content (major, secondary and micro nutrients) compared to plants treated with 100% NPK alone (Table 2). Based on an eight year field experiment, Sherchan *et al.* [40] found that finger millet yield and plant NPK uptake increased when potato, as the preceding crop, was supplied with 100% FYM or when 50% of N was supplied by lantana (*Lantana camara*, a weedy green manure).

Application of integrated nutrient management practices can reduce the amount of inorganic fertilizer used for finger millet without compromising yield. It was found that application of 50% recommended N through FYM + 50% recommended NPK fertilizer can produce a slightly higher yield than 100% of recommended NPK fertilizer alone [35] (Table 2). The authors also claimed that the benefit/cost ratio was higher with the above treatment than the traditional farmer practices and in par with the recommended NPK at 100%. Sankar *et al.* [29] also found that application of FYM at 10 t ha^{-1} + 50% recommended NPK fertilizer produced a much higher yield than the recommended NPK application at 100% (Table 2).

2.5.2. Effects of Manure + NPK on Soil Carbon

In the arid and semi-arid regions of the tropics and subtropics, soil organic carbon (C) is a limiting factor (<0.5%) [41], thus the retention capacity of nutrients is low, especially N. Therefore, improvement of the soil carbon pool through different organic manures helps to improve soil fertility and sustain yields. Based on long-term field experiments, Srinivasarao *et al.* [30,33] and Hemalatha and Chellamuthu [32] found that application of FYM along with 100% NPK inorganic fertilizer increased the grain yield of finger millet as well as the soil organic C level. Also, Srinivasarao *et al.* [30,33] found a strong correlation between soil C levels and a sustainable yield index (an approach to evaluate the sustainability of long-term cropping systems), highlighting the importance of maintaining the soil C pool in order to attain sustainable yields. It appears that sustainable finger millet production is achievable if appropriate amounts of both inorganic fertilizers and organic materials are applied together. On the other hand, application of balanced NPK along with low amounts of FYM is an alternative solution to maintain the soil C levels under limited manure availability [33]. However, application of FYM to maintain the soil C level may not be economical in the short term in the absence of compensating for C sequestration [38]. In this situation application of inorganic fertilizer may be economical for farmers.

2.5.3. Other Benefits and Challenges of Combining Manure with Inorganic Fertilizers

The application of organic manure can minimize the negative effect of continuous application of inorganic fertilizer to finger millet. Organic manure helps to maintain soil C levels [30,32,33], which minimizes N losses from the cropping system while increasing the sustainability of the system. Furthermore Hemalatha and Chellamuthu [32] found that application of FYM with NPK increased the cation exchange capacity (CEC) of the soil, possibly due to buildup of soil humus by FYM. Based on a simulation model used to explore the long-term impact of different fertilizer management scenarios for maize-millet systems in Nepal, Matthews and Pilbeam [38] observed that application of FYM reduced the decline in rates of soil C and N compared to the application of inorganic fertilizer alone (Table 2). Based on a two-year field study conducted under irrigated conditions, Kumara *et al.* [34] found that application of organic fertilizer helped to improve the microbial biomass and maintain soil pH at a neutral level compared to the application of inorganic fertilizer alone (Table 2).

It is also important to consider the extra cost involved in purchasing manure and its transportation, because in reality it may be more economical to apply inorganic fertilizer rather than organic fertilizer [58]. On the other hand, farmers in some nations are in favor of applying inorganic fertilizer rather than organic fertilizer when government subsidies are available for inorganic fertilizers (e.g., India), and also due to the ease of application, ease of transportation, ready availability, and consistency of results.

2.6. Alternative Sources of Organic Fertilizer: By-Products, Biofertilizers, and Green Manures

The availability of organic fertilizer is becoming a limiting factor for farmers, thus alternative and local organic fertilizer sources need to be explored to meet demand. Research has been conducted to evaluate the possibility of using different organic byproducts as organic fertilizers in finger millet production: composted coirpith, a by-product of the coconut coir industry [48,49,59]; neem oil cake, a by-product of bio-fuel production from neem trees [45,60]; *Pongamia* and mahua cake, by-products of biofuel production from *Pongamia* and mahua legume trees, respectively [45]; and pressmud, a by-product of industrial sugar production from sugarcane [49]. Parasuraman and Mani [48] reported that FYM can be substituted with composted coirpith, wherein the finger millet yield under recommended NPK + composted coir pith (at 12.5 t ha^{-1}) was in par with NPK + FYM (at 12.5 t ha^{-1}). Shivakumar *et al.* [45] found that application of neem cake equivalent to 100% N, along with the recommended FYM, increased finger millet yield (12.8%) and available NPK in soil compared to the addition of inorganic NPK fertilizer + FYM alone (Table 2). However, the experiment was conducted for only one season, whereas long term trials are needed in order to evaluate the organic fertilizer effect on soil. Subbiah *et al.* [60] also claimed that neem cake treated with $(NH_4)_2SO_4$ and urea significantly increased grain yield and NP uptake of finger millet. Based on a two-year field study, Rangaraj *et al.* [49] found that application of different agro-industrial wastes (composted coirpith, FYM, and pressmud) could improve finger millet yield, soil fertility (macro and micro nutrients), soil microbial population (bacteria, fungi, and actinomycetes), and soil chemical and physical properties (Table 2).

Green manures and bio-fertilizers are also becoming valuable organic sources in finger millet production. Research conducted on green manure is mainly focused on *Gliricidia* (a leguminous tree fodder) [8,44], which is rich in nutrients and decomposes rapidly [8]. Different bio-fertilizer products have been tested in finger millet such as *Azotobacter* [8,43,61,62], *Azospirillium* [61], phosphorus solubilizing bacteria (PSB) [8,62], and mycorrhizae fungi [61,63]. Based on a three year field study at Odisha, India, Dass *et al.* [8] found that finger millet supplied with 50% of the recommended inorganic fertilizers, *Gliricidia* green leaf manure (2.5 t ha^{-1}), and *Azotobacter* and PSB, produced the highest grain yield (3.95 t ha^{-1}) compared to 1.76 t ha^{-1} using the farmers' traditional practice (2 t ha^{-1} FYM + 17 kg ha^{-1} P_2O_5 + 12 kg ha^{-1} K_2O); the combined organic treatment also increased soil moisture, organic C, and NPK content (Table 2). Furthermore, the study found that treatments with *Gliricidia* (5 t ha^{-1}) combined with the above farmers' practice increased the available P and K in the soil, compared to the farmers' traditional practice alone. Based on a two-year field study, Vijaymahantesh *et al.* [44] also found that greater finger millet yield can be achieved by combining FYM (25% N), *Gliricidia* (25% N) and urea (50% N) compared to 100% N added using urea alone (Table 2). Based on a three-year field experiment, Sridhara *et al.* [43] found that application of *Azotobacter*, $ZnSO_4$, and gypsum along with recommended NPK fertilizer enhanced the growth, yield parameters, yield and profitability of finger millet compared to the recommended fertilizer application alone (Table 2). Based on pot experiments with soil as the growing medium, Ramakrishnan and Bhuvaneswari [61] found that finger millet treated with *Azospirillium* + arbuscular mycorrhizal (AM) fungi + PSB increased plant growth and N, P uptake. Uptake of macro (N, P) and micronutrients (Zn, Cu) by plants was also enhanced when finger millet was treated with AM fungi compared to non-inoculated plants [63]. Based on a single year field experiment, Apoorva *et al.* [62] found that application of *Azotobacter* and PSB along with fertilizer (based on soil testing) and FYM (10 t ha^{-1}) increased finger millet yield by 2000 kg ha^{-1} compared to the recommended fertilizer application alone.

3. Crop Rotations

Finger millet based crop rotations or relay cropping are common cropping practices in South Asian countries, involving maize-millet [39,40], potato-millet [46], and groundnut-millet [30,34]. In Africa (eastern Uganda), finger millet based crop rotations are beans-cassava-cowpeas-groundnuts-cotton, beans-cotton-cowpeas, and beans-cotton-maize [24]. Crop rotation is important as residual fertility from the previous crop contributes to the next crop. It was observed that finger millet benefits more from residual fertilizer from the previous crop when the fertilizer is supplied as organic fertilizer compared to inorganic fertilizer [38,40,58] (Table 2). Based on a well-planned study conducted in eastern Uganda, Ebanyat *et al.* [53] found that finger millet yields following legume crops (cowpea, green gram, groundnut, mucuna, pigeonpea, and soybean) were higher compared to continuous finger millet cropping. However, the N benefits derived from the legume crop residues decreased as the season progressed [53]. Although, there are N benefits to finger millet following legume crops, farmers were reluctant to use some of the legumes in their crop rotations, as they were not aware of their potential marketability or usefulness as fodder (mucuna) [53]. However, the residual N benefit to finger millet was shown to be low when the previous crop was a non-legume [39]. In particular, the authors found that recovery of N applied to maize by the following finger millet crop was only <3%, possibly due to most of the applied N being taken up by maize combined with a slow N turnover rate from maize residues. Therefore, selection of appropriate crops in finger millet based crop rotations is very important in order to utilize the residual nutrients and to obtain N credits for finger millet from the previous crop.

4. Intercropping/Mixed-cropping

Intercropping/mixed-cropping of finger millet with different legume crops such as pigeon pea, soybean, green gram, horsegram, common bean, and groundnut are also common farming practices [64–68] (Figure 1). As legumes fix atmospheric nitrogen through symbiotic N fixation, finger millet obtains N benefits from neighboring legumes. Based on a two-year field experiment in West Bengal, India, Maitra *et al.* [66] claimed that finger millet yield increased under intercropping with pigeon pea and groundnut rather than sole cropping with finger millet. However, it was reported that intercropping is beneficial only under low fertilizer input systems [22]. However, intercropping of finger millet with grain legumes can reduce legume yields due to competition, thus transplanting of finger millet after a legume is established can be beneficial [22]. Also the application of balanced inorganic fertilizer is important to minimize competition under intercropping. Maitra *et al.* [66] reported that application of NPK at a rate of 60:13.3:25 kg ha^{-1} maximized productivity and net return under finger millet-grain legume (pigeon pea and groundnut) cropping systems.

Other than nutrient benefits, intercropping has been shown to enhance finger millet yield through disease control [9,69]. Based on a multi season field trial in Western Kenya, it was observed that intercropping of finger millet with green manure legumes (leaves of *Desmodium*, a ground-cover legume) increased finger millet yield compared to mono-cropping, mainly by controlling pests associated with finger millet (*Striga hermonthica* and cereal stem borer) [9]. The authors also evaluated the economic benefits of intercropping of finger millet over monocropping, wherein the former resulted in greater economic returns although the labor cost was higher under intercropping [9]. Similarly intercropping of finger millet with mungbean has been shown to reduce *Cercospora* leaf spot and leaf curl disease [69].

One of the key factors of successful intercropping is proper plant density, which depends on the plant species as well as the particular varieties used [65,67]. Padhi *et al.* [65] reported that intercropping of early duration pigeonpea with finger millet at a row ratio of 2:4 had greater productivity and economic return than a medium duration variety. In a study conducted in Bangalore, India, intercropping of finger millet with pigeonpea at an 8:2 row ratio and field bean at 8:1 also resulted in better yield and a higher net return over sole cropping [67].

5. Varietal Effects

Finger millet has high genetic diversity [3]. All finger millet varieties do not respond to nutrients in the same manner. Genotypic variability among different finger millet cultivars has been reported for responsiveness to N and P [54]. Gupta *et al.* [6] evaluated the N use efficiency (ratio of grain yield to N supply) and N utilization efficiency (ratio of grain yield to total N uptake) of three finger millet genotypes under different N inputs (0, 20, 40, 60 kg N ha^{-1}, and 7.5 t FYM ha^{-1}) under pot conditions. They found that there was genotypic variability among the finger millet genotypes' responses to different N inputs, wherein some varieties were highly responsive to N. Therefore, identification of genotypes with higher N use efficiency and N utilization efficiency especially under low available soil N levels will benefit farmers who cannot afford N fertilizer or who do not have access to N fertilizer sources.

With respect to biofertilizers, finger millet plants treated with different strains of arbuscular mycorrhizal (AM) fungi showed significantly different effects on plant growth and yield [63]. Furthermore, the authors found variability for plant growth responses to inoculation with AM fungi by different finger millet varieties. Combined, these results highlight the importance of considering the finger millet cultivar effect as well as the strain effect of AM fungi.

Unfortunately, there is little literature available regarding development of new finger millet varieties with high yield potential under low or high nutrient input conditions, although more papers have appeared in recent years [3]. For example, in Nepal there are only three finger millets varieties that have been released after 1990 for commercial purposes [70]. Finger millet varieties suitable for different seasons and different parts of India are listed in Hegde and Gowda [22].

6. Research Gaps and Recommendations for Improving Nutrient Management in Finger Millet

Based on our literature review it is clear that limited attention has been paid to soil nutrient management issues related to finger millet compared to the other cereals like maize, rice, and wheat. Some of the research reports are restricted to a single growing season (non-replicated), lack appropriate controls, and/or the controls are not well defined (e.g., "farmer practice"). Furthermore, the majority of studies have been conducted in South Asian countries while there is a paucity of research in African countries where finger millet is a major crop. As summarized in this review, finger millet farmers have available to them different nutrient management options (Figure 3). In general, in the context of subsistence farmers, an integrated nutrient management approach appears to have the best potential to reduce the yield gap between potential yield and actual yield of finger millet farmers. This section highlights specific gaps and recommendations as follows:

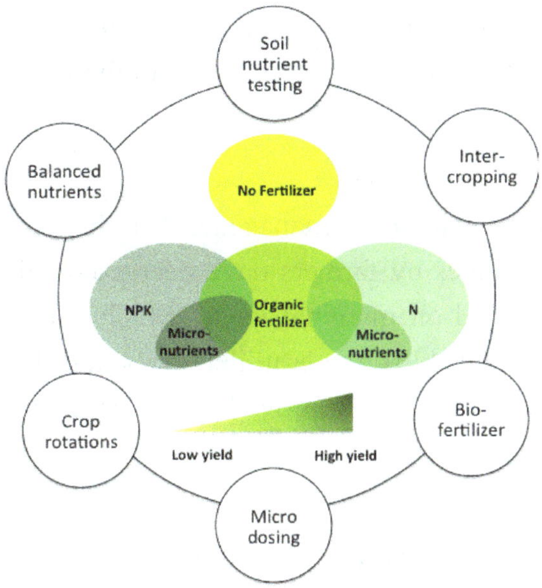

Figure 3. Nutrient management strategies that can be applied to maximize finger millet yield in farmers' fields. The internal Venn diagram (green) demonstrates the effects of integrated nutrient management on finger millet yield. The outer ring represents other nutrient management options that can be incorporated. Finger millet yield can be maximized through application of balanced NPK, farmyard manure (FYM), and micronutrients together, compared to no fertilizer, N alone or farmyard manure alone. Furthermore, a variety of management strategies can be applied together based on resource availability under different cropping systems.

6.1. Nitrogen Management

It is clear that one of the key nutrients that limits finger millet yield is N, since most of the soils under finger millet growing areas have medium to low soil N availability (75–330 kg N ha^{-1}) (Table 2). However, being a highly mobile nutrient, N is easily lost from cropping systems. There is a lack of information available regarding N fertilizer recovery by finger millet. Therefore, research needs to be conducted to evaluate and improve N use efficiency rather than a simple focus on increasing the amount of inorganic N applied. This is also important, as N is expensive for many subsistence farmers, particularly in Africa where there is a lack of fertilizer subsidies.

6.2. Phosphorus Management

Phosphorus is limited in most Asian and African soils [71]. One of the major constraints related to soil P is its low availability for plant uptake [72]. Little research has been conducted to evaluate the importance of P in finger millet and how to improve P availability in finger millet cropping systems. As P availability depends on soil pH (optimum pH of 6.5), it is important to maintain soil pH at a favourable level for nutrient uptake rather than adding extra fertilizer. Therefore, research on how to increase P solubility in soil using different P solubilizing bacteria, endophytes, and mycorrhizae may be beneficial, in addition to efforts to breed varieties that secrete organic acids from roots. Finding alternatives to inorganic P is very important such as different animal manures and green manures, as they may be locally available at low cost for subsistence finger millet farmers.

6.3. Potassium Management

There has been limited attention paid to K management in finger millet. Lack of K fertilizer application and removal of crop residues by farmers have decreased soil K levels [73]. Most of the finger millet growing areas are located on marginal lands, which suffer from drought stress. As K improves the drought resistance of plants [74], finger millet can benefit from K fertilizer. Furthermore, application of K along with N fertilizer can also improve N use efficiency [75].

6.4. Micronutrients

Most micronutrient studies on finger millet have focused on Zn and B. There is a need to evaluate the effects of other micronutrients (Mn, Cu, Mo) and secondary macronutrients (e.g., Fe) on finger millet growth and yield. As micronutrients are expensive compared to the major macronutrients, research on micronutrient seed treatments may be beneficial.

6.5. Organic Manure

A significant amount of research has been conducted to evaluate the importance of FYM and alternative sources of organic matter on finger millet growth (Table 2). However, most of the studies are restricted to Asia, whereas there is a need to explore different sources of organic manure for finger millet farmers in Africa, as soil and climatic conditions are unfavorable for crop growth in most of Sub-Saharan Africa.

6.6. NPK + FYM + Soil Testing

Typically, farmers in South Asia and Africa apply N as the sole fertilizer to their crops, as N fertilizer is subsidized by the government. However, an optimal nutrient balance is necessary to obtain higher yields of finger millet. The current literature suggests that application of NPK along with micronutrients and FYM (7.5–12.5 t ha^{-1}) increases finger millet yield. As soil fertility varies from field to field, fertilizer recommendations based on a soil test will be ideal in order to maximize yield while enhancing fertilizer use efficiency. Best-management fertilizer practices, which involve the identification of the right source, right place, right timing, and right application method [76] will also lead to more efficient management of fertilizers in finger millet systems.

6.7. Green Manures, Organic Byproducts, and Bio-fertilizers

Where fertilizer subsidies are not available, the application of alternative organic fertilizers may be beneficial for subsistence finger millet farmers due to the high costs associated with inorganic fertilizers. As discussed above, different organic options can be used to minimize the amount of inorganic fertilizer required (Table 2). Possible microbial bio-fertilizer options for finger millet include

Azospirillum, PSB, *Trichoderma*, *Bacillus*, arbuscular mycorrhizal fungi (AMF), and plant growth-promoting rhizobacteria (PGPR).

6.8. Crop Rotations and Intercropping

Subsistence farmers already grow a diversity of crops to meet their needs. Inclusion of carefully selected legume crops in finger millet based crop rotations can help to minimize inorganic N fertilizer requirements, as symbiotically fixed N can be transferred from previous legumes to a subsequent finger millet crop. Similarly, intercropping of finger millet with legumes also improves the overall yield and sustainability of the system, however selection of suitable legume crops and optimization of finger millet:legume sowing densities are critical to minimize competition.

6.9. Fertilizer Micro-Dosing and Split Applications

Micro-dosing is a strategy wherein fertilizers are applied in small quantities close to the seed or plant by digging a small hole [77]. Compared to the traditional farmer method of broadcasting fertilizers, micro-dosing minimizes the amount and cost of fertilizer as it is targeted to where roots are positioned, thus preventing leaching. Micro-dosing is an appropriate strategy to increase the yield and fertilizer use efficiency under marginal lands with low moisture availability [78]. Micro-dosing can also be practiced with organic fertilizers in areas limited to manure. In East and Sothern Africa, it was found that the grain yield of finger millet can be increased by 20%–40% by micro-dosing N fertilizer at a rate of 20 kg N ha^{-1} compared to much higher rates using traditional broadcasting [77]. As N is vulnerable to losses especially through leaching and volatilization, split application of inorganic N fertilizer (and other mobile nutrients) is an alternative management approach. The concept is to synchronize nutrient supply with plant demand, resulting in increased nutrient use efficiency. As a recommended practice, N fertilizer can be applied at the time of planting and again at 25–30 days after planting.

6.10. System of Ragi Intensification (SRI method)

With the success of the system of rice intensification (SRI), a similar method has also been introduced to finger millet, which in India is known as ragi, and hence this strategy uses the same acronym (SRI). This method involves transplanting seedlings at the two-leaf stage with some soil attached to the root, at a distance of 30 × 30 cm in a square pattern. Weeding between the rows is performed three times at an interval of 10–15 days by using a cycle hoe or hand weeder [79]. Application of the SRI method has been shown to improve finger millet grain yield significantly while reducing the cost. Therefore, the SRI method may hold significant potential for subsistence finger millet farmers in Asia and Africa.

6.11. Varietal Breeding

As already noted, there appear to be limited efforts to breed new varieties of finger millet that are adapted to low or high nutrient conditions. The development of high-yielding finger millet varieties for the arid and semiarid regions is a high priority. In particular, breeding is required to adapt finger millet to deficiencies of specific nutrients, and on the opposite end of the spectrum, to adapt this crop to high doses of synthetic fertilizers, analogous to the selection of semi-dwarfs in rice and wheat during the Green Revolution. Given recent changes in rainfall, there is a particular need to improve the yields of short duration varieties of finger millet grown under high or low input conditions. There are also opportunities to breed finger millet to be more compatible with companion crops (e.g., legumes) that improve nutrient availability to finger millet. Unfortunately, multiple years are required to release a new variety, since station trials, multi location trials, and adaptive research trials need to be conducted to compare a new variety with previously recommended varieties [80]. Therefore, a long-term funding commitment is required to enable developing countries to develop high-yielding, locally-adapted finger millet varieties. The yield evaluation of currently available local millet varieties is also important in order to provide appropriate varietal recommendations to local farmers. As an example, large numbers of Nepalese finger millet cultivars have been evaluated for yield and agro morphological characteristics [81]. Local farmers in Nepal select their own seeds for the next growing seasons using extensive seed selection procedures, and extension work can play a positive supporting role as has been previously shown [82].

7. Conclusions

Although the world's food supply depends on approximately 150 plant species, the majority of cereal based calories comes from three major sources: maize, wheat, and rice. However, due to its high nutritional quality, finger millet still plays a vital role in supplying staple food mainly to poor peoples of the world, especially in Sub-Saharan Africa and South Asia. Due to its longer storability, finger millet provides food security for poor people in these regions. Land degradation due to poor crop management practices, and low land availability for cultivation due to increased population have limited global finger millet production. Therefore, there is a need to improve finger millet productivity in order to improve the nutritional status of vulnerable poor rural people. Farmer-friendly proper nutrient management practices along with rational cropping systems can play a key role in achieving this goal. It is hoped that this review will help to better inform agricultural extension officers and other groups who make recommendations to subsistence farmers.

Acknowledgements

We thank Vijay Bhosekhar (University of Guelph, Canada) for providing an initial framework for this review, and Kirit Patel (Canadian Mennonite University, Winnipeg, Canada) for inspiring our interest in finger millet. We thank Travis Goron (University of Guelph) for his helpful comments, Jaclyn Clark (University of Guelph) for editorial assistance, and our Nepalese partner organization, LI-BIRD, especially Kamal Khadka, for generously providing photos of finger millet cropping systems. This research was supported by a grant to MNR from the CIFSRF program, jointly funded by the International Food Development Centre (IDRC, Ottawa) and the Canadian Department of Foreign Affairs, Trade and Development.

Author Contributions

Both Malinda S. Thilakarathna and Manish N. Raizada conceived of the manuscript. Malinda S. Thilakarathna analyzed the literature and wrote the manuscript, and Manish N. Raizada edited the manuscript. Both authors read and approved the final manuscript.

Conflict of Interest

The authors declare no conflict of interest.

References

1. Kerr, R.B. Lost and found crops: Agrobiodiversity, indigenous knowledge, and a feminist political ecology of sorghum and finger millet in Northern Malawi. *Ann. Assoc. Am. Geogr.* **2014**, *104*, 577–593.
2. Pokharia, A.K.; Kharakwal, J.S.; Srivastava, A. Archaeobotanical evidence of millets in the Indian subcontinent with some observations on their role in the Indus civilization. *J. Archaeol. Sci.* **2014**, *42*, 442–455.
3. Goron, T.L.; Raizada, M.N. Genetic diversity and genomic resources available for the small millet crops to accelerate a new green revolution. *Front. Plant Sci.* **2015**, *6*, 157.
4. NRC (National Research Council). *Lost Crops of Africa*, 1st ed.; National Academy Press: Washington, DC, USA, 1996.
5. Dida, M.M.; Wanyera, N.; Dunn, M.L.H.; Bennetzen, J.L.; Devos, K.M. Population structure and diversity in finger millet (*Eleusine coracana*) germplasm. *Trop. Plant Biol.* **2008**, *1*, 131–141.
6. Gupta, N.; Gupta, A.K.; Gaur, V.S.; Kumar, A. Relationship of nitrogen use efficiency with the activities of enzymes involved in nitrogen uptake and assimilation of finger millet genotypes grown under different nitrogen inputs. *Sci. World J.* **2012**, 1–10.
7. Chandrashekar, A. Finger millet *Eleusine coracana*. *Adv. Food Nutr. Res.* **2010**, *59*, 215–262.
8. Dass, A.; Sudhishri, S.; Lenka, N.K. Integrated nutrient management to improve finger millet productivity and soil conditions in hilly region of Eastern India. *J. Crop Improv.* **2013**, *27*, 528–546.
9. Midega, C.A.O.; Khan, Z.R.; Amudavi, D.A.; Pittchar, J.; Pickett, J.A. Integrated management of *Striga hermonthica* and cereal stemborers in finger millet (*Eleusine coracana* (L.) Gaertn.) through intercropping with *Desmodium intortum*. *Int. J. Pest Manag.* **2010**, *56*, 145–151.
10. Department of Agriculture and Cooperation. *State of Indian Agriculture 2012–2013*; Ministry of Agriculture, Government of India: New Delhi, India, 2013.
11. Department of Agriculture and Cooperation. *State of Indian Agriculture 2013–2014*; Ministry of Agriculture, Government of India: New Delhi, India, 2014.
12. Agri-business Promotion and Statistics Divisions. *Statistical Information on Nepalese Agriculture 2013/2014*; Agri Statistics Section, Ministry of Agricultural Development, Government of Nepal: Kathmandu, Nepal, 2014.

13. Central Statistics Agency. *Report on Area and Production of Major Crops. Agricultural Sample Survey 2010–2011*; The Fedaral Democratic Republic of Ethiopia: Addis Ababa, Ethiopia, 2011; Volume 1.
14. Central Statistics Agency. *Report on Area and Production of Major Crops. Agricultural Sample Survey 2011–2012*; The Fedaral Democratic Republic of Ethiopia: Addis Ababa, Ethiopia, 2012; Volume 1.
15. Central Statistics Agency. *Report on Area and Production of Major Crops. Agricultural Sample Survey 2012–2013*; The Fedaral Democratic Republic of Ethiopia: Addis Ababa, Ethiopia, 2013; Volume 1.
16. Central Statistics Agency. *Report on Area and Production of Major Crops. Agricultural Sample Survey 2014–2015*; The Fedaral Democratic Republic of Ethiopia: Addis Ababa, Ethiopia, 2015; Volume 1.
17. Shobana, S.; Krishnaswamy, K.; Sudha, V.; Malleshi, N.G.; Anjana, R.M.; Palaniappan, L.; Mohan, V. Finger millet (Ragi, *Eleusine coracana* L.). A review of its nutritional properties, processing, and plausible health benefits. *Adv. Food Nutr. Res.* **2013**, *69*, 1–39.
18. Verma, V.; Patel, S. Value added products from nutri-cereals: Finger millet (*Eleusine coracana*). *Emirates J. Food Agric.* **2013**, *25*, 169–176.
19. Singh, P.; Raghuvanshi, R.S. Finger millet for food and nutritional security. *Afr. J. Food Sci.* **2012**, *6*, 77–84.
20. Rurinda, J.; Mapfumo, P.; van Wijk, M.T.; Mtambanengwe, F.; Rufino, M.C.; Chikowo, R.; Giller, K.E. Comparative assessment of maize, finger millet and sorghum for household food security in the face of increasing climatic risk. *Eur. J. Agron.* **2014**, *55*, 29–41.
21. Devi, P.B.; Vijayabharathi, R.; Sathyabama, S.; Malleshi, N.G.; Priyadarisini, V.B. Health benefits of finger millet (*Eleusine coracana* L.) polyphenols and dietary fiber: A review. *J. Food Sci. Technol.* **2014**, *51*, 1021–1040.
22. Hegde, B.R.; Gowda, L. Cropping systems and production technology for small millets in India. In Proceedings of the First International Small Millets Workshop, Bangalore, India, 29 October–2 November 1986; pp. 209–236.
23. Samarajeewa, K.B.D.; Horiuchi, T.; Oba, S. Finger millet (*Eleucine corocana* L. Gaertn.) as a cover crop on weed control, growth and yield of soybean under different tillage systems. *Soil Tillage Res.* **2006**, *90*, 93–99.
24. Tenywa, J.S.; Nyende, P.; Kidoido, M.; Kasenge, V.; Oryokot, J.; Mbowa, S. Prospects and constraints of finger millet production in Eastern Uganda. *Afr. Crop Sci. J.* **1999**, *7*, 569–583.
25. Rao, B.K.R.; Krishnappa, K.; Srinivasarao, C.; Wani, S.P.; Sahrawat, K.L.; Pardhasaradhi, G. Alleviation of multinutrient deficiency for productivity enhancement of rain-fed soybean and finger millet in the semi-arid region of India. *Commun. Soil Sci. Plant Anal.* **2012**, *43*, 1427–1435.
26. Wu, W.; Ma, B. Integrated nutrient management (INM) for sustaining crop productivity and reducing environmental impact: A review. *Sci. Total Environ.* **2015**, *512–513*, 415–427.
27. Chen, X.-P.; Cui, Z.-L.; Vitousek, P.M.; Cassman, K.G.; Matson, P.A.; Bai, J.-S.; Meng, Q.-F.; Hou, P.; Yue, S.-C.; Römheld, V.; Zhang, F.-S. Integrated soil-crop system management for food security. *Proc. Natl. Acad. Sci. USA* **2011**, *108*, 6399–6404.

28. Drinkwater, L.E.; Snapp, S.S. Nutrients in agroecosystems: Rethinking the management paradigm. *Adv. Agron.* **2007**, *92*, 163–186.
29. Sankar, G.R.M.; Sharma, K.L.; Dhanapal, G.N.; Shankar, M.A.; Mishra, P.K.; Venkateswarlu, B.; Grace, J.K. Influence of soil and fertilizer nutrients on sustainability of rainfed finger millet yield and soil fertility in semi-arid Alfisols. *Commun. Soil Sci. Plant Anal.* **2011**, *42*, 1462–1483.
30. Srinivasarao, C.; Venkateswarlu, B.; Singh, A.K.; Vittal, K.P.R.; Kundu, S.; Chary, G.R.; Gajanan, G.N.; Ramachandrappa, B.K. Critical carbon inputs to maintain soil organic carbon stocks under long-term finger-millet (*Eleusine coracana* [L.] Gaertn.) cropping on Alfisols in semiarid tropical India. *J. Plant Nutr. Soil Sci.* **2012**, *175*, 681–688.
31. Hemalatha, S.; Chellamuthu, S. Effect of long term fertilization on phosphorous fractions under finger millet-maize cropping sequence. *Madras Agric. J.* **2011**, *98*, 344–346.
32. Hemalatha, S.; Chellamuthu, S. Impacts of long term fertilization on soil nutritional quality under finger millet: Maize cropping sequence. *J. Environ. Res. Dev.* **2013**, *7*, 1571–1576.
33. Srinivasarao, C.; Venkateswarlu, B.; Lal, R.; Singh, A.K.; Kundu, S.; Vittal, K.P.R.; Ramachandrappa, B.K.; Gajanan, G.N. Long-term effects of crop residues and fertility management on carbon sequestration and agronomic productivity of groundnut–finger millet rotation on an Alfisol in southern India. *Int. J. Agric. Sustain.* **2012**, *10*, 230–244.
34. Kumara, O.; Naik, T.B.; Ananadakumar, B.M. Effect weed management practices and fertility levels on soil health in finger millet-groundnut cropping system. *Int. J. Agric. Sci.* **2014**, *10*, 351–355.
35. Basavaraju, T.B.; Gururaja Rao, M.R. Integrated nitrogen management in finger millet under rainfed conditions. *Karnataka J. Agric. Sci.* **1997**, *10*, 855–856.
36. Pushpa, H.M.; Gowda, R.C.; Naveen, D.V; Bhagyalakshmi, T.; Hanumanthappa, D.C. Influence of long term fertilizer application on root biomass and nutrient addition of finger millet. *Asian J. Soil Sci.* **2013**, *8*, 67–71.
37. Kumara, O.; Naik, T.B.; Palaiah, P. Effect of weed management practices and fertility levels on growth and yield parameters in finger millet. *Karnataka J. Agric. Sci.* **2007**, *20*, 230–233.
38. Matthews, R.B.; Pilbeam, C. Modelling the long-term productivity and soil fertility of maize/millet cropping systems in the mid-hills of Nepal. *Agric. Ecosyst. Environ.* **2005**, *111*, 119–139.
39. Pilbeam, C.J.; Gregory, P.J.; Tripathi, B.P.; Munankarmy, R.C. Fate of nitrogen-15-labelled fertilizer applied to maize-millet cropping systems in the mid-hills of Nepal. *Biol. Fertil. Soils* **2002**, *35*, 27–34.
40. Sherchan, D.P.; Pilbeam, C.J.; Gregory, P. Response of wheat-rice and maize/millet systems to fertilizer and manure applications in the mid-hills of Nepal. *Exp. Agric.* **1999**, *35*, 1–13.
41. Srinivasarao, C.; Wani, S.P.; Sahrawat, K.L.; Rego, T.J.; Pardhasaradhi, G. Zinc, boron and sulphur deficiencies are holding back the potential of rainfed crops in semi-arid India: Experiences from participatory watershed management. *Int. J. Plant Prod.* **2008**, *2*, 89–99.
42. Ahiwale, P.H.; Chavan, L.S.; Jagtap, D.N. Effect of establishment methods and nutrient management on yield attributes and yield of finger millet (*Eleusine coracana* G.). *Adv. Res. J. Crop Improv.* **2011**, *2*, 247–250.

43. Sridhara, C.J.; Mavarkar, N.S.; Naik, S.K. Yield maximization in Ragi under rainfed condition. *Karnataka J. Agric. Sci.* **2003**, *16*, 220–222.
44. Vijaymahantesh; Nanjappa, H.V.; Ramachandrappa, B.K. Effect of tillage and nutrient management practices on weed dynamics and yield of fingermillet (*Eleusine coracana* L.) under rainfed pigeonpea (*Cajanus cajan* L.) fingermillet system in Alfisols of Southern India. **2013**, *8*, 2470–2475.
45. Shivakumar, B.C.; Girish, A.C.; Gowda, B.; Kumar, G.C.V.; Mallikarjuna, A.P.; Thimmegowda, M.N. Influence of pongamia, mahua and neem cakes on finger millet productivity and soil fertility. *J. Appl. Nat. Sci.* **2011**, *3*, 274–276.
46. Saravanane, P.; Nanjappa, H.V.; Ramachandrappa, B.K.; Soumya, T. Effect of residual fertility of preceding potato crop on yield and nutrient uptake of finger millet. *Karnataka J. Agric. Sci.* **2011**, *24*, 234–236.
47. Govindappa, M.; Vishwanath, A.P.; Harsha, K.N.; Thimmegowda, P.; Jnanesh, A.C. Response of finger millet (*Eluesine coracana* L.) to organic and inorganic sources of nutrients under rainfed condition. *J. Crop Weed* **2009**, *5*, 291–293.
48. Parasuraman, P.; Mani, A.K. Growth, yield and economics of rice (*Oryza sativa*-finger millet (*Eleusine coracana*) crop sequence as influenced by coirpith and farmyard manure with and without inorganic fertilizers. *Indian J. Agron.* **2003**, *48*, 12–15.
49. Rangaraj, T.; Somasundaram, E.; Amanullah, M.M.; Thirumurugan, V.; Ramesh, S.; Ravi, S. Effect of agro-industrial wastes on soil properties and yield of irrigated finger millet (*Eleusine coracana* L. Gaertn) in coastal soil. *Res. J. Agric. Biol. Sci.* **2007**, *3*, 153–156.
50. Roy, D.K.; Chakraborty, T.; Sounda, G.; Maitra, S. Effect of fertility levels and plant population on yield and uptake of nitrogen, phosphorus and potassium in finger millet (*Eleusine coracana*) in lateritic soil of West Bengal. *Indian J. Agron.* **2001**, *46*, 707–711.
51. Rao, K.L.; Rao, C.P.; Rao, K.V. Response of finger miller (*Eleusine-coracana* L. Gaertn) cultivars to nitrogen under rain-fed conditions. *Indian J. Agron.* **1989**, *34*, 302–306.
52. Subramanian, K.S.; Kumaraswami, K. Effect of continuous cropping and fertilization on the response of finger millet to phosphorus. *J. Ind. Soc. Soil Sci.* **1989**, *37*, 328–332.
53. Ebanyat, P.; de Ridder, N.; de Jager, A.; Delve, R.J.; Bekunda, M.A.; Giller, K.E. Impacts of heterogeneity in soil fertility on legume-finger millet productivity, farmers' targeting and economic benefits. *Nutr. Cycl. Agroecosys.* **2010**, *87*, 209–231.
54. Bhoite, S.V.; Nimbalkar, V.S. Response of finger millet cultivars to nitrogen and phosphorus under rain-fed condition. *J. Maharashtra Agric. Univ.* **1996**, *20*, 189–190.
55. Kang, B.T.; Osiname, O. Micronutrient problems in tropical Africa. *Fertil. Res.* **1985**, *7*, 131–150.
56. Maury, A.N.; Verma, K.P. Zn tolerance by Finger millet (*Eleusine coracana* L.) under interactions of certain heavy metals and NPK. *Geobios (Jodhpur)* **1997**, *24*, 138–141.
57. Ramachandrappa, B.K.; Sathish, A.; Dhanapal, G.N.; Babu, P.N. Nutrient management strategies for enhancing productivity of dryland crops in Alfisols. *Indian J. Dryl. Agric. Res. Dev.* **2014**, *29*, 49–55.

58. Pilbeam, C.J.; Tripathi, B.P.; Munankarmy, R.C.; Gregory, P.J. Productivity and economic benefits of integrated nutrient management in three major cropping systems in the mid-hills of Nepal. *Mt. Res. Dev.* **1999**, *19*, 333–344.
59. Parasuraman, P. Response of farming studies in rainfed finger millet (*Eleusine coracana*) under erratic monsoon conditions of North-Western agroclimatic zone of Tamil Nadu. *Indian J. Agron.* **2002**, *47*, 384–389.
60. Subbiah, S.; Ramanathan, K.M.; Francis, H.J.; Sureshkumar, R.; Kothandaraman, G.V. Influence of nitrogenous fertilizers with and without neem cake blending on the yield of finger millet (*Eleucine coracana* Gaertn.). *J. Ind. Soc. Soil Sci.* **1982**, *30*, 37–43.
61. Ramakrishnan, K.; Bhuvaneswari, G. Effect of inoculation of am fungi and beneficial microorganisms on growth and nutrient uptake of *Eleusine coracana* (L.) Gaertn. (Finger millet). *Int. Lett. Nat. Sci.* **2014**, *13*, 59–69.
62. Apoorva, K.B.; Prakash, S.S.; Rajesh, N.; Nandin, B. STCR approach for optimizing integrated plant nutrient supply on growth, yield and economics of finger millet (*Eleusine coracana* (L.) Garten.). *EJBS* **2010**, *4*, 19–27.
63. Tewari, L.; Johri, B.N.; Tandon, S.M. Host genotype dependency and growth enhancing ability of VA-mycorrhizal fungi for *Eleusine coracana* (finger millet). *World J. Microbiol. Biotechnol.* **1993**, *9*, 191–195.
64. Adipala, E.; Okoboi, C.A.; Ssekabembe, C.K.; Ogenga-Latigo, M.W. Foliar diseases and yield of finger millet [*Eleusine coracana* (L.)] under monocropping and intercropping systems. *Discov. Inov.* **1994**, *6*, 301–306.
65. Padhi, A.K.; Panigrahi, R.K.; Jena, B.K. Effect of planting geometry and duration of intercrops on performance of pigeonpea-finger millet intercropping systems. *Indian J. Agric. Res.* **2010**, *44*, 43–47.
66. Maitra, S.; Ghosh, D.C.; Sounda, G.; Jana, P.K.; Roy, D.K. Productivity, competition and economics of intercropping legumes in finger millet (*Eleusine coracana*) at different fertility levels. *Indian J. Agric. Sci.* **2000**, *70*, 824–828.
67. Mal, B.; Padulosi, S.; Ravi, S.B. *Minor millets in South Asia: Learning from IFAD-NUS Project in India and Nepal*; Bioversity International, Maccarese, Rome, Italy; The M.S. Swaminathan Research Foundation, Chennai, India, 2010.
68. Pradhan, A.; Rajput, A.S.; Thakur, A. Yield and economic of finger millet (*Eleusine coracana* L . Gaertn) intercropping system. *Int. J. Curr. Microbiol. App. Sci.* **2014**, *3*, 626–629.
69. AICRP. *Research Achievements of AICRPs on Crop Sciences*; Indian Council of Agricultural Research Directorate of Information and Publications of Agriculture: Krishi Anusandhan Bhavan, New Delhi, India, 2007.
70. LI-BIRD. Constraints and opportunities for promotion of finger millet in Nepal. 2014, pp. 1–4. Available online: http://www.dhan.org/smallmillets/docs/report/Constraints_and_Opportunities_for_Promotion_of_Finger_Millet_in_Nepal.pdf (accessed on 17 May 2015).
71. Ryan, J.; Ibrikci, H.; Delgado, A.; Torrent, J.; Sommer, R.; Rashid, A. Significance of phosphorus for agriculture and the environment in the West Asia and North Africa region. In *Advances in Agronomy*; Elsevier Inc.: Amsterdam, The Netherlands, 2012; Volume 114, pp. 91–153.

72. Lynch, J.P.; Brown, K.M. Topsoil foraging—An architectural adaptation of plants to low phosphorus availability. *Plant Soil* **2001**, *237*, 225–237.
73. Timsina, J.; Singh, V.K.; Majumdar, K. Potassium management in rice-maize systems in South Asia. *J. Plant Nutr. Soil Sci.* **2013**, *176*, 317–330.
74. Cakmak, I. The role of potassium in alleviating detrimental effects of abiotic stresses in plants. *J. Plant Nutr. Soil Sci.* **2005**, *168*, 521–530.
75. Sangakkara, R.; Amarasekera, P.; Stamp, P. Growth, yields, and nitrogen-use efficiency of maize (*Zea mays* L.) and mungbean (*Vigna radiata* L. Wilczek) as affected by potassium fertilizer in tropical South Asia. *Commun. Soil Sci. Plant Anal.* **2011**, *42*, 832–843.
76. Ryan, J.; Sommer, R.; Ibrikci, H. Fertilizer best management practices: A perspective from the dryland West Asia-North Africa region. *J. Agron. Crop Sci.* **2012**, *198*, 57–67.
77. CGIAR Annual Progress Report. CGIAR Research Program on Dryland Cereals. Performance monitoring report for calendar year 2013. Available online: http://www.cgiar.org/resources/crp-documents/ (accessed on 17 May 2015).
78. Mgonja, M.; Audi, P.; Mgonja, A.P.; Manyasa, E.; Ojulong, H. Integrated blast and weed management and microdosing in finger millet. *A HOPE Project Manual for Increasing Finger Millet Productivity*; International crops research institute for the Semi-Arid Tropics, Patancheru, Hyderabad, Andhra Pradesh, India, 2013.
79. Mukherjee, K.; Gahoi, S.; Sinha, B. Promotion of SRI-Millet: Reopening a closed chapter. *Intl. J. Agric. Innov. Res.* **2014**, *3*, 812–816.
80. Geetha, K.; Mani, A.K.; Suresh, M.; Vijayabaskaran, S. Identification of finger millet (*Eleusine coracana* gaertn.) variety suitable to rainfed areas of north western zone of Tamil Nadu. *Indian J. Agric. Res.* **2012**, *28*, 60–64.
81. Bajracharya, S.; Prasad, R.C.; Budhathoki, S.K. Participatory crop improvement of Nepalese finger millet cultivars. *Nepal Agric. Res. J.* **2009**, *9*, 12–16.
82. Baniya, B.K.; Tiwari, R.; Chaudhary, P.; Shrestha, S.K.; Tiwari, P.R. Planting materials seed systems of finger millet, rice and taro in Jumla, Kaski and Bara districts of Nepal. *Nepal Agric. Res. J.* **2005**, *6*, 39–48.

6

A Combined Field/Laboratory Method for Assessment of Frost Tolerance with Freezing Tests and Chlorophyll Fluorescence

Franz-W. Badeck and Fulvia Rizza *

CRA-GPG—Council for Agricultural Research and Economics, Genomics Research Centre, Via San Protaso 302, Fiorenzuola d'Arda (PC) 29017, Italy; E-Mail: franz-werner.badeck@entecra.it

* Author to whom correspondence should be addressed; E-Mail: fulvia.rizza@entecra.it

Academic Editor: Tristan Coram

Abstract: Recent progress in genotyping allows for studies of the molecular genetic basis of cold resistance in cereals. However, as in many other fields of molecular genetic analysis, phenotyping for high numbers of genotypes is still a major bottleneck. The use of chlorophyll fluorescence measurements as an indicator for freezing stress is a well established and rapid method for evaluation of frost tolerance. In order to extend the applicability of this technique beyond plants grown under controlled conditions in growth chambers and sacrificed for the test, here we study its applicability for leaves harvested from field trials during winter and subjected to freezing tests. Such an approach allows for simultaneous studies of the advancement of cold hardening and other components of winter survival apart from frost tolerance. It is shown that cutting or senescence of cut leaves does not have adverse effects on the outcome of subsequent freezing stress tests. The time requirements for field sampling and laboratory testing on high numbers of genotypes allow for the application of the proposed approach for genotyping/phenotyping studies.

Keywords: phenotyping; genotyping; frost tolerance; hardening; freezing test; chlorophyll fluorescence; field-laboratory method; cereals

1. Introduction

In mid-latitudes, winter survival is an important factor that influences the productivity of autumn sown cereals. Successfully surviving winter depends on capacities to resist a series of stresses provoked by environmental conditions in winter, such as soil heaving, physiological drought, ice encasement, anoxy, diseases like snow molds and freezing damage [1]. Freezing tolerance (FT), *i.e.*, the resistance to cold stress, plays a major role in winter survival [2]. Rizza *et al.* [3] conservatively estimated that at least 50% of winter hardiness was associated with the level of FT in several European field tests on barley.

FT is a complex trait with polygenic inheritance [4]. Genetic variability in FT has been shown to be related to allele combinations of vernalization loci (e.g., [3,5–7]). Further dissection of the genetic determinants of FT as related to photoperiod (*PPD*) genes, CBF transcription factors, variation in copy number, polymorphisms of relevant genes and discovery of new QTLs (e.g., *FR-H3* [8]) are needed to refine the predictability of FT. Thus further progress in studies of association mapping and subsequent use in marker assisted selection (MAS) depends on fast phenotyping methods applicable to high numbers of genotypes.

Traditional phenotyping through electrolyte leakage tests and/or survival screening after exposure to a range of freezing stress temperatures are cost and labor intensive. For example, Fisk *et al.* [8] reported a cost of $90 per line for hardening and freezing under controlled conditions with subsequent survival analyses. Chlorophyll fluorescence as a rapid and reproducible alternative has been shown to produce results that correlate well with those obtained using other well-established methods [3,7,9–13]. Measurements of F_v/F_m provide an indirect assessment of damage, *i.e.*, they have an indicator function. Unlike studies of the direct effects of low temperatures on the efficiencies of various components of the photosynthetic apparatus [14], assessments of frost damage measuring F_v/F_m take F_v/F_m as an indicator of the breakdown of compartmentalization due to membrane damage [15] through the subsequent decline in maximum quantum yield of photosystem II photochemistry. FT characterized at the leaf level (either through F_v/F_m measurements after recovery or through electrolyte leakage tests) is not identical with FT at the whole plant level, *i.e.*, the capacity to survive and eventually resprout. However, previous studies have shown that both are correlated [3]. Subsequently, FT assessed at the leaf level is addressed as FT. Note that this approach aims at establishing a relative ranking of genotypes with respect to FT applicable for biodiversity and genomics studies. Studies that aim at distinguishing the direct and indirect effects of low freezing temperatures on the photosynthetic apparatus as well as characterization of whole plant FT in absolute terms require application of other experimental approaches.

At any given stage of plant development, FT depends on the previous environmental conditions to which the plants were exposed, because FT increases upon exposure to low non-lethal temperatures, due to an adaptive process termed cold hardening, and decreases due to dehardening when plants are exposed to mild temperatures [1,16]. Thus, the FT of a given genotype will vary between sites and instants. Therefore, it is not possible to determine FT with observations in the field, whereas winter survival as a complex response to different stresses can be comprehensively assessed. In addition, the frequency distribution of winter stress conditions at inter-annual time scales may include a low fraction of winters that are suited for the purpose of FT evaluation (so called "test winters" or "differential winters", *sensu* Levitt [2]). Especially in environments characterized by winters with aleatory severe frost events, many experimental years will not result in stress-conditions suitable for discrimination between more or less resistant cultivars. It will be excluded during mild winters that do not damage even the most sensitive genotypes and during winters with extremely cold episodes when all genotypes are severely damaged [8]. Controlled and repeatable growth and hardening conditions in growth chamber experiments and artificial freezing stress at defined temperatures are thus needed for quantification of FT at defined levels of hardening. For tests performed under controlled conditions, plants can be grown and acclimated in containers that occupy relatively little space for some weeks only.

Besides the evaluation of FT for plants grown under controlled conditions, it remains important to evaluate FT and/or survival probability at a given state of hardening and potential supplementary stresses in the field, excluding subsequent additional stresses [17]. To this end, several methodological approaches have been proposed and tested in the literature. One modification of the field methods is the "provocation pot method." Plants are grown and wintered in boxes, placed at different heights above ground and transferred into a greenhouse after exposure to a strong frost event or other stresses. Subsequently, plant survival is determined after regeneration [18]. With several variants of field-laboratory methods (FLM) coined "provocation tests," plants grown in the field or in pots that are placed outside during winter are transferred to the laboratory and exposed to frost using freezers [16,19,20].

In combination with these field-laboratory methods, high-throughput phenotyping techniques are needed to study the FT after hardening under field conditions as well as to allow for combined assessment of frost hardiness during winter and winter survival. Harvesting of single leaves from field trials and subsequently exposing them to freezing temperatures in controlled laboratory freezing tests combined with chlorophyll fluorescence analysis is potentially a viable solution [21,22]. Such an approach is only minimally invasive in the sense that the plant from which the leaf is harvested for a freezing test is not killed. In addition, the high number of individuals present even in small experimental field plots allows for repeated harvesting with sufficient sampling size in the course of the winter. Furthermore, laboratory freezing tests increase the capacity for evaluation of frost resistance, especially in environments that are characterized by winters with aleatory severe frost events only. The laboratory freezing test allows for provocation of damaging low temperatures within the background of a state of hardening obtained under ambient weather conditions.

Analogous applications of chlorophyll fluorescence analyses on cut leaves exposed to stress have been reported for heat stress in laboratory and greenhouse experiments on tomato [23], cotton (for screening genetic diversity) [24], and wheat [25], as well as rice (for chilling sensitivity) [26]. Chlorophyll fluorescence can also be measured to study the response of the photosynthetic apparatus to non-lethal low-temperature stress through, e.g., effects on ϕPSII, diverse quenching mechanisms [14], and several parameters that can be derived from fast fluorescence transients [20,27].

However, here we focus on the use of chlorophyll fluorescence as an indicator for lethal damage applied in the context of combined field-laboratory methods. This approach needs to be scrutinized for some potential pitfalls. Firstly, it must be determined whether subjecting detached leaves to the freezing test introduces a bias. Secondly, the time horizon up to which leaf senescence induced by detaching the leaf from the plant does not impact repeated measurements during recovery from freezing stress needs to be determined. Detaching leaves from the plant leads to physiological changes, e.g., switching from export of carbohydrates to storage as fructans in gramineae leaves, a change through which sink limitation of photosynthesis is avoided for up to several days after cutting [28]. However, in the context of using PSII chlorophyll fluorescence as an indicator for leaf damage, cutting-induced changes that lead to senescence, defined as reduction in chlorophyll content and photosynthetic capacity, remobilization of nitrogen from the leaf [29] that could potentially interfere with the assessment of FT, need to be excluded.

Therefore, following the development of FLM approaches described above and coupled with the use of chlorophyll fluorescence as an indicator, we further investigate the feasibility of using such an approach when phenotyping FT in large panels of genotypes. To this end we test the following hypotheses:

- that leaf damage assessed with chlorophyll fluorescence measurements after freezing tests does not differ between leaves left attached to the living plant and cut leaves; and
- that the onset of leaf senescence in detached leaves is sufficiently late as to not interfere with measurements of recovery after freezing tests.

These tests are reported with the objective of demonstrating the reliability and sensitivity of measurements of chlorophyll fluorescence in leaves detached in the field and evaluating the feasibility of application to discriminate genotypes with different FT in cereals grown under various environmental conditions during winter (see Table 1 for a list of the experiments). Some applications are reported that illustrate the utility of field sampling combined with laboratory freezing tests to address questions that cannot be approached with pure growth chamber studies or measurements taken exclusively in the field.

Table 1. Abbreviations for the experiments; studied species; growth conditions: GC = plants grown in growth chambers, field = plants grown in the field and leaves cut for subsequent freezing tests and measurements; Treatments: i = measurements on attached leaves, c = measurements on cut leaves, FACE = Free Air Carbon Dioxide Enrichment; Stress temperature = minimum temperature during freezing stress tests (°C); number of genotypes studied. See Table S1 in the Electronic Supplementary Materials (ESI) for more information on the field trials.

Experiment	Species	Growth Conditions	Treatments	Stress Temperature	Number of Genotypes
E1	*H. vulgare*	GC	E1i, E1c, E1i, E1c	−13 and no stress test	31
E2	*H. vulgare*	GC	E2i, E2c, E2i, E2c	−13 and no stress test	31
E3	*H. vulgare*	field	E3	−14	30
E4	*A. sativa*	field	E4	No stress test	34
E5	*A. sativa*	field	E5	No stress test	46
E6	*T. aestivum*	field	E6 FACE, ambient & elevated CO_2	No stress test	1
E7	*T. durum*	field	E7a, E7b ambient CO_2	No stress test	1
E8	*T. durum*	field	E8 FACE, ambient & elevated CO_2	−14	12
E9	*H. vulgare*	field	E9	−14	55

2. Results and Discussion

The combination of hardening in the field, harvesting of leaves from field plots and subsequent testing for FT builds on the assumptions that cut leaves are not subject to changes in the capacities and functioning of their photosynthetic apparatus, due to the immediate effects of cutting. In addition, F_v/F_m, the maximum quantum yield of photosystem II used as indicator of damage, should not change due to incipient senescence for at least four days after cutting, a time period that exceeds the length of a full freezing test experiment. Thus, it needs to be shown that cutting leaves for testing FT does not have an additional effect due to cutting.

2.1. Effects of Test Conditions

Before addressing the effects of cutting, the dynamics of the manifestation of damage symptoms needs to be described. Damage symptoms in the field are affected by conditions at harvest (hardening stage, degree of freezing stress in preceding nights, light intensity, other potential causes of damage). Factors that cause variability not related to the freezing damage itself can be minimized under controlled conditions in growth chamber experiments by testing F_v/F_m in intact leaves before and after freezing stress, as well as after a recovery period. At recovery, damage is revealed in the presence of light, and differences in FT are more evident because the difference in PS II maximum yield between damaged and non-damaged leaves is greater relative to the measurements taken directly after the end of the stress treatment [11,21].

2.1.1. Effect of Test Conditions in Absence of Freezing Stress: Attached *vs.* Cut Leaves (Effect of Cutting) in Growth Chamber Experiments

When F_v/F_m was measured for attached and cut leaves (E1, E2) that were not subjected to a freezing stress, at the time at which stressed leaves were measured after the recovery period (Figure 1, filled triangles), both attached and cut leaves had high F_v/F_m values typical for intact non-stressed PS II. Thus, in plants grown under controlled conditions within growth chambers, leaf senescence effects on PS II efficiency in detached leaves did not interfere with freezing tests. The implicit hypothesis that this also indicates there is no membrane damage should be tested with further studies.

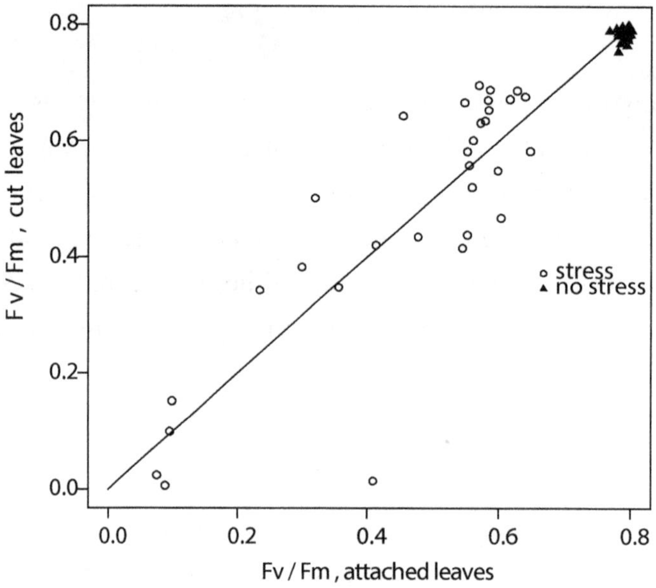

Figure 1. FT of 31 barley genotypes (E1, E2) assessed with chlorophyll fluorescence after freezing tests and recovery. Varietal mean F_v/F_m of leaves still attached to the plant *vs.* those cut prior to the freezing test for experiments with a stress temperature of −13 °C. F_v/F_m measured on non-stressed leaves in E2 in parallel to the stressed leaves after time elapsed for the freezing test and 24 h of recovery is shown with filled triangles.

2.1.2. Freezing Tests on Attached and Cut Leaves Lead to Consistently Similar Results in Growth Chamber Experiments

Performing freezing stress tests on attached and cut leaves in parallel (experiments E1 and E2) led to highly significant correlations ($p < 0.001$) between the varietal averages for FT with correlation coefficients of 0.81 for E1 and 0.66 for E2. The mean FT did not differ significantly between attached and cut leaves (paired *t*-test, $p = 0.577$). The mean coefficient of variation was about 0.1 for non-damaged varieties and up to 0.8 for varieties close to their LT_{50}. Thus, for measurements on a variety under identical conditions, the correlation coefficients indicate a good correspondence in FT between tests done on attached leaves and tests done on cut leaves. The ranking for FT for the average of experiments E1 and E2 was highly similar when assessed on attached or on cut leaves (Figure 1, open circles). The same conclusion was drawn from the comparison of cut leaves and leaves that remained attached to the plant exposed to heat stress for a test completed within two hours of cutting [23].

2.1.3. Daylight during Recovery Accelerates the Development of Damage Symptoms E1, E2

Recovery of attached leaves from stress in a 10/14 h day/night regime resulted in faster development of the damage symptoms (Figure 2) as compared to recovery in 24 h of darkness. However, the average genotype FT under the two recovery conditions were highly significantly correlated ($p < 0.001$) with correlation coefficients of 0.82 (E1) and 0.74 (E2). The measurements taken immediately after the end of the freezing stress treatment on the plants subsequently recovering in the light and in the dark significantly correlated with correlation coefficients of 0.67 (E1) and 0.78 (E2), and the correlation between measurements taken post-stress and after recovery was >0.78. Thus, it can be concluded that recovery in a day/night light regime accelerates the expression of damage symptoms but does not distort the differences between genotypes.

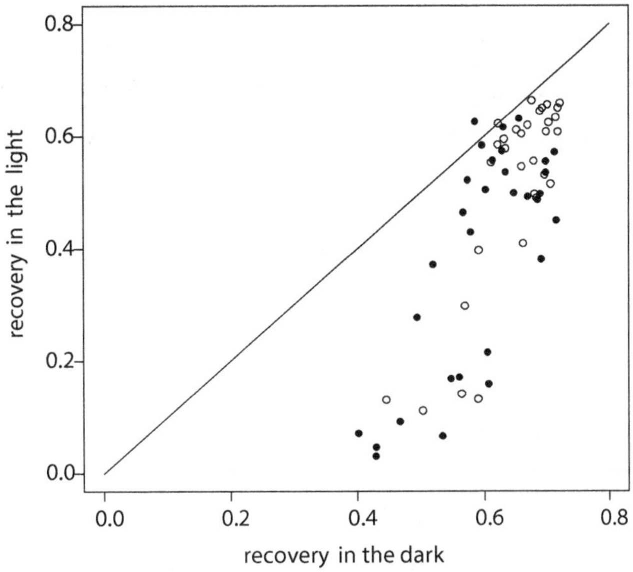

Figure 2. Comparison of FT of attached leaves as assessed with chlorophyll fluorescence subsequent to a freezing test after 24 h recovery in complete darkness *vs.* recovery in a day/night light regime for freezing test performed at −13 °C (E1 open circles and E2 closed circles).

These tests confirm the role of chlorophyll fluorescence emitted by photosystem II as an indicator of frost damage that manifests more rapidly when the recovery proceeds in the light. The primary frost damage presumably consists of damage to membranes and consequent membrane leakage [15]. Then, the length of the exposure to the effects of membrane leakage and the facilitation of spread of damage by recovery in the light would determine the speed with which damage to PS II manifests, and thus improve discrimination between genotypes. Measurements of F_v/F_m after recovery in a 24 h day/night cycle discriminate between damaged and non-damaged leaves more effectively than measurements taken immediately after the end of the freezing treatment or after recovery in the dark and are thus better suited to discriminate between genotypes. Therefore, it is important to apply a defined protocol to the freezing and testing sequence (see also [30] for sensitivity to the length of exposure to freezing temperatures, [10] for sensitivity to temperature during the recovery period, [11,20] for increasing discriminatory power of F_v/F_m after recovery).

2.1.4. Effect of Cutting in Absence of Artificial Freezing Stress in Leaves Sampled in the Field

A condition for the application of freezing tests on leaves cut from plants in the field is that traumatic effects of cutting on the photosynthetic functions that are used for diagnosis of freezing damage can be excluded.

Barley: Leaves of the same barley genotypes tested with experiments E1 and E2 (except one genotype, Pamina) were cut in a field trial in January 2012 (E3), and F_v/F_m was subsequently monitored on leaves not subject to a freezing test for three days (Table 2). F_v/F_m slightly increased over time, indicating that the functioning of PS II was not impaired by cutting and keeping the leaves in Falcon tubes in a humid environment. Leaves sampled from the same experiment in December and February had an average F_v/F_m of 0.76 and 0.72 two hours after sampling and of 0.78 and 0.72 four hours after sampling, respectively. Leaf conditions are subject to preceding freezing stress and high light stresses in the field [22]. The variability between varieties is not an effect of cutting, as can be concluded from the differences of F_v/F_m measurements in hours subsequent to cutting taken at different sampling dates that include plants at different levels of stress experienced in the field prior to cutting.

Table 2. Average Photosystem II maximal quantum yield (F_v/F_m) after transfer of cut leaves of 30 barley genotypes acclimated in the field to the laboratory (E3). $n = 180 = 6$ leaves per genotype × 30 genotypes. F_v/F_m mean with lowest and highest genotype mean in parentheses. Sd = Standard deviation of the full set of all single measurements. Significant differences between sampling times tested with Wilcoxon rank sum test.

Time after Sampling	F_v/F_m	Sd	Differences
2 h	0.68 (0.58, 0.72)	0.061	c
4 h	0.68 (0.57, 0.72)	0.063	c
1 day	0.70 (0.58, 0.74)	0.062	b
2 days	0.68 (0.59, 0.74)	0.069	c
3 days	0.72 (0.55, 0.76)	0.077	a

Are these observations on barley representative of other cereal species? In order to respond to this question, post-cutting monitoring was also performed on oat, bread and durum wheat genotypes.

Oat: After two hours of dark acclimation, the average F_v/F_m of 34 oat genotypes (E4) was 0.536, and after five hours it had risen to 0.584. Four genotypes were damaged by frosts occurring in the field prior to the sampling. Two hours after sampling on dark adapted leaves, the average F_v/F_m of 46 oat genotypes (E5) was 0.694 and had risen after four hours after sampling to 0.718.

Wheat: After 0.75 h, 4 h, and 15.25 h in the dark, the F_v/F_m leaves of the bread wheat variety Bologna (E6) sampled on 31 January, 2012 of elevated CO_2 (FACE) and control treatments (Table 3) did not differ. The cut leaves, kept in closed falcons to avoid dessication, recovered from slight non photochemical quenching as indicated by the increase in F_v/F_m in the course of time. As high values of F_v/F_m at the maximum usually exhibited by healthy, unstressed leaves (0.80 to 0.82) were recorded, there is no indication of detrimental effects on the functioning of photosystem II when keeping leaves for 15 h after cutting them from the plant. These results are in accordance with constancy of high F_v/F_m reported by Sharma *et al.* [25] for a spring wheat variety up to two hours after cutting.

Table 3. Bologna FACE (E) and ambient control (A) plots. $n = 8$ leaves per treatment. F_v/F_m mean with lowest and highest genotype mean in parentheses. Sd = standard deviation. Significant differences between sampling times tested with Wilcoxon rank sum test.

Treatment	Time after Sampling	F_v/F_m	Sd	Differences
A	0.75 h	0.76 (0.73–0.79)	0.020	c
A	4 h	0.79 (0.77–0.81)	0.017	b
A	15.25 h	0.81 (0.80–0.82)	0.006	a
E	0.75 h	0.76 (0.70–0.81)	0.040	c
E	4 h	0.79 (0.75–0.84)	0.028	bc
E	15.25 h	0.81 (0.79–0.83)	0.014	ab

The measurements made on the durum wheat variety Claudio (E7) also sampled on 31 January 2012 produced similar results. After 0.75 h, 4 h, and 15.25 h in the dark, the F_v/F_m of leaves cut from plants sown at two different sowing dates did not differ (Table 4).

Table 4. Claudio first sowing date (S1) and second sowing date (S2) plots. $n = 9$ leaves per sowing date, of which 5 were kept in the light and 4 in darkness after the measurement at 4 h after cutting. F_v/F_m mean with lowest and highest genotype mean in parentheses. Sd = standard deviation. Significant differences between sampling times tested with Wilcoxon rank sum test.

Treatment	Time after Sampling	F_v/F_m	Sd	Differences
S1	0.75 h	0.73 (0.68–0.78)	0.036	c
S1	4 h	0.77 (0.75–0.80)	0.015	b
S1	15.25 h	0.81 (0.80–0.83)	0.010	a
S2	0.75 h	0.75 (0.70–0.80)	0.034	c
S2	4 h	0.77 (0.75–0.80)	0.019	b
S2	15.25 h	0.82 (0.81–0.83)	0.010	a

The cut leaves recovered from slight reduction in the maximum yield of photosystem II as indicated by the increase in F_v/F_m in the course of time. As high values of F_v/F_m at the maximum usually exhibited by healthy, unstressed leaves (0.80 to 0.82) were recorded, there is no indication of detrimental effects on the functioning of photosystem II when keeping leaves for 15 h after cutting them from the plant. No significant differences were detected between plants kept under PPFD of 150 μmol $m^{-2}\cdot s^{-1}$ or those kept in the dark between the second and the last measurement.

In order to assess the integrity of the photosynthetic apparatus on even longer time scales, leaves of the durum wheat variety Claudio were cut in the field on 31 January 2012 (H1), 3 February (H2), 8 February (H3), and 9 February (H4a and H4b) when plants were under snow cover and on 22 February 2012 (H5b) after snow melt. Monitoring of F_v/F_m for leaves kept in the dark confirmed the recovery of maximum capacity during the first hours after cutting and revealed that the integrity of PS II was conserved for at least 4 days after cutting (DAC, Figure 3a) with F_v/F_m values close to 0.8. Between days 3 and 6 after cutting, a slow decline in F_v/F_m is seen with values that remain above 0.7. From day 6 after cutting onwards, a rapid decline occurs, indicating leaf senescence. Also after snow melt, F_v/F_m remained high at an average of 0.77 three to five hours after sampling. Leaves cut at 10 February 2013, kept under low light (PPFD: 150 μmol $m^{-2}\cdot s^{-1}$) and low temperature

(3 °C day, 1 °C night) maintained PS II integrity for even longer (Figure 3b). Beyond PS II, the full photosynthetic machinery also remained intact and worked at unchanged capacity until DAC 11, as evidenced by the electron transport rates measured based on chlorophyll fluorescence (Figure 4). At DAC 11, between-leaf variability had started to increase, and at DAC 19, ETR was reduced by 25%.

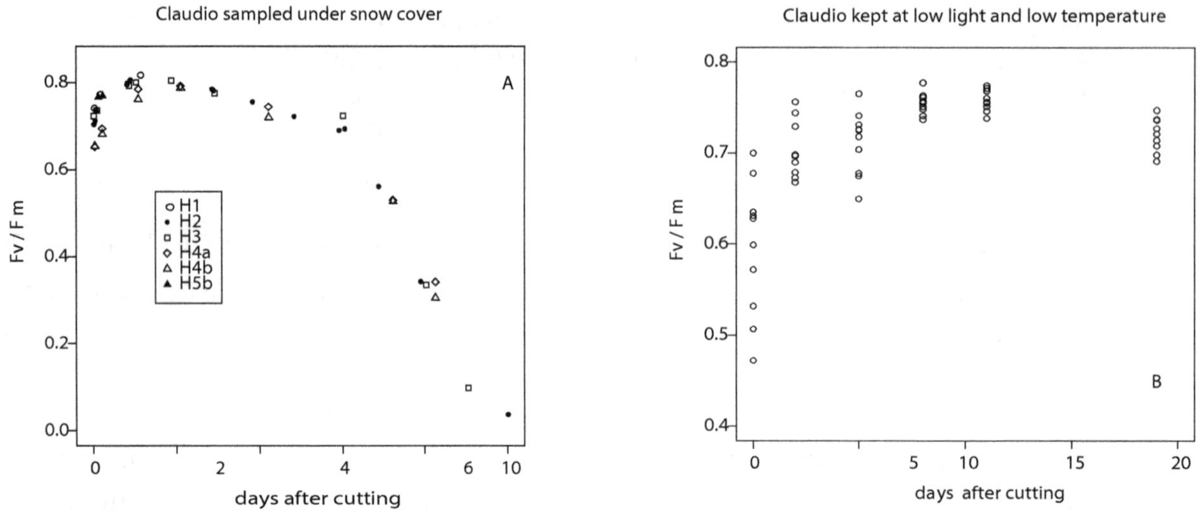

(a) (b)

Figure 3. (**a**) Time course of F_v/F_m of leaves of the durum wheat variety Claudio cut in the field when plants were under snow cover and kept in the dark. Different symbols denote harvests in January/February 2012 at different times (H1, H2, H3, H4a, H4b, H5b). H4a leaves under snow cover, H4b from a plot with snow removed the day before cutting, H5b after snow melt; (**b**) Time course of F_v/F_m of leaves of the durum wheat variety Claudio cut in the field when plants where under snow cover (10 February 2013) and kept in the light at low temperature (E7).

In summary, the tests of PS II maximum quantum yield and ETR show that leaves cut in the field and kept in Falcon tubes maintain fully functional photosynthetic machinery for several days when kept in the dark at room temperature or when kept in low light at low non-freezing temperatures. In any case, a slow decrease in photosynthetic capacities did not start before the 40 h maximally applied for the full length of freezing tests, including the subsequent recovery phase. Thus, cut cereal leaves can be regarded as a viable system for FT tests, as interference of leaf senescence or damage due to cutting with the integrity of the photosynthetic machinery, which is used as an indicator, cannot be detected.

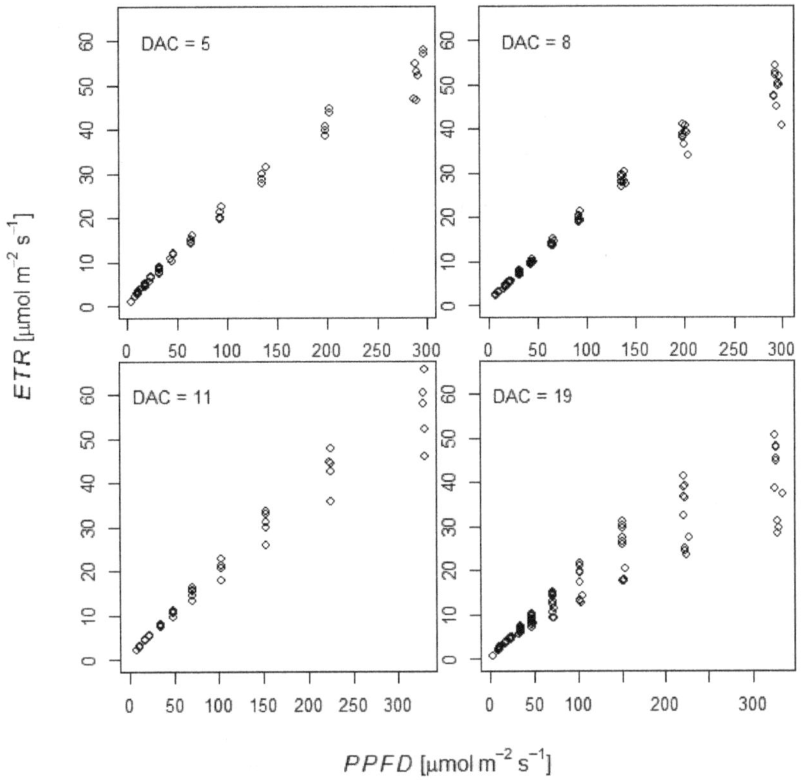

Figure 4. Response of electron transport rate (ETR) to Photosynthetically active Photon Flux Density (PPFD) of the durum wheat variety Claudio at 5, 8, 11 and 19 days after cutting (DAC), respectively (E7).

2.2. Applications of the Field Sampling/Laboratory Test Methods

2.2.1. Seasonal Monitoring of Frost Hardiness

Seasonal monitoring of FT of barley leaves harvested in the field (E3) and subjected to laboratory freezing tests at −14 °C indicated an increase in FT from early winter (December) to January (Figure 5). Frost stress treatments had more severe effects in December due to the short time for acclimation and earlier growth stage of the plants with respect to the other sampling dates. The F_v/F_m of the leaves measured before the freezing stress test also revealed differences between the sampling dates (see Section 2.1.4).

The mean relative ranks of the 30 barley varieties determined for FT measured in the chamber experiments (E1, E2) and the field-laboratory test (E3) were correlated with $r = 0.70$ ($p < 0.001$). This correlation coefficient is similar to those between cut and attached leaves within the single chamber experiments described in Section 2.1.1 (0.81 and 0.66). However, the results obtained for the single harvesting dates differed in the degree to which they led to a FT ranking similar to the one obtained with growth chamber experiments. For the sampling dates in December and February, the varietal ranking correlated with the the ranking obtained within E1 and E2, while the January sampling led to weak, non-significant correlations ($r = 0.09$ to 0.24) with the other experiments. *A priori,* it is not expected that the ranking of FT remains constant throughout the winter season in the field because FT is not a static trait and it changes with nutrition, moisture, plant age, the duration and intensity of low temperatures, light intensity [1] and daylength [31]. In addition to the resulting state of hardening/dehardening, damage to the plants under field conditions and very slowly reversable reduction of F_v/F_m can also effect the results of subsequent freezing tests [16]. In the January sampling, F_v/F_m measured prior to the freezing test correlated weakly ($r = 0.3$) but significantly ($p < 0.001$) with F_v/F_m measured after recovery, while the two measurements did not correlate for the December and the February sampling. This correlation, together with the slow recovery of high F_v/F_m (see Section 2.1.4), indicates an unidentified stress to which the plants were subjected in the field prior to sampling.

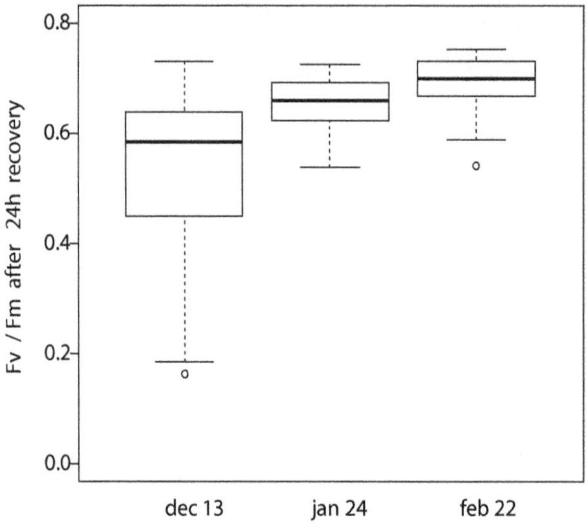

Figure 5. FT of barley varieties assessed in winter 2011/2012 on leaves harvested from experimental plots at CRA-GPG, Fiorenzuola d'Arda, Italy. Freezing stress at −14 °C. Box plots represent measurement of 30 barley varieties (E3).

In conclusion, for seasonal monitoring of FT, two peculiarities need to be taken into account: (1) the minimum temperature applied during the freezing tests used to produce good discrimination of genotype FT varies throughout the season; (2) damage or only slowly reversable downregulation of F_v/F_m present at the time of sampling impact on FT screening and need to be taken into account.

2.2.2. Effect of Elevated CO_2 on Frost Tolerance of Durum Wheat

Free Air Carbon dioxide Enrichment (FACE) techniques have been developed in order to study the effect of elevated CO_2 on plant performance under field conditions (E8). If the relevance of FT for growth, survival and production is to be evaluated in the frame of such an experimentation, evidently field sampling for evaluation of FT obtained in different growth stages is the appropriate method, as a parallel growth chamber experiment would not simulate the full set of potential interactions that may play a role in the field. A freezing test at −14 °C performed with leaves of 12 durum wheat genotypes cut from a FACE experiment [32] at three dates (9 December 2011, 3 January 2012 and 25 February 2012) resulted in highly significant effect (ANOVA) from the sampling date. On average, the plants showed a progressive increase in FT during the winter months. The average F_v/F_m increased from 0.28 in December to 0.69 in January and 0.74 in February (see Table S3 for results by genotype). The FT of the tested genotypes differed significantly, but no significant effect of elevated CO_2 or genotype by treatment interaction was found.

2.2.3. Phenotyping/Genotyping Study under Field Hardening Conditions

The association of FT of barley genotypes with the haplotypes of the vernalization loci *VRN-H1* and *VRN-H2* were studied by Rizza *et al.* [3] with freezing tests using barley seedlings grown in growth chambers under controlled conditions. Freezing tests after hardening under optimal (four weeks, 3/1 °C day/night) and suboptimal (shorter, *i.e.*, three weeks or one week, or at higher temperatures, *i.e.*, 12/7 °C day/night) hardening conditions revealed significant differences between the haplotypes AA, AB, BA and BB, where the abbreviations AA, BB, AB and BA indicate the *vrn-H1*/*Vrn-H2* (winter), *Vrn-H1*/*vrn-H2* (spring), *vrn-H1*/*vrn-H2* (facultative), and *Vrn-H1*/*Vrn-H2* (spring) haplotypes, respectively, with reference to the alleles of the varieties Nure (AA, winter) and Tremois (BB, spring). The genotypes carrying the haplotype AB were most resistant, especially when suboptimal hardening conditions were applied. Throughout the test, the order of resistance was AB > AA > BA > BB. It remains to be determined whether these results, obtained with laboratory studies, are representative of field conditions. The same set of barley genotypes was grown in the field, with leaves harvested and subsequently exposed to a freezing test (E9). The results (Figure 6, Table S2) are highly similar to the growth chamber/laboratory test studies. The FT difference between haplotypes is significant, with the order of the median response corresponding to the order described above. The FT of the genotype mean in the field-laboratory experiment and in the growth chamber experiment is correlated with $r = 0.61$ ($p < 0.001$).

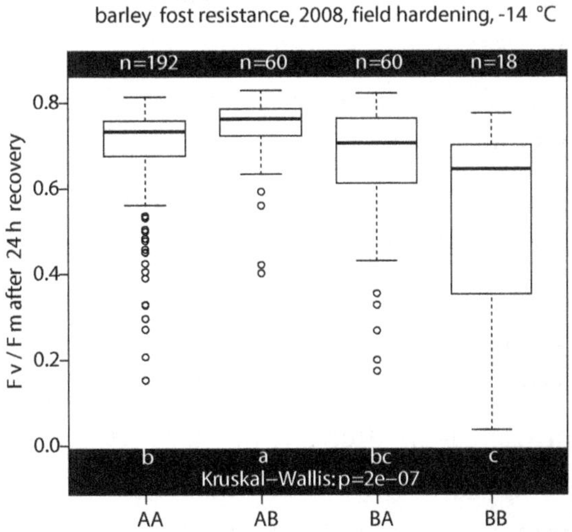

Figure 6. FT of 55 barley varieties sampled in the field in February 2008, assessed with F_v/F_m after freezing tests and 24 h of recovery. n = number of measurements = 6 replicates × N, with N = number of genotypes. AA, AB, BA and BB refer to the Nure (A) and Tremois (B) alleles for the *VRN-H1* and *VRN-H2* vernalization genes [6].

These results confirm the applicability of the combined field sampling/laboratory testing approach as scrutinized in the previous sections for phenotyping large sets of genotypes as required for combined phenotyping/genotyping studies.

3. Experimental Section

For growth chamber experiments (Table 1; E1, E2) barley plants were grown from seed at 20/15 °C day/night temperatures for one week at 10 h daylength and 200 μmol $m^{-2} \cdot s^{-1}$ PPFD. Subsequently, they were transferred to hardening conditions of 3/1 °C day/night for 4 weeks of hardening. Two series of independent freezing tests were carried out at freezing temperatures of −13 °C ± 1 °C. For these experiments, E1 and E2, either intact plants (E1i, E2i) or cut leaves (E1c, E2c) were subjected to the freezing treatment [11]. F_v/F_m was measured immediately before applying the freezing stress (pre stress), immediately after the end of the freezing stress (post stress), and after 24 h of recovery. In order to assess the role of light in the expression of post-stress damage, leaves were kept in darkness or in the light during the recovery period. Eventual rapid senescence was tested by comparing the results after recovery to measurements done on non-stressed plants kept under growth chamber conditions.

Plants from barley (E3, E9) and oat (E4, E5) field-trials and a bread and durum wheat FACE (Free Air Carbon dioxide Enrichment) experiment (E6, E7, E8) [32] were used for field sampling of the last fully developed leaves and then kept in the dark (see Table 1 and Table S1 for further information on the field trials). F_v/F_m was monitored after different lengths of dark acclimation to study reversion of initial reduction in the intrinsic maximal yield of photosystem II. Samples from E3 were taken at three sampling dates in December, January and February. For experiment E6, leaves were sampled on 31 January 2012 from the soft wheat variety Bologna in FACE and control rings, and from the durum wheat variety Claudio in the field surrounding the FACE experiment (E7). Claudio was sampled from plots with sowing dates 23 October 2011 (E7a) and 30 October 2011 (E7b). Five leaves of 34 oat genotypes each were cut in the field on 25 January 2011 (E4). Three leaves of 46 oat genotypes each were cut in the field on 19 December 2011 (E5) at 12:30 on a sunny day. Samples from E9 were taken on 13 February 2008. Information on sowing dates of the field experiments and freezing temperatures during nights preceding field sampling are reported with Table S1 in the Electronic Supplementary Materials.

Immediately after cutting, leaves were kept in the dark in Falcon tubes under the external weather conditions during the sampling time. Successively, F_v/F_m was measured after further dark acclimation at room temperature for at least half an hour (as detailed for the individual experiments). In some experiments, a second measurement was taken after one or more additional hours in the dark to follow reversion in the intrinsic maximal yield of photosystem II.

The freezing treatments were simulated in the dark in a temperature test cabinet (Vötsch VT 3050V, Weiss Technik, Magenta (Milano, Italy) as described in [11]: temperature was gradually reduced (2 $°C \cdot h^{-1}$) to −3 °C, and plants were kept at this temperature for 16 h. Subsequently, temperature was gradually lowered (2 $°C \cdot h^{-1}$) to the freezing temperature indicated for each experiment. Plants were kept at this temperature for 16 h. The temperature was then gradually raised to 1 °C at 2 $°C \cdot h^{-1}$. Plants were kept for 1 h at 1 °C to enable assessment of leaf-tissue damage. Freezing tolerance was quantified by measuring chlorophyll fluorescence using the last fully expanded leaf. F_v/F_m in a dark adapted state (Butler and Kitajima, 1975) was assessed by using a PAM-2000 fluorometer (Walz, Effeltrich, Upper Franconia, Germany). The measuring modulated light for determination of initial F_0 was sufficiently low (<0.1 μmol $m^{-2} \cdot s^{-1}$). Saturating white light of about 10,000 μmol $m^{-2} \cdot s^{-1}$ was used for determination of F_m. F_v/F_m measurements were performed according to [11] before and immediately after the freezing treatment and after 24 h of subsequent recovery under the same conditions as used for growth prior to the hardening treatment in the growth chamber experiments (20 °C day, 15 °C night). Two methods were employed with leaves in Petri dishes or in Falcon tubes. No differences were found, thus the use of Falcon tubes was preferred because of the required shorter time.

All statistical analyses were performed with R [33]. Paired t-tests and correlation analysis were used to assess differences in means and correlation between FT of different treatments. A Kruskal-Wallis test was applied to test for the effect of haplotypes of vernalization genes on FT and subsequent non-parametric multiple contrast test with Tukey contrasts (R package nparcomp) to test for between group differences. Box and whisker plots were used to visualize distributions of results where the line within the box stands for the median, the box range includes the second and third quartile and the whiskers are located at the maximum and minimum values or at 1.5 times the interquartile range from the box if more extreme values are present.

4. Conclusions

Tests on the applicability of a protocol for combined field sampling and laboratory freezing tests showed that (a) leaves cut from plants in the field and kept in Falcon tubes in the laboratory retained full photosynthetic competence for time periods that exceed the time applied for the complete freezing test and subsequent fluorescence screening after a 24 h recovery period (see Section 2.1.2). Thus, interference from senescence with the vital functions used as indicators of frost damage in cereal leaves is highly unlikely; (b) freezing tests performed on leaves attached to or detached from the plant lead to similar rankings of genotypes subjected to the tests (see Section 2.1.3). We did not find indications that the proposed method introduces a bias in the comparative ranking of FT across time or between genotypes. Homogeneity of the freezing conditions within freezing cabinets and potential effects of changes in microclimatic conditions for leaves enclosed in a Falcon tube, as opposed to leaves exposed to the air currents in the freezing cabinet, lead to differences in the absolute levels of damage to PS II that should be further studied as they are also relevant for tests done with whole plants.

Fields of application of the combined field sampling and laboratory freezing test method are illustrated with several examples concerning (a) monitoring of the seasonal course of hardening for FT obtained under field conditions (see Section 2.2.1) (b) study of FT within experiments that are genuinely performed as field experiments (example FACE study, Section 2.2.2) and (c) use for genotyping studies (see Section 2.2.3) that aim at measuring additional traits in plants that continue to grow after cutting single leaves for the freezing tests or that target other traits determined within a field experiment in parallel.

Measuring F_v/F_m for leaves cut from field grown plants one or several times prior to the freezing stress treatment, immediately after the freezing stress and after recovery, provides information beyond the FT ranking. The first pre-stress measurement monitors the maximal quantum yield of PSII as modulated by the environmental conditions in the field and traces eventual reversible or non-reversible reduction of F_v/F_m. Subsequent pre-stress measurements can be added to monitor the reversability. As the artificial freezing test is applied in darkness, the post-stress measurement characterizes the transition between pre-stress and the fully damaged state when F_v/F_m is generally not yet or only minimally affected by damage [11], which is analogous with observations on chilling effects [27]. The measurement after recovery integrates the factors that lead to irreversible damage. The use of 24 h of recovery under a day/night light regime is sufficient, as results obtained with this measurement were highly correlated with the results of assessments made after 2 days ($r > 0.97$) and five days ($r > 0.86$) of recovery [11].

The time requirement for the fluorescence measurements was determined for a total of 4900 measurements done within several experiments. On average, 5.2 measurements were performed within 1 min. The measurements were done by two people, one putting the leaves into the clamp of the PAM2000 instrument and one controlling the measured signals and taking notes that serve to check and amend the output in PAM2000 files. Thus, e.g., for a large set of 200 genotypes replicated 5 times, the total of 1000 measurements to be done for a post-recovery assessment results in a time requirement of 3 h and 12 min per 2 experimenters. If pre-stress and post-stress measurements are taken as well, then the total time requirement is nine and a half hours per person.

Acknowledgments

We thank the three anonymous reviewers for valuable comments on the manuscript. The current research was partially financed by MIPAAF through the AGROSCENARI and FAO-RGV projects and the Duco project by the "Fondazione in rete per la ricerca agroalimentare" within the AGER program. We thank Donata Pagani and Flavio Astesano for their excellent technical support.

Conflict of Interest

The authors declare no conflict of interest.

References

1. Saulescu, N.N.; Braun, H.J. Cold tolerance. In *Application of Physiology in Wheat Breeding*; Reynolds, M.P., Ortiz-Monasterio, J.I., McNab, A., Eds.; CIMMYT: Mexico, DF, Mexico City, 2001; pp. 111–123.
2. Levitt, J. *Response of Plants to Environmental Stresses: Chilling, Freezing, and High Temperature Stresses*, 2nd ed.; Academic Press: New York, NY, USA, 1980.
3. Rizza, F.; Pagani, D.; Gut, M.; Prasil, I.T.; Lago, C.; Tondelli, A.; Orru, L.; Mazzucotelli, E.; Francia, E.; Badeck, F.W.; *et al*. Diversity in the Response to Low Temperature in Representative Barley Genotypes Cultivated in Europe. *Crop Sci.* **2011**, *51*, 2759–2779.
4. Thomashow, M.F. Plant cold acclimation: Freezing tolerance genes and regulatory mechanisms. *Annu. Rev. Plant Physiol. Plant Mol. Biol.* **1999**, *50*, 571–599.
5. Fowler, D.B.; Limin, A.E. Interactions among factors regulating phenological development and acclimation rate determine low-temperature tolerance in wheat. *Ann. Bot.* **2004**, *94*, 717–724.
6. Francia, E.; Rizza, F.; Cattivelli, L.; Stanca, A.M.; Galiba, G.; Toth, B.; Hayes, P.M.; Skinner, J.S.; Pecchioni, N. Two loci on chromosome 5H determine low-temperature tolerance in a "Nure" (winter) × "Tremois" (spring) barley map. *Theor. Appl. Genet.* **2004**, *108*, 670–680.
7. Akar, T.; Francia, E.; Tondelli, A.; Rizza, F.; Stanca, A.M.; Pecchioni, N. Marker-assisted characterization of frost tolerance in barley (Hordeum vulgare L.). *Plant Breed.* **2009**, *128*, 381–386.

8. Fisk, S.P.; Cuesta-Marcos, A.; Cistue, L.; Russell, J.; Smith, K.P.; Baenziger, S.; Bedo, Z.; Corey, A.; Filichkin, T.; Karsai, I.; *et al.* FR-H3: A new QTL to assist in the development of fall-sown barley with superior low temperature tolerance. *Theor. Appl. Genet.* **2013**, *126*, 335–347.
9. Clement, J.M.A.M.; van Hasselt, P.R. Chlorophyll fluorescence as a parameter for frost hardiness in winter wheat. A comparison with other hardiness parameters. *Phyton* **1996**, *36*, 29–41.
10. Herzog, H.; Olszewski, A. A rapid method for measuring freezing resistance in crop plants. *J. Agron. Crop Sci. (Z. Acker Pflanzenbau)* **1998**, *181*, 71–79.
11. Rizza, F.; Pagani, D.; Stanca, A.M.; Cattivelli, L. Use of chlorophyll fluorescence to evaluate the cold acclimation and freezing tolerance of winter and spring oats. *Plant Breed.* **2001**, *120*, 389–396.
12. Thalhammer, A.; Bryant, G.; Sulpice, R.; Hincha, D.K. Disordered Cold Regulated15 Proteins Protect Chloroplast Membranes during Freezing through Binding and Folding, But Do Not Stabilize Chloroplast Enzymes *in Vivo*. *Plant Physiol.* **2014**, *166*, 190–201.
13. Mishra, A.; Mishra, K.B.; Höermiller, I.I.; Heyer, A.G.; Nedbal, L. Chlorophyll fluorescence emission as a reporter on cold tolerance in Arabidopsis thaliana accessions. *Plant Signal. Behav.* **2011**, *6*, 301–310.
14. Dai, F.; Zhou, M.; Zhang, G. The change of chlorophyll fluorescence parameters in winter barley during recovery after freezing shock and as affected by cold acclimation and irradiance. *Plant Physiol. Biochem.* **2007**, *45*, 915–921.
15. Steponkus, P.L.; Webb, M.S. Freeze-induced dehydration and membrane destabilisation in plants. In *Water and Life*; Somero, G.N., Osmond, C.B., Bolis, C.L., Eds.; Springer: Berlin, Germany, 1992; pp. 338–362.
16. Rapacz, M.; Tyrka, M.; Kaczmarek, W.; Gut, M.; Wolanin, B.; Mikulski, W. Photosynthetic acclimation to cold as a potential physiological marker of winter barley freezing tolerance assessed under variable winter environment. *J. Agron. Crop Sci.* **2008**, *194*, 61–71.
17. Prášil, I.T.; Prášilová, P.; Marík, P. Comparative study of direct and indirect evaluations of frost tolerance in barley. *Field Crops Res.* **2007**, *102*, 1–8.
18. Prášil, I.T.; Rogalewicz, V. Accuracy of wheat winterhardiness evaluation by a provocation method in natural conditions. *Genetika Šlechtení (Genet. Plant Breed. Praha)* **1989**, *25*, 223–230.
19. Koch, H.-D.; Lehmann, C.O. Resistenzeigenschaften im Gersten- und Weizensortiment Gatersleben. 7. Prüfung der Frostresistenz von Wintergersten im künstlichen Gefrierversuch. *Kulturpflanze* **1966**, *14*, 263–282.
20. Rapacz, M. Chlorophyll a fluorescence transient during freezing and recovery in winter wheat. *Photosynthetica* **2007**, *45*, 409–418.
21. Rapacz, M.; Wozniczka, A. A selection tool for freezing tolerance in common wheat using the fast chlorophyll a fluorescence transient. *Plant Breed.* **2009**, *128*, 227–234.
22. Witkowski, E.; Waga, J.; Witkowska, K.; Rapacz, M.; Gut, M.; Bielawska, A.; Luber, H.; Lukaszewski, A.J. Association between frost tolerance and the alleles of high molecular weight glutenin subunits present in Polish winter wheats. *Euphytica* **2008**, *159*, 377–384.
23. Willits, D.H.; Peet, M.M. Measurement of chlorophyll fluorescence as a heat stress indicator in tomato: Laboratory and greenhouse comparisons. *J. Am. Soc. Hortic. Sci.* **2001**, *126*, 188–194.

24. Wu, T.T.; Weaver, D.B.; Locy, R.D.; McElroy, S.; van Santen, E. Identification of vegetative heat-tolerant upland cotton (*Gossypium hirsutum* L.) germplasm utilizing chlorophyll fluorescence measurement during heat stress. *Plant Breed.* **2013**, *133*, 250–255.
25. Sharma, D.K.; Fernández, J.O.; Rosenqvist, E.; Ottosen, C.O.; Andersen, S.B. Genotypic response of detached leaves *vs.* intact plants for chlorophyll fluorescence parameters under high temperature stress in wheat. *J. Plant Physiol.* **2014**, *171*, 576–586.
26. Sthapit, B.R.; Witcombe, J.R.; Wilson, J.M. Methods of selection for chilling tolerance in Nepalese rice by chlorophyll fluorescence analysis. *Crop Sci.* **1995**, *35*, 90–94.
27. Strauss, A.J.; Krüger, G.H.J.; Strasser, R.J.; Heerden, P.D.R.V. Ranking of dark chilling tolerance in soybean genotypes probed by the chlorophyll a fluorescence transient O-J-I-P. *Environ. Exp. Bot.* **2006**, *56*, 147–157.
28. Housley, T.L.; Pollock, C.J. Photosynthesis and Carbohydrate-Metabolism in Detached Leaves of *Lolium-Temulentum* L. *New Phytol.* **1985**, *99*, 499–507.
29. Fischer, A.M. The Complex Regulation of Senescence. *Crit. Rev. Plant Sci.* **2012**, *31*, 124–147.
30. Min, K.; Chen, K.; Arora, R. Effect of short-term *vs.* prolonged freezing on freeze-thaw injury and post-thaw recovery in spinach: Importance in laboratory freeze-thaw protocols. *Environ. Exp. Bot.* **2014**, *106*, 124–131.
31. Mahfoozi, S.; Limin, A.E.; Fowler, D.B. Influence of vernalization and photoperiod responses on cold hardiness in winter cereals. *Crop Sci.* **2001**, *41*, 1006–1011.
32. Badeck, F.-W.; Rizza, F.; Maré, C.; Cattivelli, L.; Zaldei, A.; Miglietta, F. Durum wheat growth under elevated CO_2: First results of a FACE experiment. In *Atti del XV Convegno Nazionale di Agrometeorologia. Nuovi Scenari Agro Ambientali: Fenologia, Produzioni Agrarie ed Avversita. Palermo, 5-6-7 Giugno 2012*; Ventura, F., Pieri, L., Eds.; PA.TRON EDITORE: Bologna, Italy, 2012; pp. 15–16.
33. R Core Team. R: A Language and Environment for Statistical Computing. Vienna: R Foundation for Statistical Computing. 2014. Available online: http://www.R-project.org/ (accessed on 21 November 2014).

7

The Potential of *Lr19* and *Bdv2* Translocations to Improve Yield and Disease Resistance in the High Rainfall Wheat Zones of Australia

Garry Rosewarne [1,2,*], David Bonnett [1,3], Greg Rebetzke [1], Paul Lonergan [1] and Philip J. Larkin [1]

[1] CSIRO Agriculture Flagship, Clunies Ross St, Black Mountain, Canberra 2601, Australia; E-Mails: d.bonnett@cgiar.org (D.B.); greg.rebetzke@csiro.au (G.R.); paul.lonergan@csiro.au (P.J.L.)

[2] Department of Environment and Primary Industries, 110 Natimuk Rd, Horsham Vic 3400, Australia

[3] International Maize and Wheat Improvement Centre, Apdo. 06600 Mexico Distrito Federal, Postal 6-6-41, Mexico

* Author to whom correspondence should be addressed; E-Mail: Garry.rosewarne@depi.vic.gov.au

Academic Editor: Bertrand Hirel

Abstract: Chromosomal translocations in wheat derived from alien species are a valuable source of genetic diversity that have provided increases in resistance to various diseases and improved tolerance to abiotic stresses in wheat. These alien genomic segments can also affect multiple traits, with a concomitant ability to alter yield potential in either a positive or negative fashion. The aim of this work was to characterize the effects on yield of two types of translocations, namely T4-derived translocations from *Thinopyrum ponticum*, carrying the leaf rust resistance gene *Lr19*, and the TC14 translocation from *Th. intermedium*, carrying the barley yellow dwarf virus resistance gene *Bdv2*, in Australian adapted genetic backgrounds and under Australian conditions. A large range of germplasm was developed by crossing donor sources of the translocations into 24 Australian adapted varieties producing 340 genotypes. Yield trials were conducted in 14 environments to identify effects on yield and yield components. The T4 translocations had a positive effect on yield in one high yielding environment, but negatively affected yield in low-yielding environments. The TC14 translocation was generally benign, however, it was associated with a negative impact on yield and reduced height in two genetic backgrounds.

The translocation was also associated with a delayed maturity in several backgrounds. The T4 translocations results were consistent with previously published data, whilst this is the first time that such an investigation has been undertaken on the TC14 translocation. Our data suggests a limited role for each of these translocations in Australia. The T4 translocations may be useful in high yielding environments, such as under irrigation in NSW and in the more productive high rainfall regions of south-eastern Australia. Traits associated with the TC14 translocation, such as BYDV resistance and delayed maturity, would make this translocation useful in BYDV-prone areas that experience a less pronounced terminal drought (e.g., south-eastern Australia).

Keywords: translocations; wheat; *Lr19*; *Bdv2*; *Thinopyrum ponticum*; *Thinopyrum intermedium*

1. Introduction

Hexaploid wheat (*Triticum aestivum* L.) has recently been estimated to have formed between 230,000 and 430,000 years ago through rare hybridization events between diploid and tetraploid progenitors [1] with domestication occurring approximately 10,000 years ago. Consequently, modern wheats developed from this original bottleneck [2]. An effective method to increase genetic diversity is to introgress chromosomal segments into the wheat genome from wild relatives. These chromosomal translocations have been used to enhance resistance to various wheat diseases. However, the incomplete homology, referred to as "homoeology", between the donor and recipient [1] chromosomes, usually prevents recombination. Any deleterious genes that are introgressed along with the useful trait can be difficult to remove. One specific translocation on the 7DL chromosome of wheat, called variously 7Ag.7DL, Agatha or T4 [3–5], from *Thinopyrum ponticum* (tall wheatgrass), contained a number of characterized genes including *Lr19* for leaf rust resistance, *Sr25* for stem rust resistance, *Sd1*, a segregation distortion gene, and a linked gene deleterious for yellow pigment endosperm colour. This T4 translocation has also been associated with higher grain yield, anthesis biomass and delayed maturity in a number of genetic backgrounds [6–8]. Interestingly, the higher yield potential is only expressed under non-moisture stress conditions where yields exceed 2.5 tonnes per hectare [8].

The average yield in the Australian wheat belt is around 2 t/ha largely due to limited water conditions. Under these conditions, the T4 translocation is likely to be unproductive. Indeed, although there is a strong emphasis in maintaining stem rust resistance in Australia, and that *Sr25* is one of the few valuable and effective resistance genes available, no Australian cultivars carry this gene. However, changing agricultural markets have seen an interest in cropping in the higher rainfall areas of Australia (400–800 mm annual rainfall). Under these more favourable conditions, it is more likely that the T4 translocation may be able to provide a yield boost. The use of this translocation is further limited by the yellow pigment locus, conferring a trait that is detrimental to wheat marketing in Australia. Knott [9] used EMS mutagenesis to produce lines with reduced yellow pigment in the flour, yet retain *Lr19* effectiveness. Two of the mutant lines were further evaluated agronomically, with Agatha 28-4 maintaining both *Lr19* and *Sr25* resistance, whilst Agatha 235-6 lost the *Sr25* resistance [3].

Another major yield constraint in parts of the higher rainfall zones of Australia is Barley Yellow Dwarf Virus (BYDV). Banks, *et al.* [10] created a series of translocations from *Th. intermedium* to introduce the BYDV resistance locus, *Bdv2*, onto the 7DL chromosome of bread wheat. These translocations were known as the "TC" series, with TC14 containing the smallest alien DNA segment. Subsequently, this translocation was crossed into various backgrounds but there has not been any definitive study on how this translocation affects yield components of wheat.

In this study, we investigated the effects of the T4 and TC14 translocations on yield in Australian genetic backgrounds under Australian conditions. Potential uses in the high rainfall zones of Australia are discussed.

2. Results

Two sets of germplasm were developed to investigate the effects of the T4 translocation series, which consisted of the original yellow floured T4 translocation and two white floured mutants, Agatha 28-4 and Agatha 235-6 translocations [3], hereafter termed T4m1 and T4m2. In the trials run from 2000 and 2001, single lines from 15 genetic backgrounds were selected from multiple backcrossed material (two to six backcrosses) that contained either the T4m1 or T4m2 translocations. The molecular marker STS-Lr19130 [11] was employed to identify translocation status. The 2006–2008 trials used a different set of T4 translocation material. Multiple selections were made from the pedigree, Seri 82/Superseri, that segregated for the original T4 translocation. Similarly, selections that segregated for the T4m1 translocation were made from a six backcrossed derived series from Condor. Several other elite CIMMYT lines either with or without the T4 translocation were also included in these trials. The TC14 translocation was bred into 12 backgrounds with three to four backcrosses prior to doubled-haploid production from randomly sampled lines. Translocation status was determined with the STS molecular marker BYAgi [12,13].

In total, there were 14 yield trials conducted over five years with average yields ranging from 0.64–6.7 t/ha. There was a large amount of data collected with 1128 plots harvested in relation to the T4 translocations and 1325 plots harvested in relation to the TC14 translocation. All trials were inspected regularly (minimum of three times) throughout the growing seasons and there were no obvious disease in any of the experiments. All germplasm used in each of the trials are listed in Supplementary Table 1 and average yields, number of plots and number of genetic backgrounds in each of the trials is listed in Table 1.

There was a large range of average yields per site due to a wide range of environments and growth conditions. There were also significant treatment effects, genetic background effects and translocation effects for both the T4 translocations and the TC14 translocation. The T4, T4m1 and T4m2 translocations had a negative impact on yield when compared to non-translocation containing lines in two of four, seven of 11, and seven of seven environments, respectively. Yield reductions as a percentage of yield for the T4 and T4m1 translocations were more pronounced in the more water stressed environments and trended towards the positive in sites with higher water availability. The T4m2 translocation had negative effects in all environments tested, whether they were under low or high water stress (Figure 1).

Table 1. Trial sites, numbers of plots and number of genetic backgrounds used in trials.

Year	Site	Average Site Yield	T4 Translocation		TC14 Translocation	
			No. Plots	Genetic Backgrounds	No. Plots	Genetic Backgrounds
2000	Ginninderra [a]	4.71	192	12	-	-
2000	Gundibindyal	6.70	120	9	-	-
2000	Moombooldool	4.27	160	11	-	-
2001	Condobolin	1.57	80	10	-	-
2001	Ginninderra [a]	6.24	80	11	-	-
2001	Gundibindyal	4.26	96	12	-	-
2001	Moombooldool	3.09	96	11	-	-
2006	Griffith [a]	4.66	23	7	177	13
2006	Gundibindyal	n.d.	6	4	90	11
2007	Gundibindyal	1.55	63	8	225	9
2007	Yanco [a]	3.44	29	8	211	9
2008	Temora	1.91	63	9	225	9
2008	Yanco	0.74	59	9	213	9
2008	Yanco [a]	3.97	61	9	211	9
Total	**14**		**1128**		**1325**	

[a] Signifies an irrigated trial.

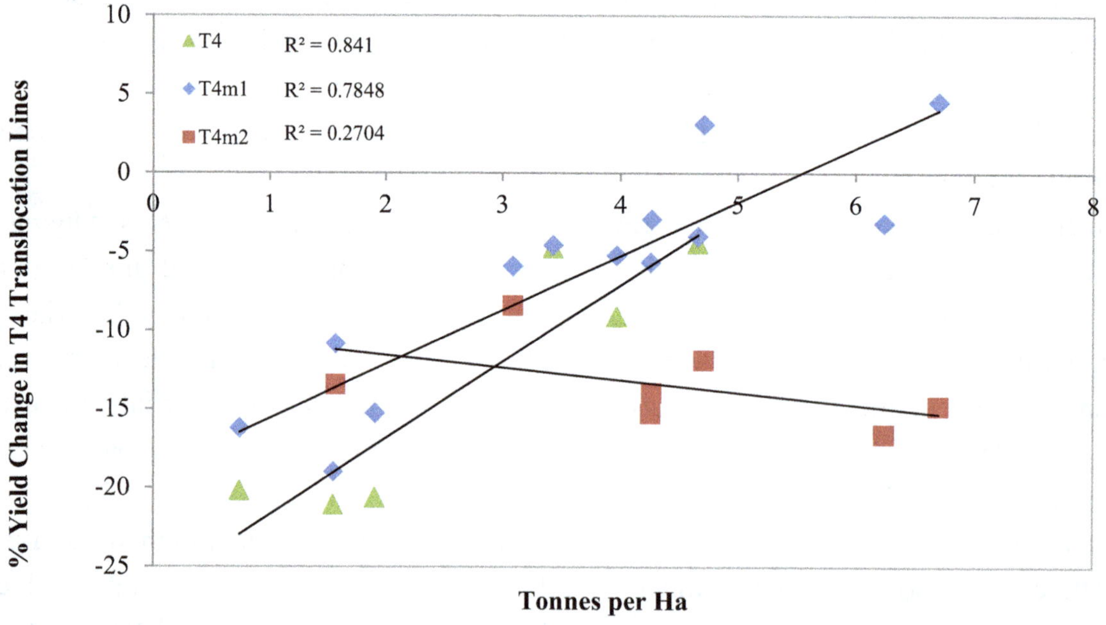

Figure 1. Effects of translocations on yield. Linear trends showing the percentage effects of the different T4 translocations against non-T4 containing lines plotted against the average site yield.

Yield components were most commonly measured for the T4m1 and T4m2 translocations (Table 2). The effects of T4 and T4m1 on yield in the different sites were similar (Figure 1). The T4 or T4m1 translocations had a significant effect on grains·m^{-2} at three of the five sites where this was measured

(Table 2) and the percentage effect of the translocation in comparison to non-T4 lines increased as site yield increased (Figure 2a). Thousand kernel weight (TKW) was significantly reduced in three of the five sites tested with T4m1 and the percentage effect of this change in T4 lines was proportionally smaller in higher yielding environments. Harvest index (HI) and height were also significantly different in approximately half of the sites where these traits were investigated for the T4m1 translocation. This compares to the T4m2 translocation which was significantly different from the recurrent parents in most environments for virtually all traits tested. T4m2 was associated with a lower level of grains·m^{-2} in two of two environments, lower TKW in one of two environments, lower HI in all five environments, lower biomass in four of five environments, later maturity in three of six environments and were significantly taller in four of six environments.

The T4m2 lines were consistently taller than the recurrent parents by an average of 7% while T4m1 were 2% shorter. Some of the translocations also affected maturity. The T4m2 containing lines were on average 3.4 days later in maturity than the recurrent parents while the other T4 translocations were no different.

In 2008, the Seri 82/Superseri segregants indicated that the T4 translocation had a highly significant effect on yield in rainfed trials where moisture limited yield. The irrigated trial did not show such an effect, neither did the T4m1 translocation in the Condor background (Table 3).

The TC14 translocation had a significant negative effect on yield in two of the six locations tested, namely Temora 2008 and the irrigated site at Yanco in 2008, both being moderately yielding sites. Maturity was significantly delayed in the TC14 lines in all four environments tested, and ranged between 1.8 and 4.0 days (Table 4). The TC14 translocation had a significant negative effect on grains·m^{-2} in one environment and in this environment, it also was associated with a significant yield reduction. TKW was unaffected by the TC14 translocation, and hectolitre and height were significantly different in one environment each.

Background genotypes had an effect on whether TC14 was associated with reduced yield. The lower yields in the irrigated site at Yanco in 2008 were most prevalent in the H45 background ($p > 0.01$), followed by Janz ($p > 0.05$) and to a lesser degree in the Westonia/Sunbrook backgrounds ($p > 0.1$). The H45 background also showed a highly significant correlation of yield reduction in the TC14 lines at Temora 2008 ($p > 0.01$) along with Camm and Westonia ($p > 0.05$) (Table 5). Maturity and height were also investigated where records were taken. At the irrigated Yanco 2008 site, the TC14 containing lines in H45 and Janz backgrounds were significantly shorter than their null counterparts($p > 0.05$) (Table 5). This data was not collected for the Temora 2008 trial.

Table 2. Effects of T4 translocations on grain yield, yield components, maturity and height.

Environment	Germplasm [a]	Yield (t/ha)	KGrain·m^{-2}	TKW (mg)	HI	Total Biomass (t/ha)	Maturity [b]	Height (cm)
Ginninderra 2000 [c,d]								
	Recurrent Parents	4.80					61.5	
	T4m1	5.24					58.3	
$H^2 = 0.81$	T4m2	4.46 **					58.3 **	
Gundibindyal 2000 [d]	Recurrent Parents	6.64					58.4	101.3
	T4m1	6.93					61.2	100.7
$H^2 = 0.89$	T4m2	5.32 **					58.6 *	116.1 *
Moombooldool 2000 [d]	Recurrent Parents	4.35			0.429	10.48	61.5	79.8
	T4m1	4.31			0.438	9.74	63.0	77.5 ***
$H^2 = 0.74$	T4m2	3.81 **			0.415 *	9.09 *	61.6 ns	87.1 ***
Condobolin 2001 [d]	Recurrent Parents	1.67	6017	28.1	0.357	4.74	67.2	63.4
	T4m1	1.56 ***	5786	26.6	0.341	4.50	68.7	61.5 *
$H^2 = 0.83$	T4m2	1.39 ***	5420 *	26.7	0.321 ***	4.42 *	65.0 ns	62.8
Ginninderra 2001 [c,d]	Recurrent Parents	6.57	17981	36.9	0.430	15.28	62.3	91.0
	T4m1	6.28	18517	34.1 ***	0.406 **	15.53	61.4	90.6
$H^2 = 0.86$	T4m2	5.49 ***	17150 *	32.3 ***	0.371 ***	14.75 ns	59.2 **	100.2 ***
Gundibindyal 2001 [c]	Recurrent Parents	4.57			0.393	11.64		89.2
	T4m1	4.23 **			0.380 *	11.18		85.4 ***
$H^2 = 0.68$	T4m2	3.82 ***			0.344 ***	11.10 *		90.8
Moombooldool 2001 [c]	Recurrent Parents	3.33			0.384	8.81	58.8	78.9
	T4m1	3.06 **			0.375 *	8.10	58.9	74.6 **
$H^2 = 0.73$	T4m2	2.98 ***			0.359 ***	8.33 **	57.6	83.0 ***

Table 2. *Cont.*

Environment	Germplasm [a]	Yield (t/ha)	KGrain·m^{-2}	TKW (mg)	HI	Total Biomass (t/ha)	Maturity [b]	Height (cm)
Gundibindyal 2007 [e]	T4 negative lines	1.92	5527	34.9	0.366	5.05	59.5	63.8
	T4	1.70 ***	5163 **	33.5	0.343 *	4.88	63.5 ns	64.3
$H^2 = 0.66$	T4m1	1.59 **	5152	31.4 **	0.371	4.22 *	62.0 ns	62.6 *
Temora 2008 [e]	T4 negative lines	2.22	6006	35.5				
	T4	1.85 ***	5150	34.9				
$H^2 = 0.73$	T4m1	1.74 ***	4844 **	35.8 *				
Yanco 2008 [c,e]	T4 negative lines	4.00	13194	30.5			69.4	82.4
	T4	3.80	13626	27.9 *			69.9	86.3 **
$H^2 = 0.81$	T4m1	3.70 *	12975 *	28.7			68.9 *	83.7
Yanco 2008[e]	T4 negative lines	0.71						
	T4	0.69						
$H^2 = 0.47$	T4m1	0.56 *						

* Represents significant at $p > 0.1$, ** at $p > 0.05$ and *** at $p > 0.01$ between the recurrent parents/T4 negative lines and either the T4, T4m1 or T4m2 lines; [a] Recurrent parents refers to the backcross parents that did not contain the translocation, T4 negative lines to a range of elite germplasm and null lines from the two T4 doubled haploid series, T4 to lines containing the original translocation, T4m1 and T4m2 to the mutated translocations as described in materials and methods [9]; [b] Maturity scores according to Zadoks *et al.* [14]; [c] irrigated trials; [d] Trials compared single lines from highly backcrossed selections that contained the translocation to the recurrent parents; [e] Trials compared a number of elite CIMMYT lines (with and without the translocation) combined with two sets of sister lines that segregated for either the T4 or T4m1 translocations; ns = not significant.

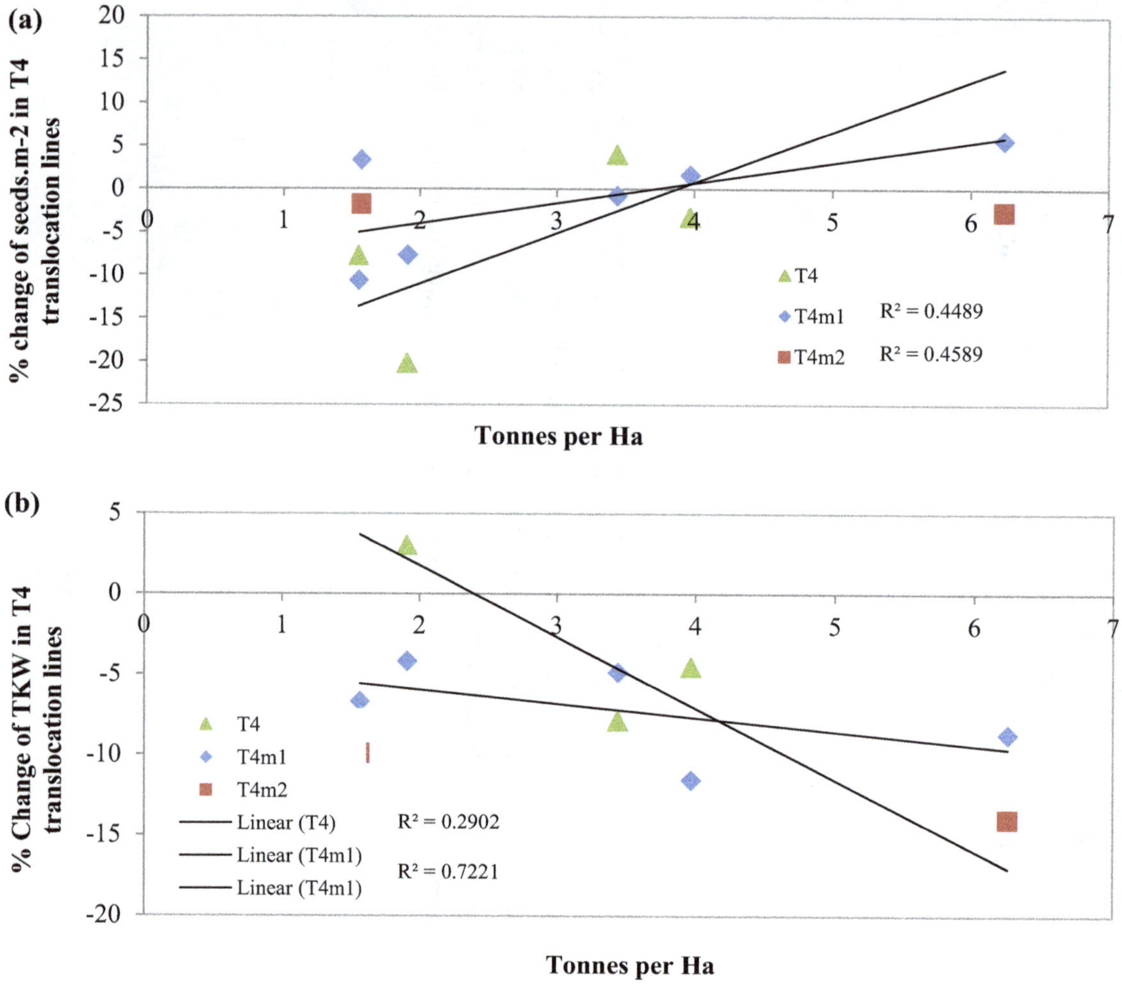

Figure 2. The percentage change attributed to the T4 translocations for (**a**) seeds·m^{-2} and (**b**) TKW. T4m2 only had two data-points.

Table 3. Effects of the T4 and T4m1 translocations on yield (t·ha^{-1}), grains·m^{-2}, thousand kernel weight (TKW) and maturity. Comparisons were made from two sets of sister lines that segregated for the original T4 translocation in the Seri 82/Superseri cross, and for the T4m1 translocation in a highly backcrossed Condor background.

Site	Germplasm	No. of lines	t·ha^{-1}	Grains·m^{-2}	TKW	Maturity [a]
Temora 2008	Seri 82/Superseri null	15	2.43	7387	32.8	
	Seri 82/Superseri T4	6	1.74 ***	5432 ***	32.0	
	Condor null	13	2.14	6389	32.9	
	Condor T4m1	10	1.93	6055	31.7	
Yanco 2008	Seri 82/Superseri null	13	0.81			
	Seri 82/Superseri T4	6	0.50 ***			
	Condor null	10	0.58			
	Condor T4m1	10	0.54			
Yanco 2008 [b]	Seri 82/Superseri null	15	3.98	13484	29.53	69.13
	Seri 82/Superseri T4	6	3.58	13355	26.73 ***	68.67
	Condor null	10	3.77	13745	27.59	69.50
	Condor T4m1	10	3.38	13634	24.87 ***	69.90

*** Represents significance at $p > 0.01$, significance was tested at $p > 0.1$, $p > 0.05$ and $p > 0.01$. [a] Maturity scores according to Zadoks *et al.* [14]; [b] irrigated trial.

Table 4. Effects of the TC14 (*Bdv2*) translocation on yield, yield components, maturity and height.

Environment	Germplasm [a]	Grain Yield (t/ha)	Grain Number (no·m^{-2})	TKW (mg)	Maturity [b]	Hectolitre Weight	Height (cm)
Griffith 2006[c]	Parents	4.87			57.4		87.6
	Null	4.55			59.1		85.7
$H^2 = 0.72$	TC14	4.62			55.1 ***		86.3
Gundibindyal 2007	Parents	1.43	4,012	36.2	60.0	80.0	57.3
	Null	1.55	4,642	34.5	60.8	79.3	60.4
$H^2 = 0.59$	TC14	1.48	4,398	34.8	59.0 *	80.0 **	60.2
Yanco 2007 [c]	Parents	3.82	11,362	33.9	63.7	79.7	76.3
	Null	3.75	11,572	32.7	63.8	80.0	75.6
$H^2 = 0.60$	TC14	3.63	11,093	33.0	61.9 *	80.1	73.0 *
Temora 2008	Parents	2.34	6,757	33.6			80.4
	Null	2.28	6,611	33.5			79.7
$H^2 = 0.46$	TC14	2.10 **	6,238	33.0			79.8
Yanco 2008	Parents	0.69					
	Null	0.67					
$H^2 = 0.49$	TC14	0.65					
Yanco 2008 [c]	Parents	4.19	13,494	31.4	66.4	79.6	83.3
	Null	4.11	14,170	29.2	68.0	78.5	81.9
$H^2 = 0.71$	TC14	3.95 ***	13,763 *	29.2	65.6 **	78.9	80.2

* Represents significance at $p > 0.1$, ** at $p > 0.05$ and *** at $p > 0.01$ between the null and TC14 containing lines; [a] Parents represents the backcross parents used to generate the null and TC14 lines, null lines are progeny that do not contain TC14, and TC14 are progeny lines with the translocation; [b] Maturity scores according to Zadoks *et al.*[14]; [c] irrigated trials.

Table 5. Association of the TC14 lines with changes in yield, maturity [14] and height in different genetic backgrounds.

Germplasm	No. of Lines [a]	2006 Griffith [b]			2007 Gundibindyal			2007 Yanco [b]			2008 Yanco [b]			2008 Temora	2008 Yanco
		Yield	Maturity	Height	Yield	Maturity	Height	Yield	Maturity	Height	Yield	Maturity	Height	Yield	Yield
Thelin/4*Camm Null	12	-	-	-	1.81	49	63	3.66	63	77	3.76	70	77	2.68	0.62
Thelin/4*Camm TC14	12	-	-	-	1.70	48	63	3.54	62	75	3.71	69	78	2.39 *	0.56
Thelin/4*Chara Null	12	-	-	-	1.26	47	61	4.14	57	75	4.52	67	85	2.33	0.45
Thelin/4*Chara TC14	12	-	-	-	1.32	44	63	3.86	55	70 *	4.33	64	82	2.28	0.46
Thelin/4*Drysdale Null	(15) 11	4.87	57	87	2.17	48	63	4.23	60	75	4.25	70	83	1.46	0.55
Thelin/4*Drysdale TC14	(4) 7	5.85	52	96 *	2.09	46 *	61	4.08	58 *	72	4.10	67 *	76	1.60	0.53
Thelin/4*H45 Null	(17) 12	4.99	67	85	1.53	52	67	3.85	66	73	4.70	70	85	2.86	0.76
Thelin/4*H45 TC14	(8) 12	4.85	64 **	85	1.53	52	67	3.66	66	73	4.33 **	70	81 **	2.66 **	0.76
Thelin/4*Janz Null	(5) 12	4.65	51	79	1.14	48	56	3.83	60	73	4.23	69	81	2.40	0.69
Thelin/4*Janz TC14	(3) 12	4.61	49	76	1.13	46 **	51 *	3.85	56 *	72	3.87 *	68	78 *	2.13	0.64
Thelin/3*Drysdale//Sunbrook Null	(19) 14	4.19	56	88	0.95	45	53	3.71	53	78	4.14	60	83	2.53	0.65
Thelin/3*Drysdale//Sunbrook TC14	(10) 13	4.29	52	86	1.15	45	56	3.43	51	70 *	4.04	60	81	2.39	0.78
Thelin/3*Westonia//Sunbrook Null	(14) 14	4.54	60	84	1.75	48	62	3.42	59	75	4.40	65	79	2.21	0.75
Thelin/3*Westonia//Sunbrook TC14	(10) 17	4.35	54	91	1.58	47	61	3.74	57	78	4.09	59	83	2.10	0.69
Thelin/4*Westonia Null	(10) 12	4.69	63	79	1.15	54	50	4.01	66	70	4.37	70	77	2.24	0.64
Thelin/4*Westonia TC14	(9) 12	4.86	59	79	1.16	52	52	4.27	65	74	4.38	69	80	1.80 *	0.54
Thelin/3*H45//Darter Null	(5)	3.72	63	91	-	-	-	-	-	-	-	-	-	-	-
Thelin/3*H45//Darter TC14	(2)	3.89	64	95	-	-	-	-	-	-	-	-	-	-	-
Chara/Ohm/Sunbrook Null	(2)	4.25	65	74	-	-	-	-	-	-	-	-	-	-	-
Chara/Ohm/Sunbrook TC14	(2)	4.94	41 **	95	-	-	-	-	-	-	-	-	-	-	-

* Represents significance at $p > 0.05$ and ** at $p > 0.01$ between the null and TC14 containing lines; [a] The number of lines in each trial are shown for Griffith 2006 as a parenthesis, and as normal text for the remainder of the trials; [b] Indicates trials that had irrigation.

3. Discussion

Extensive genetic material was developed and thoroughly tested in yield trials in Australia. Previous research with the T4 translocation indicated that it drove a significant yield boost under high-moisture conditions, but had a negative impact on yield when water limited production to under 2 $t \cdot ha^{-1}$ [6–8]. This study identified similar phenomenon with the white floured T4m1 translocation, but not with the T4m2 translocation (Figure 1). In our trials, the original T4 translocation was only tested in relatively low yielding environments, and although it did not show a yield boost, its relative yield effect in relation to average site yield trended in a similar fashion to that of T4m1. Both T4m1 and T4m2 were developed through an EMS mutagenesis screen with selection based upon white flour colour [9]. Zhang and Dubcovsky [15] surmised that the different mutations in the original translocation were responsible for the white flour colour in these two lines but it was not clear which regions within the translocation were affected. These mutations did not affect leaf rust resistance however T4m1 retained the effective *Sr25* locus whilst T4m2 did not [3], indicating this latter mutation line may have an abnormal makeup following the mutagenic treatment. Physical mapping placed *Sr25* and the yellow flour colour locus (*PSY-E1*) together at the telomeric end of the translocation [16–18]. This large T4 translocation obviously contains many loci affecting different traits in wheat, the data presented in Figure 1 suggests that the T4m2 translocation not only lost yellow pigmentation and *Sr25* resistance, but appears to no longer provide a yield increase under non-water stress conditions.

Simplistically, grain yield can be determined from the two major yield components of $grains \cdot m^{-2}$ and grain weight [19]. The correlations between yield and $grains \cdot m^{-2}$ with both the T4 and T4m1 translocations indicate that $grains \cdot m^{-2}$ impacted on yield, and this impact overrode a negative correlation between yield and kernel weight (Figure 2). Interestingly, the T4m2 lines had neither correlation, further supporting the lack of a genetic effect of this mutated translocation on yield. Previous work has shown that under high yielding conditions, the increased yield potential conferred by the T4 translocation was mainly due to an increase in $grains \cdot m^{-2}$, grains per spike and harvest biomass [6,7]. Singh, Huerta-Espino, Rajaram and Crossa [6] also observed a higher kernel weight but a lower harvest index in the T4 material. Harvest index appeared altered most significantly in the T4m2 containing lines, and is likely to be related to a significant change in maturity.

Significant drivers of wheat yield are optimal height and maturity for any given environment. The evidence presented herein indicates that T4 and T4m1 had no effect on maturity but did reduce height, while T4m2 lines were taller and later. Previous work provided conflicting results with Singh, Huerta-Espino, Rajaram and Crossa [6] indicating that T4 lines were significantly later in maturity and taller, whereas Reynolds, Calderini, Condon and Rajaram [7] indicated that there was no difference in maturity conferred by the presence of T4. Knott [3] developed the original white floured mutants of T4m1 and T4m2 and found that they conferred a mild delay in maturity, averaging one day, although significance tests were not presented. Knott's study also indicated that there was no difference in height. The average delay in flowering of the T4m2 lines of 3.4 days could have a significant impact on yield under the conditions in which we tested them. The trial sites in southern NSW generally have a tight finish brought on by hot, dry weather conditions. Later flowering would result in grain-filling under less than optimal conditions and is evidenced by the consistently lower harvest index. The effect of T4m2 on TKW was only investigated in one environment, and showed a highly significant reduction, and this could also be attributed to grain-filling under sub-optimal conditions.

The genetic background may have affected the expression of the yield reduction associated with the T4 translocation. The highly significant lower yield in the T4 containing lines in moisture-stressed environments was only observed in the Seri 82/Superseri series. It seems unlikely that the Condor background was better adapted to these environments as the null lines yielded lower than the corresponding Seri 82/Superseri null lines. It would be interesting to trial these two series in higher yielding environments. This potential background genetic interaction needs to be considered when deploying such material in a breeding program.

A negative effect associated with the TC14 translocation on yield appeared to be both environment and genotype dependent. Moderately yielding 2008 trials were the only ones to show this negative effect. September and October were drier and hotter than long term averages in 2008 (http://www.bom.gov.au/climate/annual_sum/2008/index.shtml), conditions which may have affected yields in the later maturing TC14 retaining lines. However, the background genotypes with the greatest yield reductions associated with TC14 were H45 and Camm. There was no evidence that TC14 was associated with changed maturity in these particular backgrounds. Furthermore, in the Drysdale background, maturity was consistently delayed in the TC14 containing lines, yet there was no significant yield difference. This evidence would suggest that later maturity did not drive yield reductions associated with the TC14 translocation. On the other hand, a height reduction was significantly associated with both a yield reduction and the presence of TC14 in some genetic backgrounds. The impact of height on yield in the 2008 season is unknown although if the shorter plants also had smaller root systems, the hot, dry spring of that year may have enabled longer rooted plants to better extract water from the soil profile, however no direct evidence of this was collected.

The resistance loci that are present on these translocations are of undoubted use in Australia. However, their deployment would need to be done judiciously. The T4m1 translocation is preferred in Australia over T4 or T4m2 due to loss of yellow flour pigmentation and the maintenance of both *Sr25* and *Lr19*. The effects on yield would limit the use of T4m1 to the more productive region where the yield effect could be advantageous. Due to the lack of use in Australia, *Lr19* is still effective against all known leaf rust pathotypes in this country, and could potentially be a useful adjunct to breeders. However, there is a caveat to this in that *Lr19* has been overcome in Mexico [20] and the Indian sub-continent [21]. This would be unlikely to happen in Australia unless *Lr19* were to become common in varieties across the wheat growing regions. It is more probable that the yield effects of this translocation would limit its use to the more productive higher rainfall zones in south-eastern Australia.

The stem rust resistance gene, *Sr25*, also has potential for Australian deployment as it is effective against a potentially devastating family of stem rust races currently evolving in western Africa, the Ug99 stem rust family. Stem rust can be of major concern in highly productive regions and this gene would not only protect varieties from endemic stem rust races, but also against the Ug99 family should an incursion into Australia occur.

The TC14 translocation also contains a useful resistance locus in the *Bdv2* gene. Barley yellow dwarf virus is a disease that has the potential to cost the Australian wheat industry $15 million annually [22] and is also associated with the more productive Australian wheat growing regions. Ayala-Navarrete, Mechanicos, Gibson, Singh, Bariana, Fletcher, Shorter and Larkin [18] recombined T4 translocations with TC14 translocations to develop the Pontin translocation series, some of which contain combinations of *Bdv2* and *Lr19* on a much shorter recombinant translocation than the T4 translocations in this current study. Other Pontin lines have *Lr19* and *Sr25* on a short translocation, or *Sr25* alone on a shorter translocation. Some Pontin recombinants have retained *Bdv2* and *Lr19* while losing the PSY-E1 marker associated with yellow flour. It is hoped that some of these will have useful combinations of genes for use in breeding, and it will be important to assess the impact they have on yield potential in different environments.

4. Experimental Section

4.1. Development of Germplasm

Germplasm was developed through a combination of traditional backcrossing and doubled-haploid production. Most of the T4-containing lines were developed by the National Cereal Rust Control Program (Cobbitty, Australia) and comprised 15 recurrent parents (B2806, Batavia, Banks, Bayonet, Condor, Cook, H45, Kiata, Lark, Lowan, Matong, Oxley, R38549H, R38568C and Vulcan) using two white-floured EMS generated T4 donor mutants Agatha 28-4 and Agatha 235-6 [9]. Herein, we have termed these donor lines as T4m1 and T4m2, respectively, maintaining consistency with the original T4 translocation. Both of these variants maintained their *Lr19* status but only the former retained *Sr25* conferred resistance. Other sets of sister lines were developed through doubled haploid methodology from Seri 82/Superseri which segregated for the original yellow floured T4 translocation and from Condor/CondorT4m1 which segregated for T4m1. The TC14 lines used Oasis 86//TC14/2*Spear as the translocation donor and either three or four backcrosses made to either elite parents or from crosses between elite parents. The recurrent parents of these lines were Brookton, Camm, Chara, Drysdale, H45, Janz, Westonia, H45/Darter, Janz/Sunbri, Drysdale/Sunbrook, Westonia/Sunbrook and Chara/Ohm//Sunbrook. All lines and pedigrees are listed in Supplementary Table 1. These doubled haploids were generated using the maize cross method under contract through the South Australian Research and Development Institute, Adelaide.

4.2. Yield Trials

Yield trials were run in two stages. Most of the T4 material was used in experiments conducted at multiple sites in the year 2000 (Ginninderra, ACT, Stockinbingal, NSW and Moombooldool, NSW), and in 2001 (Ginninderra, Gundibindyal, NSW and Condobolin, NSW). A second set of experiments contained up to 19 Condor-derived lines and up to 23 Seri/Super Seri derived lines that had differential statuses for the T4m1 and the T4 translocations, respectively. These experiments were also used for the TC14 translocation. The second set of experiments were conducted between 2006 and 2008 at Griffith, Gundibindyal, Yanco and Temora (all in NSW). Details of trial sites are listed in Table 1. The trials conducted in 2000 and 2001 were alpha-lattice designs, the 2006, 2007 and 2008 experiments were partially replicated, row-column designs.

All trials were sown in 10-row plots with 18 cm spacing and a 50 cm gap between plots. Plots were 8 m in length. Sowing rate was 100 kg/ha with 105 kg/ha Starter 15® fertilizer applied at sowing while irrigated experiments were broadcast with an additional 40–80 kg/ha urea at late booting. Where trials were irrigated, seed was sown on a full soil-water profile with an additional irrigation at late booting. Dryland sites were sown under the prevailing weather conditions. Sites were monitored for foliar diseases, but as none were observed, no fungicide treatments were required. Maturity scores were taken according to Zadoks, Chang and Konzak [14] when most of the lines were at midear emergence (Growth Stage 55). Heights were also scored on plots. Thousand kernel weight was determined from 250 random grains and hectoliter weight calculated on a chondrometer designed in CSIRO that weighs the volume of 0.5 L of grain.

4.3. Statistical Analysis

Data were analyzed statistically after first checking for normality and error variance heterogeneity across environments. Hereafter, a two-stage, mixed model approach was utilized [23] using ASReml [24]. Each experiment was analyzed separately with the best spatial models being determined after first fitting the experimental design and then modelling the residual variation with autoregressive row and column terms. Significant spatial effects were then identified and residuals assessed before determinations made to the need for fitting of other (e.g., linear) effects [25]. For each experiment, the best linear unbiased estimates (BLUEs) and their weights [26] were used as inputs in the subsequent across-experiment analyses. Individual entry and translocation group means were obtained for each experiment and SED and LSDs obtained for pairwise comparison. Broad-sense heritability (H^2) was calculated on a single-plot and genotype-mean basis with standard errors estimated after Holland, *et al.* [27].

5. Conclusions

In conclusion, the T4 and T4m1 translocation results confirmed previous reports that indicate they provide a yield boost under high yielding conditions, yet are detrimental to yield under low yielding conditions. However, the T4m2 translocation would appear to be of little use as the yellow flour mutation also rendered *Sr25* and a yield locus ineffective. The TC14 translocation appeared to have a specific interaction with reduced height in two genetic backgrounds that were correlated with lower yields. Overall, the TC14 translocation appeared to delay flowering, but more detailed analysis showed this was also dependent on genetic background and did not appear to be associated with a yield penalty. BYDV resistance conferred by TC14 should make if useful in higher yielding environments where this disease is prevalent.

Acknowledgments

We would also like to acknowledge the dedicated technical support of Bernie Mickelson and staff at the CSIRO Ginninderra Experiment Station, Canberra ACT.

Author Contributions

Garry Rosewarne conducted trials in 2006–2008, collated and interpreted all data and wrote manuscript. David Bonnett conducted all trials in 2000–2001. Greg Rebetzke conducted trial design and statistical analysis, Paul Lonergan oversaw field trial operations, Philip J. Larkin originally conceived experiments and developed TC14 genetic material.

Conflicts of Interest

The authors declare no conflict of interest.

References

1. Marcussen, T.; Sandve, S.R.; Heier, L.; Spannagl, M.; Pfeifer, M.; Jakobsen, K.S.; Wulff, B.B.H.; Steuernagel, B.; Mayer, K.F.X.; Olsen, O.-A. Ancient hybridizations among the ancestral genomes of bread wheat. *Science* **2014**, *345*, doi: 10.1126/science.1250092.
2. Trethowan, R.M.; Mujeeb-Kazi, A. Novel germplasm resources for improving environmental stress tolerance of hexaploid wheat. *Crop Sci.* **2008**, *48*, 1255–1265.
3. Knott, D.R. The effect of transfers of alien genes for leaf rust resistance on the agronomic and quality characteristics of wheat. *Euphytica* **1989**, *44*, 65–72.
4. Sharma, D.; Knott, D.R. The transfer of leaf-rust resistance from *agropyron* to *triticum* by irradiation. *Can. J. Genet. Cytol.* **1966**, *8*, 137–143.
5. Somo, M.; Chao, S.; Acevedo, M.; Zurn, J.; Cai, X.; Marais, F. A genomic comparison of homoeologous recombinants of the *Lr19* (T4) translocation in wheat. *Crop Sci.* **2014**, *54*, 565–575.
6. Singh, R.P.; Huerta-Espino, J.; Rajaram, S.; Crossa, J. Agronomic effects from chromosome translocations 7DL.7AG and 1BL.1RS in spring wheat. *Crop Sci.* **1998**, *38*, 27–33.
7. Reynolds, M.P.; Calderini, D.F.; Condon, A.G.; Rajaram, S. Physiological basis of yield gains in wheat associated with the *Lr19* translocation from agropyron elongatum. *Euphytica* **2001**, *119*, 139–144.
8. Monneveux, P.; Reynolds, M.P.; Aguilar, J.G.; Singh, R.P.; Weber, W.E. Effects of the 7DL.7AG translocation from lophopyrum elongatum on wheat yield and related morphophysiological traits under different environments. *Plant Breed.* **2003**, *122*, 379–384.
9. Knott, D.R. Mutation of a gene for yellow pigment linked to *Lr19* in wheat. *Can. J. Genet. Cytol.* **1980**, *22*, 651–654.
10. Banks, P.M.; Larkin, P.J.; Bariana, H.S.; Lagudah, E.S.; Appels, R.; Waterhouse, P.M.; Brettell, R.I.S.; Chen, X.; Xu, H.J.; Xin, Z.Y., *et al.* The use of cell culture for subchromosomal introgressions of barley yellow dwarf virus resistance from thinopyrum intermedium to wheat. *Genome* **1995**, *38*, 395–405.
11. Prins, R.; Groenewald, J.Z.; Marais, G.F.; Snape, J.W.; Koebner, R.M.D. AFLP and STS tagging of *Lr19*, a gene conferring resistance to leaf rust in wheat. *Theor. Appl. Genet.* **2001**, *103*, 618–624.
12. Stoutjesdijk, P.; Kammholz, S.J.; Kleven, S.; Matsay, S.; Banks, P.M.; Larkin, P.J. PCR-based molecular marker for the *Bdv2 Thinopyrum intermedium* source of barley yellow dwarf virus resistance in wheat. *Aust. J. Agric. Res.* **2001**, *52*, 1383–1388.
13. Ayala-Navarrete, L.; Bariana, H.S.; Singh, R.P.; Gibson, J.M.; Mechanicos, A.A.; Larkin, P.J. Trigenomic chromosomes by recombination of *Thinopyrum intermedium* and *Th. ponticum* translocations in wheat. *Theor. Appl. Genet.* **2007**, *116*, 63–75.
14. Zadoks, J.C.; Chang, T.T.; Konzak, C.F. A decimal code for the growth stages of cereals. *Weed Res.* **1974**, *14*, 415–421.
15. Zhang, W.; Dubcovsky, J. Association between allelic variation at the phytoene synthase 1 gene and yellow pigment content in the wheat grain. *Theor. Appl. Genet.* **2008**, *116*, 635–645.

16. Prins, R.; Marais, G.F.; Marais, A.S.; Janse, B.J.H.; Pretorius, Z.A. A physical map of the thinopyrum-derived *Lr19* translocation. *Genome* **1996**, *39*, 1013–1019.
17. Prins, R.; Marais, G.F.; Pretorius, Z.A.; Janse, B.J.H.; Marais, A.S. A study of modified forms of the *Lr19* translocation of common wheat. *Theor. Appl. Genet.* **1997**, *95*, 424–430.
18. Ayala-Navarrete, L.I.; Mechanicos, A.A.; Gibson, J.M.; Singh, D.; Bariana, H.S.; Fletcher, J.; Shorter, S.; Larkin, P. The pontin series of recombinant alien translocations in bread wheat: Single translocations integrating combinations of *Bdv2*, *Lr19* and *Sr25* disease-resistance genes from *Thinopyrum intermedium* and *Th. ponticum*. *Theor. Appl. Genet.* **2013**, *126*, 2467–2475.
19. Slafer, G.A. Differences in phasic development rate amongst wheat cultivars independent of responses to photoperiod and vernalization. A viewpoint of the intrinsic earliness hypothesis. *J. Agric. Sci.* **1996**, *126*, 403–419.
20. Huerta-Espino, J.; Singh, R.P. First report of virulence to wheat with leaf rust resistance gene *Lr19* in Mexico. *Plant Dis.* **1994**, *78*, 640.
21. Bhardwaj, S.C.; Prashar, M.; Kumar, S.; Jain, S.K.; Datta, D. *Lr19* resistance in wheat becomes susceptible to *Puccinia triticina* in India. *Plant Dis.* **2005**, *89*, 1360–1360.
22. Murray, G.M.; Brennan, J.P. *The Current and Potential Costs from Diseases of Wheat in Australia*; Grains Research and Development Corporation: Kingston, Australia, 2009.
23. Cullis, B.R.; Thomson, F.M.; Fisher, J.A.; Gilmour, A.R.; Thompson, R. The analysis of the NSW wheat variety database. 2. Variance component estimation. *Theor. Appl. Genet.* **1996**, *92*, 28–39.
24. Gilmore, A.R.; Gogel, B.J.; Cullis, B.R.; Thompson, R. *Asreml User Guide Release 3.0*; VSN International Ltd.: Hemel Hempstead, UK, 2001.
25. Cullis, B.; Gogel, B.; Verbyla, A.; Thompson, R. Spatial analysis of multi-environment early generation variety trials. *Biometrics* **1998**, *54*, 1–18.
26. Smith, A.; Cullis, B.; Thompson, R. Analyzing variety by environment data using multiplicative mixed models and adjustments for spatial field trend. *Biometrics* **2001**, *57*, 1138–1147.
27. Holland, J.B.; Nyquist, W.E.; Cervantes-Martínez, C.T. Estimating and interpreting heritability for plant breeding: An update. In *Plant Breeding Reviews*; John Wiley & Sons, Inc.: Raleigh, CO, USA, 2003; pp. 9–112.

8

Analysis of Temporal Variation of Soil Salinity during the Growing Season in a Flooded Rice Field of Thessaloniki Plain-Greece

Emanuel Lekakis [1], Vassilis Aschonitis [2,*], Athina Pavlatou-Ve [3], Aristotelis Papadopoulos [1] and Vassilis Antonopoulos [3]

[1] ELGO-DIMITRA, Soil Science Institute of Thessaloniki, Thermi 57001, Greece; E-Mails: elekakis@agro.auth.gr (E.L.); gendirpap@nagref.gr (A.P.)

[2] Department of Life Sciences and Biotechnology, University of Ferrara, Via L.Borsari 46, Ferrara 44121, Italy

[3] Department of Hydraulics, Soil Science & Agricultural Engineering, School of Agriculture, Aristotle University of Thessaloniki, Thessaloniki 54124, Greece; E-Mails: ave@agro.auth.gr (A.P.-V.); vasanton@agro.auth.gr (V.A.)

* Author to whom correspondence should be addressed; E-Mail: schvls@unife.it

Academic Editor: Steve Robinson

Abstract: The effects of regional water management practices (WMPs) on the soil salinity of a representative rice field under Mediterranean conditions (Thessaloniki plain, Greece) were investigated. The temporal variation of soil salinity parameters in the soil solution and in the exchangeable phase was monitored at and below the root zone (15–20 and 35–40 cm) during the growing season. The comparative analysis (ANOVA for $p = 0.05$) of the measurements before and after the growing season showed that: (a) for the soil solution of the 15–20 cm layer, Ca^{2+}, Mg^{2+}, K^{+}, HCO_3^{-} and EC were significantly reduced, Na^{+} remained constant and Cl^{-} increased, while in the 35–40 cm layer no significant differences were detected to all parameters except for Cl^{-} which was increased; (b) for the exchangeable cations Ca^{2+}, Mg^{2+} and K^{+} no significant differences were found, while exchangeable Na^{+} and ESP were significantly increased in both soil layers during the short period of soil drying before harvest. The final values of Na^{+} and ESP were quite low to indicate soil degradation hazard. Overall the results showed adequate performance of

WMPs to preserve a good soil salinity status but with the cost of high water consumption, exceeding 2000 mm.

Keywords: soil salinity; suction cups; saturation extract; exchangeable cations; water management

1. Introduction

More than 10% of irrigated lands worldwide are affected by salinity problems [1,2]. Soil degradation by salinity has become one of the major threats and it is expected to be intensified by the imminent climate change and the increase in irrigation demand [1,3–5]. One of the most interesting cases to investigate soil salinity components is the case of rice-fields. Rice cropping is often held responsible for soil sodification [6], while on the other hand irrigated rice cropping is practiced to reclaim saline-sodic soils in many parts of the world [7,8].

Continuous flooding conditions is the common practice for rice irrigation, especially in lowland plains [9], leading to increased water inflows for the attainment of an adequate ponding depth. The quality and amount of irrigation water, the evapotranspiration, the soil hydraulic properties and the drainage conditions of lowland fields are the main regulators of soil salinity.

Long-term submerged soils undergo physical changes which lead to the formation of a highly compacted layer below the root-zone [10], called "plow sole layer" and is considered as the major factor controlling infiltration rate and drainage in paddy rice fields [11–14]. Furthermore, infiltration can potentially be blocked by air entrapment due to surge flooding and shallow water table [15]. Boivin *et al.* [15] noted that if the downward water transfer is blocked under flooded conditions, then solutes supplied by the irrigation water may accumulate in the top soil. Wopereis *et al.* [16] demonstrated that the salinity level of the topsoil of rice-fields was decreased, because salts from the topsoil were gradually transferred downward by infiltration while infiltration blocked the upward transport of salts from the water table. Häfele *et al.* [17] indicated that the practice of wet soil-tillage (puddling), which is commonly used in Asian countries, leads to an increase of the puddled soil layer depth and new salts from the subsoil may be brought into the soil solution. Therefore, drainage control of rice fields is of primary significance in order to avoid soil salinization and sodification. In lowland plains with low permeable soils and high water table, drainage of rice soils is restricted and in many cases it is supported by controlled overland discharge [4]. In this case, any salt deposition-accumulation on soil surface is partially subjected to dissolution and flows out of the system by runoff.

There is a great number of studies dealing with the chemistry and physics of paddy rice in waterlogged soils [15,18–24], but there is still a lack of knowledge concerning solute dynamics under these conditions. The development and application of models which describe salinity components (anions-cations dynamics) in rice fields, remain a great challenge even though many have been developed to describe the water, nutrients (nitrogen, phosphorus) and pesticides transport under these conditions [12,25–37]. Methods based on water budget approaches in combination with the monitoring of salinity are common tools to analyze soil salinity components in rice fields [38–41].

The aim of this study is to investigate the changes of soil salinity during the growing season in a representative rice field of Thessaloniki plain in Greece, under the effects of regional water management practices (WMPs). To this aim, the temporal variation of salinity parameters at and below the root zone (15–20 and 35–40 cm) was monitored in the soil solution and in the exchangeable phase of the soil during one growing period. The comparative analysis of the measurements before and after the growing season was used to provide a discussion on the sustainability of water consumption in rice fields under Mediterranean conditions.

2. Materials and Methods

2.1. Study Site

Rice (*Oryza sativa* L.) was cultivated using the regional conventional practices under flooding conditions in an experimental field of 13 × 20 m during the growing season of 2011. The field is located at the southeastern region of Axios River plain (40°35′ N, 22°41′ E, 1 m above sea level) near Thessaloniki city in Northern Greece. This area is representative of the broader area of Axios River characterized by an extensive irrigation—drainage network, where rice is the dominant crop. Rice is being cultivated in the region for more than 50 years. A rotation program with three consecutive years of rice crop interrupted by one or two years of maize, fodder, or cotton, is the common practice in the area [5,10]. The soils in the region are mostly silty clay, poorly drained, classified as Typic Xerofluvents [42] under Mediterranean climate conditions, according to Soil Survey Staff [43]. A large part of the land is near and below the sea level with permanent high water table (1–1.5 m from the soil surface) controlled by the adjacent drainage ditches. Pumps are used to lead the drainage water to the sea. During the non-flooding period, the water level in the drainage ditches is approximately 1.2–1.5 m below the soil surface, while during flooding season, it is approximately 1.0–1.2 m below the soil surface [5,44].

2.2. Agricultural Practices

Soil treatment consisted of moldboard plowing at the end of November 2010 followed by land leveling at the end of April 2011 and harrowing for seedbed preparation simultaneously with basal fertilization. Sowing was performed on 16 May with seeding density at 200 $kg \cdot ha^{-1}$. The total fertilization rate was at 176 $kg\ N \cdot ha^{-1}$. The amount of 96 $kg\ N \cdot ha^{-1}$, in the form of urea, was incorporated in the soil before sowing on 12 May. Another two surface applications of 40 $kg\ N \cdot ha^{-1}$ were performed on 17 June and 12 July (60.8 $kg\ N \cdot ha^{-1}$ in the form of ammonium nitrogen and 19.2 $kg\ N \cdot ha^{-1}$ in the form of nitrate nitrogen). The fertilizer consisted of 19% ammonium nitrogen in the form of ammonium sulfate and 6% nitrate nitrogen. Herbicide (active ingredients benzofenap and clomazone) was applied on 7 June. Irrigation water originates form River Axios and is transferred by gravity in the fields using open channels' network. Irrigation in the experimental field started on the 14 May. The maximum ponding depth during the growing season of rice was approximately 10 cm with a few intermissions in irrigation water supply. Irrigation stopped on 8 September and flooding conditions ceased on 13 September. Rice was harvested on 13 October.

2.3. Water Balance Components

Water balance components were computed and analyzed in a previous study by Aschonitis *et al.* [37] using the GLEAMS-PADDY model [28], which was modified to include algae growth in rice fields. The model was applied for the same experimental field at the same growing season, in order to assess the uptake of major nutrients by algae in the floodwater, the inflow by irrigation and the losses by runoff of N, P, Ca, Mg, K, Na.

The rice crop evapotranspiration was determined using regional modified crop coefficients, which are functions of leaf area index LAI [45,46], multiplied by the reference crop evapotranspiration, estimated by the ASCE standardized Penman-Monteith method [47]. LAI measurements were performed using the destructive planimetric technique and meteorological data were obtained by the station at the experimental farm of the Aristotle University of Thessaloniki (40°32′ N, 23°00 E, 16 m above sea level). Measurements of irrigation water inflow, surface runoff and ponding depth were used for the calibration of the model. The water balance components of irrigation inflow, precipitation, surface runoff (above the bunds of 10 cm height), drainage below the root zone, and actual crop evapotranspiration, were estimated by the model at 2260, 158, 520, 432 and 1350 mm, respectively [37]. The daily variation of ponding depth, irrigation water amount and precipitation are given in Figure 1a, while the daily variation of surface runoff, drainage and actual crop evapotranspiration are given in Figure 1b.

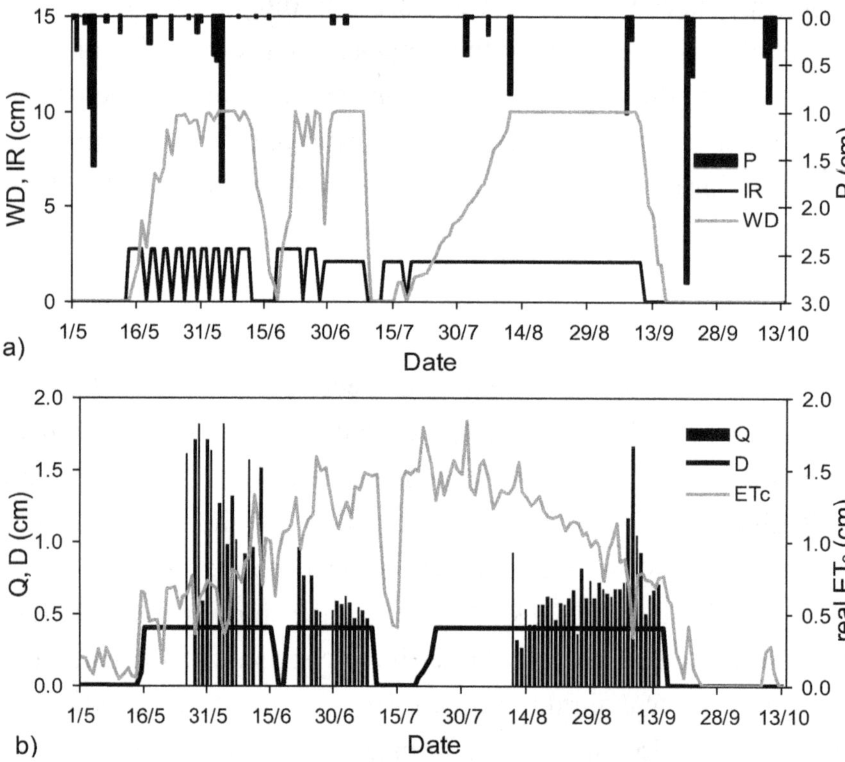

Figure 1. (**a**) Daily variation of ponding depth—WD, irrigation water amount—IR, and precipitation—P; (**b**) daily variation of surface runoff—Q, drainage—D, and actual crop evapotranspiration—ET_c as computed by the modified GLEAMS-PADDY model [37].

2.4. Measurements of Soil and Water Quality Parameters

For the collection of the soil solution at and below the root zone, three pairs of ceramic porous cups (SPS 200, SDEC France, ∅63 mm) were installed in the rice-field at two soil layers (15–20 cm and 35–40 cm), before the flooding season, on 10 May. Distance between the pairs of porous cups was 5 m. For the installation of the suction cups, a hole was opened to the intended sampling depth. The soil was collected from representative layers in the hole and it was preserved in order to measure the initial soil physicochemical characteristics of the experimental field. The soil physicochemical properties are given in Table 1.

Table 1. Soil physicochemical characteristics at 15–20 and 35–40 cm soil layers and mean water quality parameters of the experimental rice field during the irrigation season.

Soil Parameters	**Soil Layer (cm)**	
	15–20 cm	**35–40 cm**
Soil Texture	Silty Clay	Silty Clay Loam
$CaCO_3$ (%)	6.16	7.19
Organic Matter (%)	2.09	1.73
[1] CEC ($cmol_c \cdot kg^{-1}$)	40.58	33.01
θ_s ($cm3 \cdot cm^{-3}$)	0.655	0.538
θ_r ($cm3 \cdot cm^{-3}$)	0.000	0.000
a (cm^{-1})	0.0085	0.00196
n	1.200	1.129
m ($m = 1 - 1/n$)	0.167	0.114
ρ_b ($g \cdot cm^{-3}$)	1.123	1.223
Irrigation water parameters	**Mean ± S.D.**	
pH	8.1 ± 0.3	
EC ($\mu S \cdot cm^{-1}$)	429 ± 83	
Ca^{2+} [3]($mmol_c \cdot L^{-1}$)	2.31 ± 0.54	
Mg^{2+} ($mmol_c \cdot L^{-1}$)	1.04 ± 0.12	
Na^+ ($mmol_c \cdot L^{-1}$)	0.65 ± 0.08	
K^+ ($mmol_c \cdot L^{-1}$)	0.03 ± 0.01	
Cl^- ($mmol_c \cdot L^{-1}$)	0.65 ± 0.48	
HCO_3^- ($mmol_c \cdot L^{-1}$)	3.17 ± 0.61	
[2] SAR $(mmol_c \cdot L^{-1})^{0.5}$	0.50 ± 0.05	

[1] Cation Exchange Capacity (CEC) is calculated as the sum of exchangeable cations; [2] SAR is calculated by the equation $Na^+/((Ca^{2+} + Mg^{2+})/2)^{0.5}$; [3]millimoles of charge per liter.

The remaining soil was used to form slurry, which was poured back into the hole to refill a section of 10–20 cm from the bottom. The suction cups were then gently pushed into the slurry to establish tight hydraulic contact between the saturated porous cups and the surrounding soil [48]. Prior to installation, the devices were placed in a container with 5% HCl and the solution was pumped three times through the porous cups. They were then rinsed with deionized water until electrical conductivity was equal to that of the deionized water [49]. The first soil solutions from the suction cups were obtained on 31 May and then they were obtained weekly until 26 September (a total of 15 samplings) in order to measure Electrical Conductivity (EC), pH, K^+, Mg^{2+}, Ca^{2+}, Na^+, HCO_3^-

and Cl^-. At the same dates, three pairs of soil samples from the flooded soil, near the porous cups, were collected from the field at soil layers 15–20 and 35–40 cm. Soil samples were air dried and sieved through a 2 mm sieve. Measurements of EC, K^+, Mg^{2+}, Ca^{2+}, Na^+, HCO_3^- and Cl^- were also performed in the saturation extracts of the soil paste. Exchangeable K^+, Mg^{2+}, Ca^{2+} and Na^+ were measured in the soil samples. Additional soil sampling was performed approximately 20 days after the end of the flooding season, on 6 October, when soil reached unsaturated conditions. This sampling was performed for comparison purposes between the final and the initial status of soil salinity parameters.

Triplicate floodwater and irrigation water samples were obtained weekly during the growing season to measure EC, pH, K^+, Mg^{2+}, Ca^{2+}, Na^+, HCO_3^- and Cl^-. Mean irrigation water quality parameters throughout the irrigation season are given in Table 1. Even though the quality of the applied irrigation water from Axios River is considered satisfactory for preventing any problems regarding salt built up in soil and element deficiencies in rice plants [50], the combination of EC and SAR (Sodium Adsorption Ratio) poses slight to moderate soil degradation risk, regarding infiltration problems, according to the international standards for the irrigation water [51].

Particle size analysis was performed by the hydrometer method [52]. Organic matter was determined by the wet oxidation method [53] and calcium carbonate ($CaCO_3$) by the volumetric calcimeter method [54]. The concentrations of Ca^{2+}, Mg^{2+}, K^+ and Na^+ in the soil solution, the saturation extract, floodwater and irrigation water were determined using ICP [55]. EC was measured by a conductivimeter [55,56] and pH by a pH-meter (pH of the soil samples was measured in the saturated soil paste). Chloride was measured via colorimetric titration with silver nitrate using a chromate indicator [57]. Concentration of bicarbonate (HCO_3^-) was determined by titration to a pH of 4.3 using diluted 0.05 N H_2SO_4 [55]. Extracted exchangeable cations Ca^{2+}, Mg^{2+}, Na^+ and K^+, by the ammonium acetate method, at pH 8.2 [58], were determined using ICP.

For the determination of the hydraulic properties, undisturbed soil samples were obtained from twelve positions in the field. Cores of 5 cm high and 5 cm diameter were collected in thin-walled metal rings from two depths of the soil profile (15–20 cm and 35–40 cm) before the flooding period. The soil cores were wetted from below to saturation and then equilibrated on ceramic tension plates at eleven successive suctions between 0–15,000 hPa, for the determination of the soil water retention curve [59] with the van Genuchten model [60]. The parameters of the van Genuchten model were calculated using the RETC code software [61]. Bulk densities, ρ_b, were measured at the same cores after drying at 105 °C (Table 1).

2.5. Analysis and Statistics

Changes in soil salinity were analyzed and discussed based upon the measurements in the soil solutions obtained from the suction cups. To compare the initial and final status of the soluble phase under unsaturated conditions before (5 May 2011) and after the growing season (6 October 2011), the saturation extracts were used. Comparative analysis was also performed for the case of exchangeable cations at the same dates. The measurements were statistically analyzed using the software package Statgraphics Centurion XV software. ANOVA with LSD test for level of significance $p < 0.05$ was used to identify significant differences between measurements.

3. Results and Discussion

3.1. Variations in Ca^{2+}, Mg^{2+}, K^{+}, Na^{+}, Cl^{-}, HCO_3^{-}, EC and pH in Soil Solution, Irrigation and Floodwater

The variation of Ca^{2+}, Mg^{2+}, K^{+}, Na^{+}, Cl^{-}, HCO_3^{-}, EC and pH in the soil solution obtained from suction cups in the 15–20 and 35–40 cm layers, the irrigation water and floodwater during the flooding season, is presented in Figure 2.

The concentration of Ca^{2+} in the soil solution tended to increase in both soil layers (Figure 2a). That increase was followed by two interruptions during the growing season, which are mainly attributed to plant uptake. Possible source of Ca^{2+} may be the dissolution of carbonate minerals due to saturated conditions in the presence of H^{+} ions. The involvement of these H^{+} in chemical reactions occurring in the rhizosphere is of prime significance due to their implications in the acquisition of many mineral nutrients [62]. Ammonium sulfate fertilizer was applied on the surface on 17 June and 12 July, dates that coincide with the onset of high concentrations of Ca^{2+} in soil solution. Furthermore, ammonium sulfate dissolves some calcium carbonate in calcareous soils [63] and dissolution of insoluble carbonate and bicarbonate under reduced conditions contribute Ca^{2+} and Mg^{2+} in the soil solution [23]. These results are in accordance with results reported in the literature [20,21,23,64]. Larson *et al.* [65] found that water-logging of two calcareous Entisols increased the availability of cations such as Ca^{2+} and Mg^{2+} in the soluble phase. Also, pCO_2 increases in the soil-rooting zone and pH decreases due to the decomposition of organic matter leading to calcite dissolution and increase of Ca^{2+} concentrations in the soil solution [66]. Ca^{2+} concentration in the floodwater and the irrigation water followed the same fluctuation during the irrigation period and remained relatively low with no significant differences (Figure 2a).

The concentration of Mg^{2+} in the soil solution slightly increased at the beginning of the growing season and remained almost constant afterwards in the 15–20 cm soil layer, while the opposite trend was observed in the underlying layer of the 35–40 cm (Figure 2b). There was no increase recorded in Mg^{2+} concentration resulting by dissolution as in the case of Ca^{2+}. A possible explanation is that the calcareous parent material of the soil is mainly limestone, containing higher levels of Ca^{2+} in relation to Mg^{2+}. The concentration of Mg^{2+} in the floodwater and the irrigation water remained relatively low and constant throughout the irrigation period (Figure 2b).

The concentration of K^{+} in the soil solution of the 15–20 cm decreased almost two-fold from a maximum value of 0.146 $mmol_c \cdot L^{-1}$ to a minimum value of 0.065 $mmol_c \cdot L^{-1}$ while in the layer of 35–40 cm remained relatively constant during the growing season (Figure 2c). The decrease of K^{+} in the upper layer can be attributed to crop uptake. It must be mentioned that the only source of K^{+} is the irrigation water where K^{+} concentration remained constant during the irrigation period (Figure 2c). The application of the modified GLEAMS-PADDY model by Aschonits *et al.* [37] showed that during the experimental period, the irrigation water provided 31.16 $kg \cdot ha^{-1}$ of K^{+} while algae uptake from the flooding water was estimated at 39.58 $kg \cdot ha^{-1}$. These results indicate that algae consumed more K^{+} than the estimated inflow. This is justified by the fact that the concentration of K^{+} in the flooding water was regulated mainly by its concentration in the irrigation water and in the upper muddy soil layer. The effects of algae uptake on the concentration of K^{+} in the flooding water are evident in Figure 2c where K^{+} decreased at the beginning until the middle of the flooding season. This period corresponds to the highest algae biomass production and growth rates estimated by the GLEAMS-PADDY model [37]. Later in the season, K^+ increased again due to:

- the decrease of algae growth and uptake because LAI exceeded the value of 2 and algal photosynthesis was restricted [37];
- the continuous irrigation water inflow with no further intermissions by the middle of July (Figure 1a) which contributed to the increase of K^+ in the flooding water; and
- evapotranspiration reaching maximum values after the middle of July (Figure 1b), which enhanced evapoconcentration.

It is clear that flooding conditions together with the rice crop and algae biomass control the availability of K^+ in the soil solution of the root zone and in the floodwater, respectively. Phillips and Greenway [20], Lu *et al.* [23] and Larson *et al.* [65] found that flooding increased the availability of soluble K^+ in the soil solution and floodwater. However, their experiments were held in laboratory conditions and involved soil water logging without considering the effects of plants or algae, which could exploit this availability on their behalf. On the other hand, Boivin *et al.* [24] observed that K^+ concentration decreased with time during flooding, in field experiments.

The concentration of Na^+ in the soil solution, at both soil layers, remained relatively constant during the irrigation period (Figure 2d). Lu *et al.* [23] also reported no significant changes of water-soluble Na^+ during a 12-week waterlogging period. Levels of Na^+ in the floodwater and irrigation water remained low and constant throughout the irrigation period (Figure 2d).

Cl^- concentration in the soil solution of both soil layers (Figure 2e) remained relatively constant during the irrigation period but decreased significantly to a minimum value of approximately 0.4 $mmol_c \cdot L^{-1}$, after irrigation ceased. These findings are in agreement with Boivin *et al.* [15] who explained this behavior by an unstable Cl-GR formation in the soil around the porous cups during solution collection, which leads to a decrease in pCO_2. The researchers noted that Cl^- should not be used as a reference element for calculating concentration ratios of water when sampled under vacuum atmosphere in partly reduced soils. Control of Cl^- in soils with low pCO_2 might occur without vacuum applied. Indeed, measured concentrations of Cl^- in saturation extracts from soil samples, at the same depths (results not shown), revealed a different tendency, with Cl^- concentration values increasing almost three-fold after the irrigation period (from approximately 1.1 to 2.9 $mmol_c \cdot L^{-1}$). The concentration of Cl^- in irrigation water and floodwater remained almost constant throughout the irrigation season (Figure 2e).

Bicarbonate in the soil solution (Figure 2f) behaved like calcium, increasing over time from 16 to approximately 23 $mmol_c \cdot L^{-1}$ in the 15–20 cm soil layer. The same fluctuation was observed in the 35–40 cm soil layer, while its concentration decreased back to the initial levels after the flooding season, in both soil layers. Bicarbonate increased in the soil solution particularly after the second surface fertilizer application. This behavior can be attributed to the production of HCO_3^- in the presence of high levels of calcium carbonate. Boivin *et al.* [15] also reported continuous increase in bicarbonate and further assumed that calcite precipitation did not occur during the irrigation period. van Asten *et al.* [8] mentioned that precipitation and dissolution of calcite control bicarbonate of the soil solution, as in this study. The decrease in bicarbonate at both soil layers at the end of the cropping period can be explained by desorption of exchangeable Ca^{2+} which leads to calcite precipitation and a subsequent decrease of bicarbonate in the soil solution [8]. When the soil profile dries up and reoxidizes, pCO_2 drops and minerals like sepiolite, calcite and magnesite are probably crystallized and accumulated in the soil profile during the fallow period [15]. Bicarbonate in the floodwater and irrigation water remained low and constant throughout the irrigation period (Figure 2f).

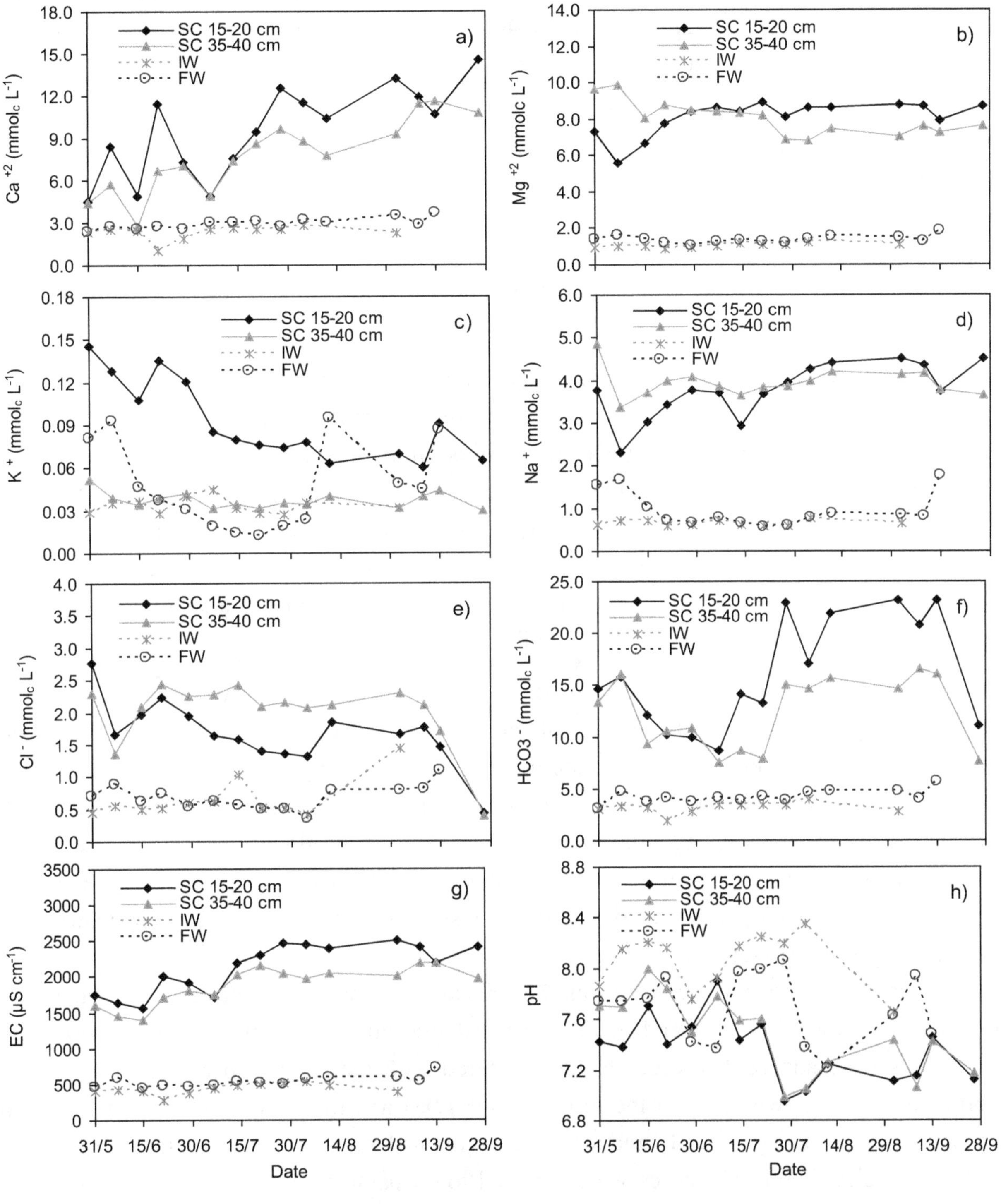

Figure 2. Variation of salinity components in the soil solution obtained from suction cups of soil layers 15–20 and 35–40 cm, in the irrigation water and in the floodwater (SC: soil solution, IW: irrigation water, FW: floodwater) (**a**) Ca^{2+} concentration; (**b**) Mg^{2+} concentration; (**c**) K^{+} concentration; (**d**) Na^{+} concentration; (**e**) Cl^{-} concentration; (**f**) HCO_3^{-} concentration; (**g**) Electrical conductivity (EC) and (**h**) pH.

During the first six weeks after the initiation of flooding, EC in the soil solution gradually increased reaching peak values in both soil layers, while afterwards it remained almost constant until the end of the cropping season (Figure 2g). That increase was mainly regulated by the increase of Ca^{2+}. The concentration of Ca^{2+} was highly correlated with the EC of the soil solution ($R = 0.818$, $p < 0.05$ at 15–20 cm and $R = 0.786$, $p < 0.05$ at 35–40 cm), while weaker correlations with Mg^{2+} ($R = 0.602$, $p < 0.05$ at 15–20 cm and $R = -0.081$, $p > 0.05$ at 35–40 cm), Na^+ ($R = 0.582$, $p < 0.05$ at 15–20 cm and $R = 0.592$, $p < 0.05$ at 35–40 cm) and HCO_3^- ($R = 0.418$, $p < 0.05$ at 15–20 cm and $R = 0.218$, $p > 0.05$ at 35–40 cm) were found. An additional factor that may have contributed to the EC increase is the soluble SO_4^{2-} originating from the two surface fertilizer applications of ammonium sulfate. The EC values of the flooding and irrigation water remained below 700 $\mu S \cdot cm^{-1}$ during the monitoring period (Figure 2g).

The pH of the soil solution (Figure 2h) remained relatively constant at the beginning of the flooding season while after 15 July a simultaneous reduction was observed in both soil layers probably because of the following reasons:

- the continuous water inflow by irrigation enhanced oxygenation; and
- when the plant reaches the tillering and internode elongation stages, the roots are at full development while stem aerenchyma starts to be fully active providing a low-resistance pathway for diffusion of O_2 within the roots, enhancing soil oxygenation [67]. In flooded soils, where lowland rice is expected to rely solely on ammonium because of the ambient reducing conditions, it has been well established that the root-induced release of O_2 in the rhizosphere, and the subsequent oxidation of the soil, can make a substantial, and sometimes major contribution to the observed rhizosphere acidification [68–70].

In addition, the root-induced precipitation of the so-called Fe plaque (Fe oxyhydroxides) in the rhizosphere of rice, promotes a significant part of the observed rhizosphere acidification [71]. Peak values of floodwater pH coincide with the ammonium sulfate surface applications. Ammonium sulfate may raise pH, especially in calcareous soils, because it dissolves some calcium carbonate [63]. Floodwater pH decreased again during August, because shading of the floodwater by the rice canopy lowered the photosynthetic activity of the aquatic biomass and thus reduced the degree of CO_2 depletion [72]. Irrigation water pH remained relatively constant with time with an average value of 8.1 during the irrigation season (Figure 2h).

For the analysis of the effects of the WMPs on the soil salinity status, soil samples were obtained on 6 October (unsaturated conditions after flooding were established in the soil by early October) and saturation extracts were analyzed in order to be compared with the saturation extracts obtained on 10 May, prior to flooding. According to the results (Table 2), Ca^{2+}, Mg^{2+}, K^+, HCO_3^- and EC of the 15–20 cm soil layer were significantly reduced, Na^+ remained constant and Cl^- increased. In the 35–40 cm layer no significant differences were detected to all parameters except for Cl^-, which was increased. The above indicate a general reduction of soil salinity of the first layer while the conditions of the second layer remained relatively constant.

Table 2. Comparison of soluble ions Ca^{2+}, Mg^{2+}, K^{+}, Na^{+}, Cl^{-}, HCO_3^{-} and EC obtained from saturation extracts before (5 May) and after (6 October) the growing season.

Parameter	Soil Layer	5 May	6 October
Ca^{2+} ($mmol_c \cdot L^{-1}$)	15–20 cm	5.58 ± 0.18a [1]	4.45 ± 0.25b
Ca^{2+} ($mmol_c \cdot L^{-1}$)	35–40 cm	3.69 ± 0.23a	3.61 ± 0.33a
Mg^{2+} ($mmol_c \cdot L^{-1}$)	15–20 cm	2.35 ± 0.02a	2.1 ± 0.07b
Mg^{2+} ($mmol_c \cdot L^{-1}$)	35–40 cm	1.76 ± 0.17a	1.83 ± 0.15a
K^{+} ($mmol_c \cdot L^{-1}$)	15–20 cm	0.13 ± 0.01a	0.08 ± 0.01b
K^{+} ($mmol_c \cdot L^{-1}$)	35–40 cm	0.05 ± 0.01a	0.04 ± 0.01a
Na^{+} ($mmol_c \cdot L^{-1}$)	15–20 cm	1.66 ± 0.13a	1.96 ± 0.13a
Na^{+} ($mmol_c \cdot L^{-1}$)	35–40 cm	1.85 ± 0.22a	2.09 ± 0.23a
Cl^{-} ($mmol_c \cdot L^{-1}$)	15–20 cm	1.07 ± 0.13b	2.40 ± 0.50a
Cl^{-} ($mmol_c \cdot L^{-1}$)	35–40 cm	1.13 ± 0.07b	3.33 ± 0.13a
HCO_3^{-} ($mmol_c \cdot L^{-1}$)	15–20 cm	5.67 ± 0.53a	3.34 ± 0.07b
HCO_3^{-} ($mmol_c \cdot L^{-1}$)	35–40 cm	3.56 ± 0.23a	3.20 ± 0.01a
EC ($\mu S \cdot cm^{-1}$)	15–20 cm	1010 ± 32a	832 ± 28b
EC ($\mu S \cdot cm^{-1}$)	35–40 cm	736 ± 42a	751 ± 72a

[1] Mean ± Standard error and results of Anova. Different letters indicate statistically significant differences before and after the flooding season at 0.05 confidence level.

3.2. Variations in Exchangeable Cations Ca^{2+}, Mg^{2+}, K^{+}, Na^{+} and ESP

The concentration of the exchangeable cations Ca^{2+}, Mg^{2+}, K^{+} and Na^{+} in the solid phase and ESP of soil layers 15–20 and 35–40 cm are presented in Figure 3. Exchangeable Ca^{2+}, Mg^{2+} and K^{+} followed an increasing trend until the middle of the flooding season and were decreased afterwards in both soil layers, reaching approximately to the initial values, when flooding ceased (Figure 3a–c). These fluctuations were more evident for Ca^{2+} and to a lesser extend for K^{+} and Mg^{2+}. The increase in exchangeable Ca^{2+} can be attributed (a) to the dissolution of insoluble calcium carbonate in the presence of H^{+} ions and (b) to the probable decomposition of organic matter which results in calcite dissolution and larger Ca^{2+} concentrations in the soil solution. The above suggestions also justify the Mg^{2+} fluctuations while K^{+} changes are considered negligible. De Datta [73] also mentioned that many soils under wet conditions may fix some NH_4^{+} in a form that is slowly replaced by cations such as Ca^{2+}, Mg^{2+} and Na^{+} but not by K^{+}. Incorporation of urea in the soil as a basal fertilizer may have enhanced NH_4^{+} fixing in the soil, which was later replaced mainly by Ca^{2+} at the beginning of flooding. Defixation occurs more rapidly under flooding in comparison to unsaturated soils [73]. Na^{+} concentration in the solid phase remained almost constant during the flooding season while after the removal of excess soil water, rapidly increased (Figure 3d). This is a possible indication that Na^{+} reoccupied exchange sites, leading to an increase of the ESP after the flooding season. The increase of Na^{+} concentration in the solid phase partly justifies the observed Cl^{-} increase in the saturation extract after irrigation ceased. As exchangeable Na^{+} increases, soluble Na^{+} is removed from the soil solution forcing to the right the equilibrium of $NaCl \leftrightarrow Na^{+} + Cl^{-}$, resulting to the maintenance of Na^{+} and the increase of Cl^{-}.

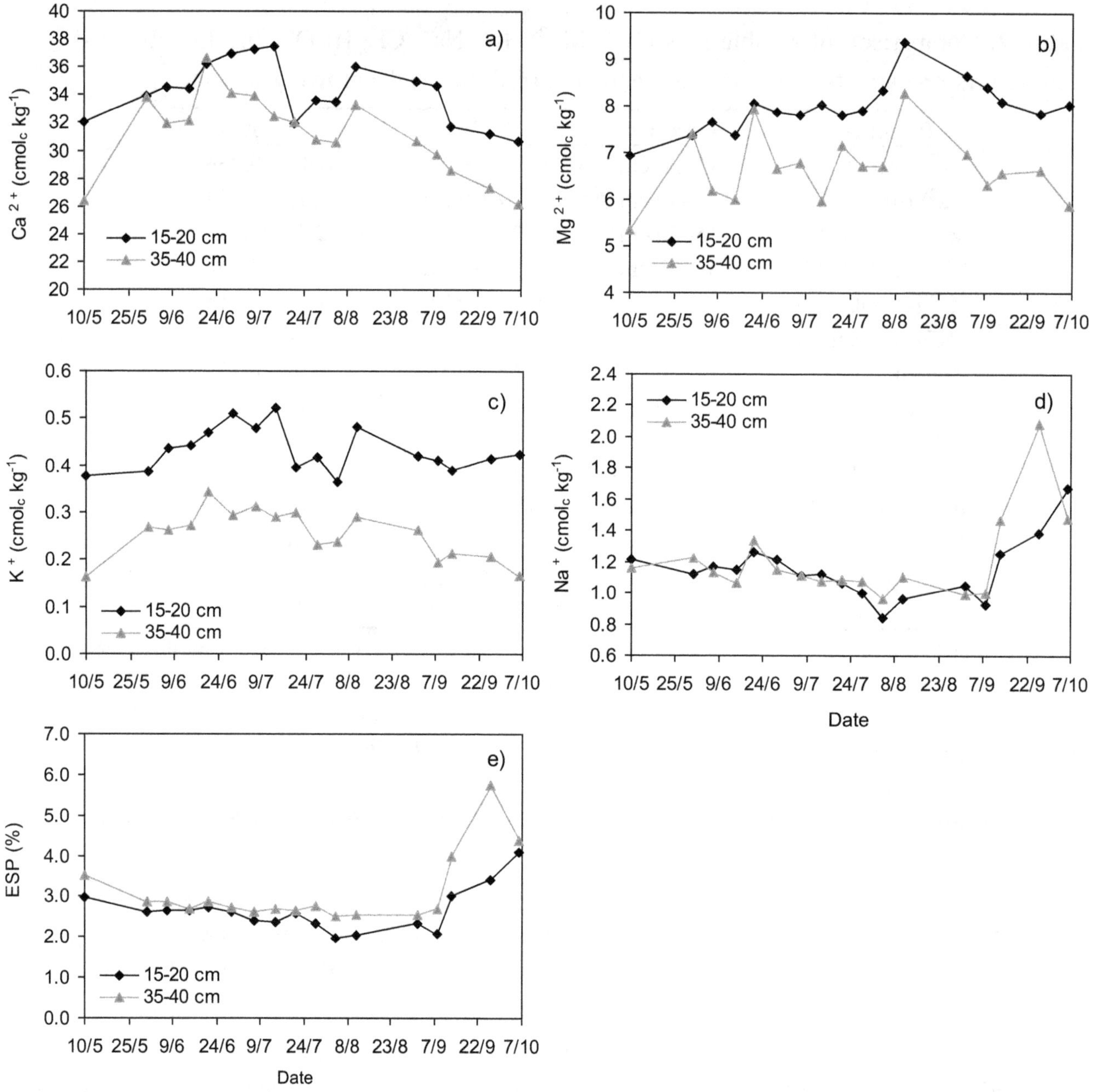

Figure 3. Fluctuation of exchangeable cations during the growing period: (**a**) for Ca^{2+}; (**b**) for Mg^{2+}; (**c**) for K^{+}; (**d**) for Na^{+} and (**e**) for ESP.

The statistical differences of exchangeable Ca^{2+}, Mg^{2+}, K^{+} and Na^{+} concentrations, before and after the cropping period, are given in Table 3. It is evident from Table 3 that statistical differences are observed only for exchangeable Na^{+} and ESP. Despite the increase of these two parameters after the growing season, their final values are quite low to indicate soil degradation.

Table 3. Comparison of exchangeable cations concentrations and ESP before (5 May) and after (6 October) the growing season.

Parameter	Soil Layer	5 May	6 October
Ca^{2+} ($cmol_c \cdot kg^{-1}$)	15–20 cm	32.04 ± 1.46a [2]	30.69 ± 0.95a
Ca^{2+} ($cmol_c \cdot kg^{-1}$)	35–40 cm	26.35 ± 1.59a	26.13 ± 1.63a
Mg^{2+} ($cmol_c \cdot kg^{-1}$)	15–20 cm	6.95 ± 0.45a	8.03 ± 0.21a
Mg^{2+} ($cmol_c \cdot kg^{-1}$)	35–40 cm	5.34 ± 0.77a	5.87 ± 0.79a
K^+ ($cmol_c \cdot kg^{-1}$)	15–20 cm	0.38 ± 0.04a	0.42 ± 0.04a
K^+ ($cmol_c \cdot kg^{-1}$)	35–40 cm	0.16 ± 0.03a	0.17 ± 0.02a
Na^+ ($cmol_c \cdot kg^{-1}$)	15–20 cm	1.21 ± 0.07b	1.67 ± 0.05a
Na^+ ($cmol_c \cdot kg^{-1}$)	35–40 cm	1.16 ± 0.15b	1.47 ± 0.16a
[1] ESP (%)	15–20 cm	2.98 ± 0.07b	4.09 ± 0.11a
[1] ESP (%)	35–40 cm	3.51 ± 0.32b	4.36 ± 0.13a

[1] Exchangeable sodium percentage = 100 × Na^+/CEC; CEC is calculated as the sum of exchangeable cations;
[2] Mean ± Standard error and results of Anova. Different letters indicate statistically significant differences before and after the flooding season at 0.05 confidence level.

3.3. WMPs Effects on the Salinity of Rice Fields

One of the major reasons that motivated this study was the debate on water consumption between the farmers and the administrative units that manage the water supply in the irrigation channels of the study area. According to the administrative agencies, the local rice farmers exploit the fact that the water cost is based on field-area/crop attributes and not to water consumption and for this reason they use excessive water that leads to quite high surface drainage losses and low water use efficiency which approximates in general 50% [5,44]. Considering that the irrigation of the experimental field was applied accordingly to the common practices followed by the farmers, the results indicate that the observed surface runoff and the respective irrigation water consumption were adequate to prevent salt built up in the soil. According to FAO classification [74], the soil of the experimental rice field is considered neither saline nor sodic, in both layers, which indicates that the long-term rice cultivation under the specific WMPs has not degraded its quality in terms of soil salinity. This also indicates that the increase of exchangeable Na^+ and ESP, which was observed in the experimental field during the short period of soil drying after flooding, may occur temporarily, but needs further investigation. Moreover, the farmers are well aware about the benefits of gypsum to prevent soil quality problems and for this reason they occasionally use it in this area when is needed.

According to the observations made by this analysis but also from other studies in the rice fields of the region, the prevention of salinity increase is not only ascribed to the adequate drainage but also to the excessive water use for surface runoff due to the following reasons:

- the continuous irrigation water inflow prevents oxygen depletion from the floodwater enhancing algae production which seems to play a crucial role on ions uptake. In the parallel experiments which were performed in the same experimental field, Aschonitis *et al.* [37] estimated the total final produced dry algal biomass at 1047 $kg \cdot ha^{-1}$ whereas algae uptake of Ca^{2+}, Mg^{2+}, Na^+ and K^+ was estimated to 62.61, 3.24, 3.35 and 39.58 $kg \cdot ha^{-1}$, respectively.

Apart from their implication in salinity components regulation, algae remove a significant part of nutrients (nitrogen and phosphorus) from the floodwater [35,37]. On the other hand, the surface runoff removed 296.3, 133.1, 87.7 and 13.2 kg·ha^{-1} of Ca^{2+}, Mg^{2+}, Na^{+} and K^{+}, respectively [37]. Furthermore, due to the continuous flooding and runoff, the quality characteristics of the runoff water are such, that it could be reused for irrigation of rice fields in this area at least once [5]. The significant contribution of controlled surface runoff in semi-arid environments has also been verified by Chen *et al.* [75]; and

- the continuous inflow of irrigation water and the attainment of a constant ponding depth in the rice-fields prevents further increase of the water temperature and evaporation [36], which lead to evapoconcentration and salt accumulation.

The generally good state of the soils in the region indicates that the regional agricultural practices are adequate to sustain the soil quality of rice-fields but with the cost of high water consumption. Amounts of water between 2000–2500 mm, depending on the drainage and the climatic conditions, are prerequisite to prevent soil degradation and to preserve rice yields. Until recently there were no significant limitations on the water supply from Axios River to support the current needs (apart from some isolated drought events), setting Thessaloniki plain one of the top regions of rice productivity ranging from 8.40–9.82 Mg·ha^{-1} grain yields [76]. Future challenges which may need to be considered in order to design better WMPs and to preserve soil quality of rice-fields are (a) the imminent climate change which may reduce the water supply of Axios River and (b) the potential expansion of rice-fields in the study area.

4. Conclusions

The comparative analysis of the measurements before and after the growing season indicated that the regional WMPs (inflows of 2260 mm irrigation plus 158 mm rainfall, continuous flooding with maximum of 10 cm ponding depth, surface runoff 520 mm, drainage below the root zone 432 mm and real crop evapotranspiration 1350 mm) significantly contributed to the removal of salts from the root zone preventing salt built up in the soil. The detailed analysis on the exchangeable cations showed that Na^{+} rapidly increased after the removal of excess soil water, possibly reoccupying exchange sites, leading to an increase in ESP. This finding is of extreme importance due to the long-term risk of increasing soil sodicity and to that aim the regional conventional management practices should additionally be followed by remediation actions (e.g., use of gypsum) in order to prevent future soil sodification.

The irrigation practice that is followed in the region is appropriate to sustain the soil quality of rice-fields but with the cost of high water consumption. The Mediterranean region is the major rice producer which covers the main rice-based food demands in Europe but the water needs to sustain this production (ranging between 2000–2500 mm) are considered extremely high. Actions to redesign and improve water management are mandatory in order to preserve rice yields. Future studies should focus on (a) the effects of the winter season to soil salinity between sequential rice growing seasons; (b) trial experiments to assess the minimum water consumption based on ratios of drainage *versus* runoff water to preserve or reduce soil salinity.

Acknowledgments

The authors wish to thank Professor Vissarion Keramidas for providing helpful comments that significantly improved this work.

Author Contributions

All authors contributed equally to this work.

Conflicts of Interest

The authors declare no conflict of interest.

References

1. Rhoades, J.D. Sustainability of irrigation: An overview of salinity problems and control strategies. In *Footprints of Humanity: Reflections on 50 Years of Water Resource Developments*. In Proceedings of the Annual Conference of the Canadian Water Resources Association, Lethbridge, AB, Canada, 3–6 June 1997.
2. Szabolcs, I. Prospects of soil salinity for the 21st Century. In Proceedings of the 15th World Congress of Soil Science (ISSS), Acapulco, Mexico City, 10–16 July 1994; Volume 1, pp. 123–141.
3. Ghassemi, F.; Jakeman, A.J.; Nix, H.A. *Salinisation of Land and Water Resources*; University of New South Wales Press Ltd.: Canberra, Australia, 1995.
4. Playán, E.; Pérez-Coveta, O.; Martínez-Cob, A.; Herrero, J.; García-Navarro, P.; Latorre, B.; Brufau, P.; Garcés, J. Overland water and salt flows in a set of rice paddies. *Agric. Water Manag.* **2008**, *95*, 645–658.
5. Litskas, V.D.; Aschonitis, V.G.; Lekakis, E.H.; Antonopoulos, V.Z. Effects of land use and irrigation practices on Ca, Mg, K, Na loads in rice-based agricultural systems. *Agric. Water Manag*. **2014**, *132*, 30–36.
6. Hammecker, C.; van Asten, P.; Marlet, S.; Maeght, J.L.; Poss, R. Simulating the evolution of soil solutions in irrigated rice soils in the Sahel. *Geoderma* **2009**, *150*, 129–140.
7. Ceuppens, J.; Wopereis, M.C.S.; Miézan, K.M. Soil salinization processes in rice irrigation schemes in the Senegal River Delta. *Soil Sci. Soc. Am. J.* **1996**, *61*, 1122–1130.
8. Van Asten, P.J.A.; van't Zelfde, J.A.; van der Zee, S.E.A.T.M.; Hammecker, C. The effect of irrigated rice cropping on the alkalinity of two alkaline rice soils in the Sahel. *Geoderma* **2004**, *119*, 233–247.
9. Sharma, P.K.; de Datta, S.K. Effects of puddling on soil physical properties and processes. In *Soil Physics and Rice*; IRRI: Los Baños, Philippines, 1985; pp. 217–234.
10. Aschonitis, V.G.; Kostopoulou, S.K.; Antonopoulos, V.Z. Methodology to assess the effects of rice cultivation under flooded conditions on van Genuchten's model parameters and pore size distribution. *Transp. Porous Med.* **2012**, *91*, 861–876.
11. Wopereis, M.C.S.; Wösten, J.H.M.; Bouma, J.; Woodhead, T. Hydraulic resistance in puddled rice soils: Measurement and effects on water movement. *Soil Till. Res.* **1992**, *24*, 199–209.

12. Wopereis, M.C.S.; Bouman, B.A.M.; Kropff, M.J.; Ten Berge, H.F.M.; Maligaya, A.R. Water use efficiency of flooded rice fields. I. Validation of the soil-water balance model SAWAH. *Agric. Water Manag.* **1994**, *26*, 277–289.
13. Chen, S.K.; Liu, C.W.; Huang, H.C. Analysis of water movement in paddy rice fields (II) simulation studies. *J. Hydrol.* **2002**, *268*, 259–271.
14. Chen, S.K.; Liu, C.W. Analysis of water movement in paddy rice fields (I) experimental studies. *J. Hydrol.* **2002**, *260*, 206–215.
15. Boivin, P.; Favre, F.; Hammecker, C.; Maeght, J.L.; Delariviére, J.; Poussin, J.C.; Wopereis, M.C.S. Processes driving soil solution chemistry in a flooded rice-cropped vertisol: Analysis of long-time monitoring data. *Geoderma* **2002**, *110*, 87–107.
16. Wopereis, M.C.S.; Ceuppens, J.; Boivin, P.; N'Diaye, A.M.; Kane, A. Preserving soil quality under irrigation in the Senegal River Valley. *Roy. Neth. Soc. Agric. Sci.* **1998**, *46*, 97–107.
17. Häfele, S.; Wopereis, M.C.S.; Boivin, P.; Diaye, A.M.N. Effect of puddling on soil desalinization and rice seedling survival in the Senegal River Delta. *Soil Till. Res.* **1999**, *51*, 35–46.
18. Cass, A.; Gusli, S.; MacLeod, D.A. Sustainability of soil structure quality in rice paddy-soybeen cropping systems in South Sulawesi, Indonesia. *Soil Till. Res.* **1994**, *31*, 339–352.
19. Favre, F.; Boivin, P.; Wopereis, M. Water movement and soil swelling in a dry, cracked vertisol. *Geoderma* **1997**, *78*, 113–123.
20. Phillips, I.R.; Greenway, M. Changes in water soluble and exchangeable ions, cation exchange capacity, and phosphorus_{max} in soils under alternating waterlogged and drying conditions. *Commun. Soil Sci. Plant Anal.* **1998**, *29*, 51–65.
21. Narteh, L.T.; Sahrawat, K.L. Influence of flooding on electrochemical and chemical properties of West African soils. *Geoderma* **1999**, *87*, 179–207.
22. Favre, F.; Tessier, D.; Abdelmoula, M.; Génin, J.M.; Gates, W.P.; Boivin, P. Iron reduction and changes in cation exchange capacity in intermittently waterlogged soil. *Eur. J. Soil Sci.* **2002**, *53*, 175–183.
23. Lu, S.G.; Tang, C.; Rengel, Z. Combined effects of waterlogging and salinity on electrochemistry, water-soluble cations and water dispersible clay in soils with various salinity levels. *Plant Soil* **2004**, *264*, 231–245.
24. Boivin, P.; Saejiew, A.; Grunberger, O.; Arunin, S. Formation of soils with contrasting textures by translocation of clays rather ferrolysis in flooded rice fields in Northeast Thailand. *Eur. J. Soil Sci.* **2004**, *55*, 713–724.
25. Inao, K.; Kitamura, Y. Pesticide paddy field model (PADDY) for predicting pesticide concentrations in water and soil in paddy fields. *Pest Sci.* **1999**, *55*, 38–46.
26. Singh, K.B.; Gajri, P.R.; Arora, V.K. Modelling the effects of soil and water management practices on the water balance and performance of rice. *Agric. Water Manag.* **2001**, *49*, 77–95.
27. Liu, C.W.; Chen, S.K.; Jang, C.S. Modelling water infiltration in cracked paddy field soil. *Hydrol. Process* **2004**, *18*, 2503–2513.
28. Chung, S.O.; Kim, H.S.; Kim, J.S. Model development for nutrient loading from paddy rice fields. *Agric. Water Manag.* **2003**, *62*, 1–17.

29. Miao, Z.; Cheplick, M.J.; Williams, M.W.; Trevisan, M.; Padovani, L.; Gennari, M.; Ferrero, A.; Vidotto, F.; Capri, E. Simulating pesticide leaching and runoff in rice paddies with the RICEWQ-VADOFT model. *J. Environ. Qual.* **2003**, *32*, 2189–2199.
30. Chowdary, V.M.; Rao, N.H.; Sarma, P.B.S. A coupled soil water and nitrogen balance model for flooded rice fields in India. *Agrc. Ecosyst. Environ.* **2004**, *103*, 425–441.
31. Watanabe, H.; Takagi, K.; Vu, S.H. Simulation of mefenacet concentrations in paddy field by improved PCPF-1 model. *Pest Manag. Sci.* **2005**, *62*, 20–29.
32. Tournebize, J.; Watanabe, H.; Takagi, K.; Nishimura, T. The development of a coupled model (PCPF-SWMS) to simulate water flow and pollutant transport in Japanese paddy fields. *Paddy Water Environ.* **2006**, *4*, 39–51.
33. Jeon, J.H.; Yoon, C.G.; Ham, J.H.; Jung, K.W. Model development for surface drainage loading estimates from paddy rice fields. *Paddy Water Environ.* **2005**, *3*, 93–101.
34. Antonopoulos, V.Z. Modelling of water and nitrogen balance in the ponded water of rice fields. *Paddy Water Environ.* **2008**, *6*, 387–395.
35. Antonopoulos, V.Z. Modelling of water and nitrogen balances in the ponded water and soil profile of rice fields in Northern Greece. *Agric. Water Manag.* **2010**, *98*, 321–330.
36. Aschonitis, V.G.; Antonopoulos, V.Z. Evaluation of the water balance and the soil and ponding water temperature in paddy-rice fields with the modified GLEAMS model. In Proceedings of the Agricultural Engineering Conference, Crete, Greece, 23–25 June 2008.
37. Aschonitis, V.G.; Lekakis, E.H.; Petridou, N.C.; Koukouli, S.G.; Pavlatou-Ve, A. Nutrients fixation by Algae and limiting factors of algal growth in flooded rice fields under semi-arid Mediterranean conditions—Case study in Thessaloniki Plain in Greece. *Nutr. Cycl. Agroecosys.* **2013**, *96*, 1–13.
38. Phogat, V.; Yadav, A.K.; Malik, R.S.; Kumar, S.; Cox, J. Simulation of salt and water movement and estimation of water productivity of rice crop irrigated with saline water. *Paddy Water Environ.* **2010**, *8*, 333–346.
39. Pochai, N.; Pongnoo, N. A numerical treatment of a mathematical model of ground water flow in rice field near marine shrimp aquaculture farm. *Proc. Eng.* **2012**, *32*, 1191–1197.
40. Singh, A.; Panda, S.N. Integrated salt and water balance modeling for the management of waterlogging and salinization. II: Application of SAHYSMOD. *J. Irrig. Drain. Eng. ASCE* **2012**, *138*, 964–971.
41. Wang, X.; Yang, J.; Yao, R.; Yu, S. Irrigation regime and salt dynamics for rice with brackish water irrigation in coastal region of North Jiangsu Province. *Trans. Chin. Soc. Agric. Eng.* **2014**, *30*, 54–63.
42. N.AG.RE.F. (National Agriculture Research Foundation-Institute of Soil Science). *Soil Map of Thessaloniki Region-Area of Gallikos and Axios Rivers*; Institute of Soil Science: Thessaloniki, Greece, 2003; pp. 137–138.
43. Soil Survey Staff. *Soil Taxonomy: A Basic System of Soil Classification for Making and Interpreting Soil Surveys*; US Department of Agriculture-Soil Conservation Service: Washington, DC, USA, 1975.

44. Litskas, V.D.; Aschonitis, V.G.; Antonopoulos, V.Z. Water quality in irrigation and drainage networks of Thessaloniki plain in Greece related to land use, water management, and agroecosystem protection. *Environ. Monit. Assess.* **2010**, *163*, 347–359.
45. Aschonitis, V.G. Modeling of Evapotranspiration, Physical Soil Properties and Water, Nitrogen and Phosphorus Balance in Flooded Rice Fields. Ph.D. Thesis, Aristotle University, School of Agriculture, Thessaloniki, Greece, 2012.
46. Aschonitis, V.G.; Papamichail, D.M.; Lithourgidis, A.; Fano, E.A. Estimation of leaf area index and foliage area index of rice using an indirect gravimetric method. *Commun. Soil Sci. Plant Anal.* **2014**, *45*, 1726–1740.
47. Allen, R.G.; Walter, I.A.; Elliott, R.; Howell, T.; Itenfisu, D.; Jensen, M.; Snyder, R.L. *The ASCE Standardized Reference Evapotranspiration Equation*; Final Report (ASCE-EWRI), ASCE: Columbus, OH, USA, 2005.
48. Tuller, M.; Islam, M.R. Field methods for monitoring solute transport. In *Soil-Water-Solute Process Characterization, an Integrated Approach*; Benedí, A., Muñoz-Carpena, J.R., Eds.; CRC Press LLC: Boca Raton, FL, USA, 2005; pp. 309–355.
49. McDonald, J.D.; Bélanger, N.; Sauvé, S.; Courchesne, F.; Hendershot, W.H. Collection and characterization of soil solutions. In *Soil Sampling and Methods of Analysis*, 2nd ed.; Canadian Society of Soil Science; Carter, M.R., Gregorich, E.G., Eds.; CRC Press LLC, Taylor & Francis Group: Boca Raton, FL, USA, 2008; pp. 179–196.
50. Tacker, P.; Vories, E.; Wilson, C., Jr.; Slaton, N. Water management. In *Rice Production Handbook*; Miscellaneous Publication 192; Slaton, N.A., Ed.; Cooperative Extension Service, University of Arkansas: Little Rock, AR, USA, 2001; pp. 75–86.
51. Ayers, R.S.; Westcot, D.W. *Water Quality for Agriculture*; Irrigation and Drainage Paper 29; FAO, U.N.: Rome, Italy, 1994.
52. Bouyoucos, G.J. Hydrometer method improved for making particle size analysis of soils. *Agron. J.* **1962**, *54*, 464–465.
53. Walkley, A.; Black, I.A. An examination of the Degtjareff method for determining soil organic matter, and a proposed modification of the chromic acid titration method. *Soil Sci.* **1934**, *37*, 29–38.
54. Allison, L.E.; Moodie, C.D. Carbonates. In *Methods of Soil Analysis. Part 2*; Black, C.A., Ed.; Agronomy Monograph. American Society of Agronomy: Madison, WI, USA, 1965; pp. 1379–1400.
55. APHA. *Standard Methods for the Examination of Water and Wastewater*, 21st ed.; Eaton, A.D., Clesceri, L.S., Rice, E.W., Greenberg, A.E. Eds.; APHA, AWWA, WEF: Washington, DC, USA, 2005.
56. Rhoades, J.D. Salinity: Electrical conductivity and total dissolved solids. In *Methods of Soil Analysis. Part 3. Chemical Methods*; Soil Science Society of America Book Series 5.3; Sparks, D.L., Page, A.L., Helmke, P.A., Loeppert, R.H., Soltanpour, P.N., Tabatabai, M.A., Johnston, C.T., Sumner, M.E., Eds.; Soil Science Society of America, American Society of Agronomy: Madison, WI., USA, 1996; pp. 417–436.
57. Chapman, H.D.; Pratt, P.F. *Methods of Analysis for Soils, Plants and Waters*; University of California Division of Agricultural Science: Riverside, CA, USA, 1961.

58. Thomas, G.W. Exchangeable cations. In *Methods of Soil Analysis. Part 2, Chemical and Microbiological Properties*, 2nd ed.; Klute, A., Ed.; American Society of Agronomy, Soil Science Society of America: Madison, WI, USA, 1982; pp. 159–165.
59. Dane, J.H.; Hopmans, J.W. Water retention and storage. In *Methods of Soil Analysis. Part 1. Physical Methods*, 3rd ed.; Dane, J.H., Topp, G.C., Eds.; Soil Science Society of America: Madison, WI, USA, 2002; pp. 671–720.
60. Van Genuchten, M.T. A closed-form equation for predicting the hydraulic conductivity of unsaturated soils. *Soil Sci. Soc. Am. J.* **1980**, *44*, 892–898.
61. Van Genuchten, M.T.; Leij, F.J.; Yates, S.R.M. *The RETC Code for Quantifying the Hydraulic Functions of Unsaturated Soils*; USDA, Agricultural Research Service: Riverside, CA, USA, 1991.
62. Hinsinger, P.; Plassard, C.; Tang, C.; Jaillard, B. Origins of root-induced pH changes in the rhizosphere and their responses to environmental constraints: A review. *Plant Soil* **2003**, *248*, 43–59.
63. Jones, C.; Brown, B.D.; Engel, R.; Horneck, D.; Orson-Rutz, K. *Factors Affecting Nitrogen Fertilizer Volatilization*; USDA, Montana State University and Montana State University Extension Bulletin: Bozeman, MT, USA, 2013; p. 8.
64. Mahrous, F.N.; Mikkelsen, D.S.; Hafez, A.A. Effect of soil salinity on the electro-chemical and chemical kinetics of some plant nutrients in submerged soils. *Plant Soil* **1983**, *75*, 455–472.
65. Larson, K.D.; Graetz, D.A.; Schaffer, B. Flood-induced chemical transformations in calcareous agricultural soils of South Florida. *Soil Sci.* **1991**, *152*, 33–40.
66. Chorom, M.; Rengasamy, P. Carbonate chemistry, pH, and physical properties of an alkaline sodic soil as affected by various amendments. *Austral. J. Soil Res.* **1997**, *35*, 149–162.
67. Armstrong, W. Aeration in higher plants. *Adv. Botan. Res.* **1979**, *7*, 225–232.
68. Begg, C.B.M.; Kirk, G.J.D.; MacKenzie, A.F.; Neue, H.U. Root-induced iron oxidation and pH changes in the lowland rice rhizosphere. *New Phytol.* **1994**, *128*, 469–477.
69. Kirk, G.J.D.; van Du, L.E. Changes in rice root architecture, porosity, and oxygen and proton release under phosphorus deficiency. *New Phytol.* **1997**, *135*, 191–200.
70. Kirk, G. *The Biogeochemistry of Submerged Soils*; John Wiley and Sons Ltd.: Chichester, West Sussex, UK, 2004; p. 297.
71. Hinsinger, P. Plant-induced changes in soil processes and properties. In *Soil Conditions and Plant Growth*; Gregory, P.J., Nortcliff, S., Eds.; John Wiley and Sons Ltd.: Chichester, West Sussex, UK, 2013; pp. 323–365.
72. Fillery, I.R.P.; Simpson, J.R.; de Datta, S.K. Influence of field environment and fertilizer management on ammonia loss from flooded rice. *Soil Sci. Soc. Am. J.* **1984**. *48*, 914–920.
73. De Datta, S.K. *Principles and Practices of Rice Production*; John Wiley and Sons, Inc. Singapore, Singapore, 1981; p. 640.
74. Abrol, I.P.; Yadav, J.S.P.; Massoud, F.I. *Salt-Affected Soils and Their Management*; FAO Soils Bulletin: Rome, Italy, 1988.
75. Chen, Y.; Zhang, G.; Xu, Y.J.; Huang, Z. Influence of irrigation water discharge frequency on soil salt removal and rice yield in a semi-arid and saline-sodic area. *Water* **2013**, *5*, 578–592.

76. Ntanos, D.A. Strategies for rice production and research in Greece. *Cah. Opt. Méditerr.* **2001**, *50*, 115–122.

9

Response of Snap Bean Cultivars to Rhizobium Inoculation under Dryland Agriculture in Ethiopia

Hussien Mohammed Beshir [1,2], Frances L. Walley [3], Rosalind Bueckert [1] and Bunyamin Tar'an [1,*]

1 Department of Plant Sciences, University of Saskatchewan, 51 Campus Drive, Saskatoon, SK S7N 5A8, Canada; E-Mails: hmb572@mail.usask.ca (H.M.B); rosalind.bueckert@usask.ca (R.B.)

2 School of Plant and Horticultural Sciences, Hawassa University, Hawassa, Ethiopia

3 Department of Soil Science, University of Saskatchewan, 51 Campus Drive, Saskatoon, SK S7N 5A8, Canada; E-Mail: flw766@mail.usask.ca

* Author to whom correspondence should be addressed; E-Mail: bunyamin.taran@usask.ca

Academic Editor: Yantai Gan

Abstract: High yield in snap bean (*Phaseolus vulgaris* L.) production requires relatively high nitrogen (N) inputs. However, little information is available on whether the use of rhizobial inoculants for enhanced biological dinitrogen fixation can provide adequate N to support green pod yield. The objectives of this study were to test the use of rhizobia inoculation as an alternative N source for snap bean production under rain fed conditions, and to identify suitable cultivars and appropriate agro-ecology for high pod yield and N_2 fixation in Ethiopia. The study was conducted in 2011 and 2012 during the main rainy season at three locations. The treatments were factorial combinations of three N treatments (0 and 100 $kg{\cdot}N{\cdot}ha^{-1}$, and *Rhizobium etli* (HB 429)) and eight snap bean cultivars. Rhizobial inoculation and applied N increased the total yield of snap bean pod by 18% and 42%, respectively. Cultivar Melkassa 1 was the most suitable for a reduced input production system due to its greatest N_2 fixation and high pod yield. The greatest amount of fixed N was found at Debre Zeit location. We concluded that N_2 fixation achieved through rhizobial inoculation can support the production of snap bean under rain fed conditions in Ethiopia.

Keywords: snap bean; *Rhizobium*; nitrogen fixation; ^{15}N; Dryland Agriculture

1. Introduction

Snap bean is a cultivar of common bean (*Phaseolus vulgaris* L.) from which immature pods are harvested and used as a vegetable for human consumption. Snap bean is characterized by its succulent and flavorful pods [1]. Snap bean production in Ethiopia has increased over time both for export and local market [2]. Ethiopia has become an important supplier of snap bean to European markets especially during the winter months when domestic snap bean production in Europe is down. At the same time the local demand and consumption of snap bean in Ethiopia have increased. To date, snap bean production in Ethiopia is mainly done by private companies, mostly business affiliates of European companies. This results in high input production systems including the heavy use of synthetic N fertilizer and irrigation that precludes most of the local small-holders farmers from export production. The burden of high input costs such as N fertilizer needs to be reduced for small-scale farmers in order to increase and sustain the production, and maximize the benefits of growing this cash crop.

Biological N_2 fixation, a key source of N for resource poor farmers, constitutes a potential solution and may play a role in sustainable common bean production in sub-Saharan Africa [3,4]. Previous report indicated that adequate N can be contributed by symbiosis between *Rhizobium* and legume crops [5]. This can potentially be a cheaper and more effective agronomic practice to improve productivity of legume crops than applied N [5,6]. The benefits of using rhizobia for N_2 fixation have been realized for grain production in chickpea [7], soybean [8], common bean [9,10], and many other grain legumes. The potential yield improvement of common bean through N_2 fixation has been reported by several authors [11–13]. However, the use of rhizobial inoculants as an alternate N source for legume vegetables including snap bean is less clear and currently is not practiced in Ethiopia. This study tested the hypothesis that N_2 fixation can support acceptable production of snap bean yield under rain fed conditions.

Earlier reports showed that effective N_2 fixation by rhizobia and further conversion into yield in common bean is affected by the cultivar [9,14–17] and agro-ecological (environmental) factors [18–20].

To date, there is no nationally registered snap bean cultivar in Ethiopia. Only few, yet to be released, snap bean cultivars are recommended by Melkassa Agricultural Research Center for local farmers in Ethiopia. Farmers in general are left to limited option to grow introduced, privately owned cultivars with high seed cost. The introduced cultivars in general require high rates of nitrogen as they have been developed under intensive production systems. We hypothesized that locally developed cultivars may have better N_2 fixation potential than introduced commercial cultivars.

Agro-ecological conditions as a cumulative effect of climate, landform and edaphic factors such as soil mineral composition, pH and soil temperature contribute significant effect on yield [21], and affect the effectiveness of rhizobia for N_2 fixation [5]. In the current experiment, three locations with different agro-ecological characteristics were selected with the hypothesis that snap bean performance for N_2 fixation and yield vary with the growing season's agro-ecology.

Assessing the potential of rhizobium inoculation as the main source of nitrogen for snap bean production will create opportunity for Ethiopian farmers to produce snap bean under low input production systems. Additionally, identification of cultivars with greatest yield and N_2 fixation potential, and identification of locations that allow greatest N_2 fixation to occur will help to maximize the benefits from N_2 fixation. The objectives of this study were to test the use of rhizobia inoculation as an alternative N source for snap bean production under rain fed conditions in Ethiopia, and to identify suitable cultivars and appropriate agro-ecology for optimum pod yield and N_2 fixation.

2. Materials and Methods

2.1. Experimental Sites

The study was conducted at three sites across different agro-ecologies in the Rift Valley of Ethiopia. The three sites were Debre Zeit (8°44′52″ N, 38°05′53″ E), Hawassa (7°4′ N, 38°31′ E) and Ziway (8°00′ N, 38°45′ E). Debre Zeit is found in a tepid to cool sub-moist agro-ecology characterized by moderate temperature and a definitive rain fall pattern between July to September (Table 1). It is situated at higher altitude in the transitional region of the Rift Valley and associated mountain ranges. The area is dominated by clay soils with high copper and cation exchange capacity, and a neutral pH (Table 2). Hawassa is in a hot to warm sub-moist humid agro-ecology with warmer temperature (Table 1). It has a longer growing season and a less definitive pattern of rain fall during the growing season (Table 1). It is a mid-highland area in the Rift Valley zone. The soil is loam characterized by slightly acidic pH and higher concentrations of micronutrients such as manganese, iron and zinc (Table 2). Ziway is found in a tepid to cool semi-arid agro-ecology with erratic rainfall and an unpredictable climate (Table 1). The area is in the Rift Valley zone with mid-altitude. It has warmer temperature. The soil is sandy loam with very high pH and relatively higher exchangeable sodium (Table 2). Ziway is located at a distance of 100 km equidistant between Debre Zeit and Hawassa.

These three locations represented distinct agro-ecologies (climate zones). But all the three climate zones are characterized by close range of temperature profile.

2.2. Experimental Design and Management

The experiment was conducted from June to September in 2011 and 2012 under rain fed conditions at the three locations. The plots were seeded on 27 June, 6 and 19 July in 2011 and on 2, 4 and 1 July in 2012 at Ziway, Debre Zeit and Hawassa, respectively.

At each location and year, eight snap bean cultivars were tested against three N treatments factorially combined in a randomized block design with three replications. Among the eight cultivars, six (Andante, Boston, Contender Blue, Lomami, Paulista and Volta) were commercial cultivars currently under production in Ethiopia. The remaining two (Melkassa 1 and Melkassa 3) were local cultivars developed and recommended by the Melkassa Agricultural Research Center (MARC) for production in Ethiopia and are currently being grown by local farmers. The three N treatments were 0 and 100 $kg{\cdot}N{\cdot}ha^{-1}$, and *Rhizobium etli* (HB 429) inoculation. The rhizobium isolate (HB 429) was developed by the National Soil Testing Center, Addis Ababa, Ethiopia. It is being used by local farmers for common bean grain production. This isolate was isolated from Hadiya District, the Southern Nations and Nationalities People Region, Ethiopia. The 100 $kg{\cdot}N{\cdot}ha^{-1}$ is the average of commonly used N fertilizer rate by commercial snap bean producers in Ethiopia.

Table 1. Average rainfall, maximum and minimum temperature during 2011 and 2012 growing seasons at Debre Zeit, Hawassa and Ziway, Ethiopia. Ten year normal climate, altitude and climate zone of each location are presented [a].

Year		Debre Zeit			Hawassa			Ziway		
		Rainfall	[b] Max. T	[c] Min. T	Rainfall	[b] Max. T	[c] Min. T	Rainfall	[b] Max. T	[c] Min. T
		mm	°C	°C	mm	°C	°C	mm	°C	°C
	July	134.6	26.9	13.5	129.6	25.7	12.8	133.7	25.8	14.8
2011	August	241.7	25.0	14.9	157.3	25.3	13.0	114.8	24.6	15.1
	September	82.6	25.0	14.9	113.3	25.7	13.3	56.2	25.5	13.3
	Annual	724.1	26.4	11.3	776.1	28.0	12.1	598.3	29.1	13.0
	July	197.4	25.0	13.5	232.5	24.9	14.7	326.3	23.2	15.0
2012	August	256.5	24.5	12.6	72.7	24.4	14.5	171.4	24.3	14.7
	September	103.0	25.6	12.5	139.8	27	15.3	136.6	27.8	9.7
	Annual	726.3	26.7	10.4	884.8	28.1	12.7	856.8	28.6	12.4
10 years	Annual Aver.	747.0	26.4	10.7	786.5	27.9	12.3	763.9	27.5	13.9
Altitude	m above sea level		1950			1700			1645	
Climate Zone [d]			Tepid to cool sub-moist			Hot to warm sub-moist humid			Tepid to cool semi-arid	

[a] Data collected by Debre Zeit Agricultural Research Center (Debre Zeit), South Agricultural Research Center (Hawassa), and Adame Tulu Agricultural Research Center (Ziway); [b] Maximum Temperature; [c] Minimum Temperature; [d] According to Ministry of Agriculture in Ethiopia [22].

Table 2. Soil physicochemical characteristics at Debre Zeit, Hawassa and Ziway, Ethiopia during the 2011 and 2012 growing seasons.

Profile Code	Debre Zeit		Hawassa		Ziway	
	2011	2012	2011	2012	2011	2012
Texture class [a]	clay	clay	loam	loam	sandy loam	sandy loam
pH-H2O (1:2.5) [b]	6.98	6.98	6.1	6.1	8.38	8.2
pH-KCl (1:2.5) [b]	5.96	6.02	5.31	5.22	7.61	7.58
EC ($ms \cdot cm^{-1}$) (1:2.5)	0.16	0.26	0.17	0.17	0.15	0.26
CEC ($cmolc \cdot kg^{-1}$ soil)	48.47	41.42	26.39	24.01	28.48	26.18
Organic Carbon (%) [c]	1.5	1.47	1.59	1.55	0.96	1.15
Nitrogen (%) [d]	0.11	0.08	0.11	0.1	0.1	0.07
Available P ($mg \cdot P \cdot kg^{-1}$ soil) [e]	19.05	18.28	21.52	40.01	19.12	36.42
Available K ($mg \cdot K \cdot kg^{-1}$ soil) [f]	131.34	117.19	515.24	808.22	646.57	717.30
Exchangeable sodium % (ESP) [f]	0.83	1.58	1.8	1.66	3.42	3.58
Micronutrients [g]						
Cu ($mg \cdot kg^{-1}$ soil)	2.04	1.47	0.3	0.39	0.33	0.32
Fe ($mg \cdot kg^{-1}$ soil)	12.46	10.64	28.96	25.93	3.13	4.58
Mn ($mg \cdot kg^{-1}$ soil)	9.27	7.82	20.76	27.03	2.7	4.63
Zn ($mg \cdot kg^{-1}$ soil)	0.86	0.86	3.61	3.78	1.08	1.5

Method: [a] Hydrometer; [b] Acid neutralization; [c] Walklay and Black; [d] Kjeldahl; [e] Olsen; [f] Ammonium acetate; [g] Instrumental (total micronutrient).

Plot size was 2 m × 2.5 m. Each plot had five rows with 0.1 m between plants within each row and 0.5 m between rows with a row length of 2 m. The distance between adjacent blocks was 1.5 m. The two outer rows were considered border rows. All samples for data collection were taken from the three internal rows. Plant population was maintained by planting two seeds per hill and thinning to one plant upon appearance of trifoliate leaves. The spacing and other management practices were adopted from commercial snap bean producers nearby the three locations. Manual weeding was used to control weed. Fungicide (Mancozeb) was applied at two-week interval from mid-vegetative growth stage to pod setting stage to protect from fungal diseases.

Temperature and rainfall data were obtained from the nearest agricultural research center at each location (Table 1). Nitrogen fixation was assessed using the ^{15}N isotope dilution method and wheat was planted as a non-nodulating control plant for ^{15}N analysis [13]. In 2011, microplots measuring 0.75 m^2 were established within the treatment plots and a urea solution was applied at a rate of 10 $kg \cdot N \cdot ha^{-1}$ labeled with 5 atom % ^{15}N excess subsequent to seedling emergence using a handheld watering can with a fine spray. The ^{15}N dilution was only successful at Debre Zeit; heavy rain at both Hawassa and Ziway shortly after ^{15}N application compromised the integrity of the ^{15}N in the microplots. The recommended rate of phosphorus fertilizer (21 $kg \cdot ha^{-1}$ phosphorus) was applied in the form of triple super phosphate at each location during seeding time.

In 2012, the ^{15}N natural abundance method was used to estimate biological N fixation and wheat was used as the non-fixing reference crop for the experiments at all locations. Wheat was planted in a row perpendicular to the inoculated snap bean plots 0.75 m from the snap bean row. Therefore, each plot had its own reference plants to minimize the impact of any natural ^{15}N variation in soils. It is assumed that non-nodulating isolines are ideal reference crops [13,23]. However, other non-fixing cereal crops including wheat and barley can be used as a reference crop for estimation of %Ndfa in common bean and other legume crops [13,24].

2.3. Data Collection

Total pod yield was determined by the weight of the pods from three central rows of each plot and converted into tonnes per hectare ($t \cdot ha^{-1}$) at optimum maturity. Pods from four randomly selected plants per plot were counted and the average was taken as pod number per plant. The dry weight of pods was determined after drying for 48 h at 70 °C.

Three randomly selected plant samples were carefully uprooted at the flowering stage and nodules were counted for total nodule number per plant root system. The diameters of all the nodules were measured using a caliper, and expressed as mean nodule diameter. Nodules were then dried in an oven for 24 h at 70 °C and weighed for determination of nodule dry weight per plant root system.

The percentage of N derived from the atmosphere (%Ndfa) was calculated by two methods, by ^{15}N isotopic dilution method in 2011 and by ^{15}N natural abundance method in 2012.

For the ^{15}N isotopic dilution method, three plants were harvested from the central row of the micro-plot at the green pod mature stage (almost all of the pods on the plant reached maturity) in 2011 at Debre Zeit. Wheat plants were also harvested at the same time. Plant samples were dried in an oven at 50 °C to constant weight. The samples were ground and 20 g subsamples were taken for further analysis. The sub-samples were reground using a ball mill and very small portions (approximately 3 mg each sample) of the fine ground samples were pelleted into 6 × 8 mm tin caps. Samples were then analyzed using a Costech ECS4010 elemental analyzer (Costech Analytical Technologies Inc., Valencia, CA, USA) coupled to a Delta V mass spectrometer with a ConFlo IV interface (Thermo Scientific, Bremen, Germany).

%Ndfa was calculated according to [25]:

$$\%\text{Ndfa} = \left(1 - \frac{atom\%\ ^{15}N\ excess\ in\ snap\ bean}{atom\%\ ^{15}N\ excess\ in\ wheat}\right) \times 100 \quad (1)$$

where "atom % excess" is the measure of the sample's ^{15}N content above the assumed atmospheric atom% ^{15}N value of 0.3663 [26].

When N fixation was estimated using the ^{15}N natural abundant method in 2012, three snap bean plants were selected randomly from inoculated snap bean plots, accompanied by several wheat plants from the nearest wheat row. The above ground plant parts were harvested and analysed. The same procedures were followed with the ^{15}N isotopic dilution method mentioned above for sample preparation and analysis, and %Ndfa was calculated as follows:

$$\%\text{Ndfa} = \left(\frac{\delta^{15}N\ of\ wheat\ -\ \delta^{15}N\ of\ snap\ bean}{\delta^{15}N\ of\ wheat - B}\right) \times 100 \quad (2)$$

where "$\delta^{15}N$" is:

$$\delta^{15}N = \left(\frac{\text{atom\%}\ ^{15}N\ sample\ -\ \text{atom\%}\ ^{15}N\ atmosphere}{\text{atom\%}\ ^{15}N\ atmosphere}\right) \times 1000 \quad (3)$$

and a B-value of −2 was assumed [24]. The largest pool of N in the environment is atmospheric N_2 and it has a constant abundance of 0.3663 atom% ^{15}N [26].

Total fixed N was calculated using the formula (4) on a per hectare (ha) basis. In this experiment, only the above ground biomass was used to calculate total N.

$$\text{Total fixed N (kg ha}^{-1}) = \left(\frac{\%\text{Ndfa}}{100}\right) \times \text{Total N (kg ha}^{-1}) \tag{4}$$

where "*%Ndfa*" is obtained from the formula (2).

2.4. Statistical Analysis

Data were analyzed using the PROC MIXED procedure of the SAS software version 9.3 [27] to determine Analysis of Variance (ANOVA). Nitrogen treatment, cultivar and location were considered as fixed effects, and replication (block), year and interactions with year were considered random. Nodule number and nodule dry weight were log transformed based on the results of Levene's test for homogeneity of variance and after inspecting the residuals. After analysis, these transformed values were back transformed for plotting graphs. The DDFM = Kr option was considered for approximating the degree of freedom for means. Treatments were compared by LSD method. Significance was declared at $p < 0.05$. The means of N treatments used for plotting the graphs were summarized from all cultivars at all locations. Similarly, the means of cultivars were summarized from all the N treatments at all the locations. The summarized mean of all cultivars and all N treatments were also used to plot the graphs for location. The means used to plot %Nfda of the cultivars in 2011 were cultivars only at Debre Zeit under inoculated treatments. The means of %Nfda and total fixed N of the cultivars in 2012 were from all locations under inoculated treatments. The %Nfda and total fixed N for location were the summary of all cultivars of inoculated treatments. Pearson correlation analysis was used to study correlations between yield and %Ndfa of the cultivars using the SAS software version 9.3 [27].

3. Results

Nitrogen treatment, cultivar and location significantly ($p < 0.05$) affected the total fresh pod yield and pod dry weight of snap bean. Application of 100 kg·N·ha^{-1} increased fresh pod yield by 42 and 17% as compared to the control (0 N·ha^{-1}) and rhizobial inoculation, respectively. Inoculation of *R. etli* (HB 429) resulted in 18% yield increase compared to the control treatment (Figure 1A). Application of 100 kg·N·ha^{-1} resulted in the highest pod dry weight (Figure 1A). Pod dry weight in rhizobia inoculation treatments was not significantly different from either 0 or 100 kg·N·ha^{-1} (Figure 1A). Melkassa 1 was the greatest in pod fresh yield and pod dry weight (Figure 1C). However, Andante was the lowest in both pod fresh yield and pod dry weight (Figure 1C). Total fresh yield and pod dry weight from Hawassa and Debre Zeit were significantly greater than the total pod fresh yield and pod dry weight from Ziway (Figure 1B).

Nitrogen treatment, cultivar, location and N treatment by location interaction significantly ($p < 0.05$) affected pod number per plant. Boston had the greatest number of pods per plant; however, it was comparable with the other cultivars, except Melkassa 1 and Melkassa 3 (Figure 2B). For the N treatment by location interaction effect, the greatest number of pods per plant was observed when 100 kg·N·ha^{-1} was applied at Hawassa (Figure 2A). The least number of pods per plant was produced at Ziway with 0 kg·N·ha^{-1}. Pod numbers in the rhizobium inoculation and 100 kg·N·ha^{-1} treatments were similar at Debre Zeit and Ziway; however, rhizobia inoculation was significantly different from the 0 kg·N·ha^{-1} at Debre Zeit (Figure 2A). The 0 kg·N·ha^{-1} treatment at Hawassa was better than 100 kg·N·ha^{-1}

application at Ziway and it was similar to 100 kg·N·ha^{-1} and rhizobia inoculation at Debre Zeit (Figure 2A). None of the N treatments significantly affected pod numbers at Ziway (Figure 2A).

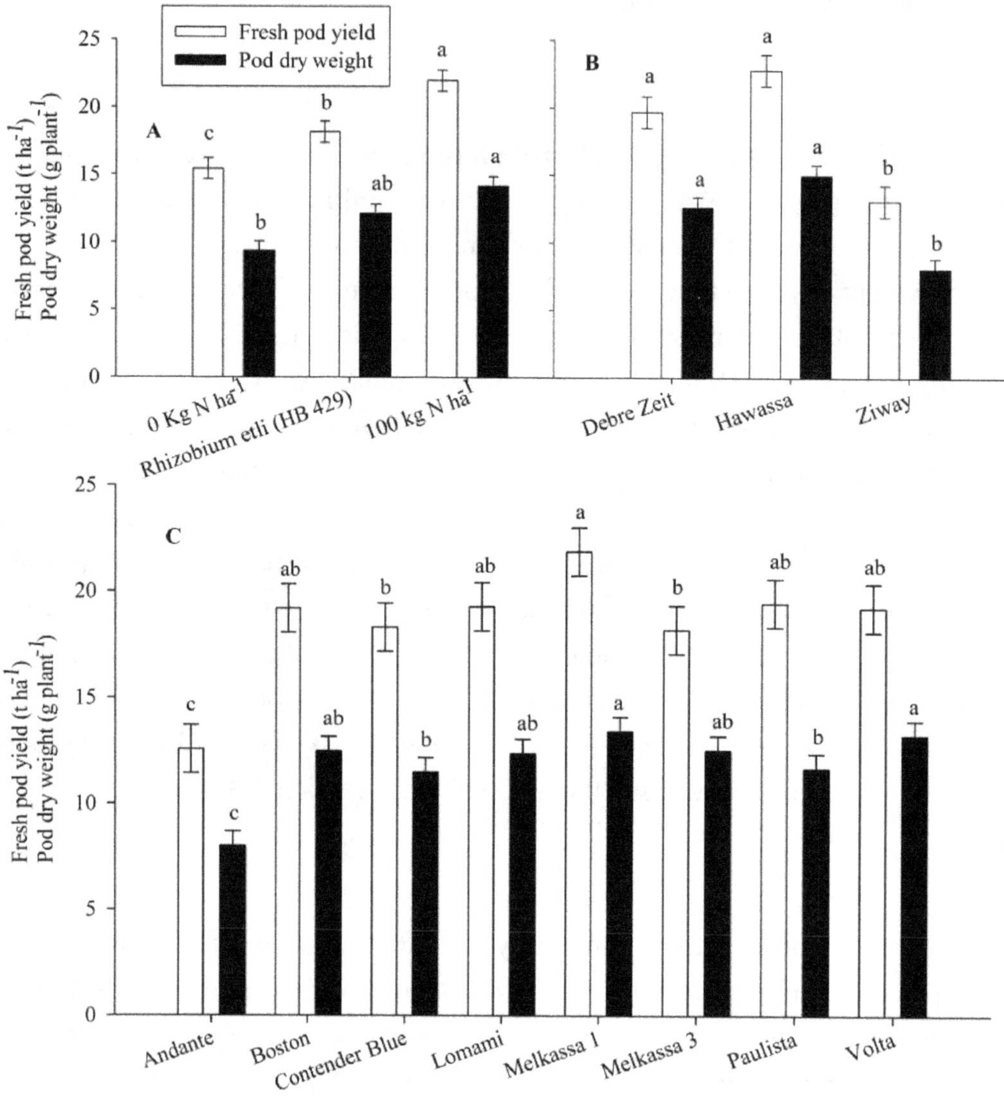

Figure 1. Fresh pod yield (t·ha^{-1}) and pod dry weight per plant of snap bean influenced by nitrogen treatment (**A**), location (**B**) and cultivar (**C**) in 2011 and 2012. The same lower letter on the bars of the same legend within panels (**A**), (**B**) or (**C**) indicates that data are not statistically significant according to LSD at $p < 0.05$.

Nodulation parameters were all influenced by the treatments in this study. Nitrogen treatment and cultivar significantly ($p < 0.05$) affected total nodule dry weight, nodule number and mean nodule diameter. Rhizobia inoculation resulted in the greatest nodule dry weight, nodule number (Figure 3A) and largest nodule diameter (Figure 4A). In contrast, N fertilizer application supressed all of these parameters. Variability across the cultivars was also observed for these nodulation parameters. Melkassa 1 produced the greatest nodule dry weight, greatest nodule number (Figure 3B) and had the largest mean nodule diameter (Figure 4C). Andante had the lowest followed by Boston for the nodulation parameters (Figures 3B and 4C). Location significantly ($p < 0.05$) affected only nodule diameter (Figure 4B) out of the three nodule parameters. Nodule diameter was larger at Debre Zeit than at Ziway (Figure 4B).

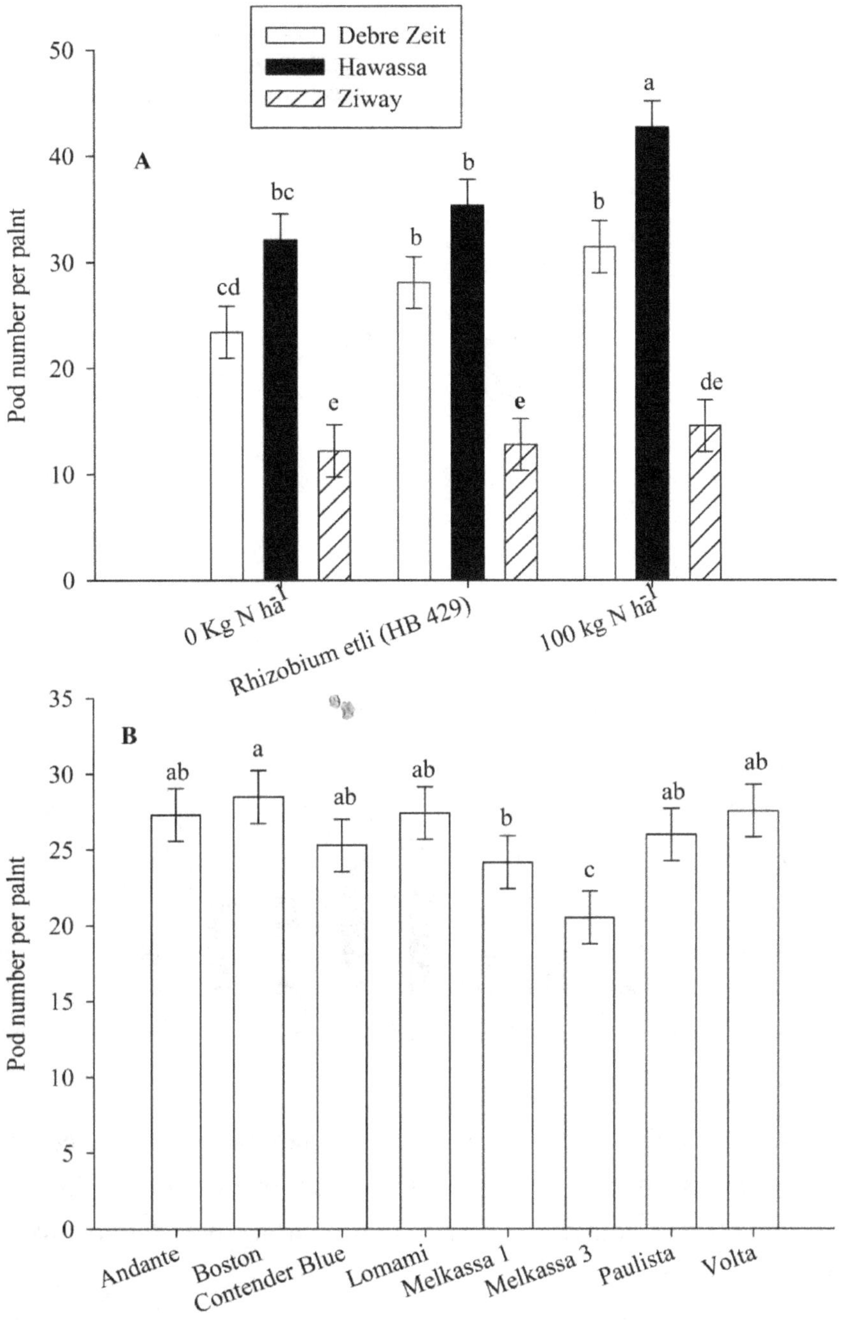

Figure 2. Pod number per plant of snap bean as a result of nitrogen treatment by location interaction (**A**) and cultivar (**B**) in 2011 and 2012. The same lower letter on the bars (among all bars in panel (**A**)) and among the bars in panel (**B**) indicates that data are not statistically significant according to LSD at $p < 0.05$.

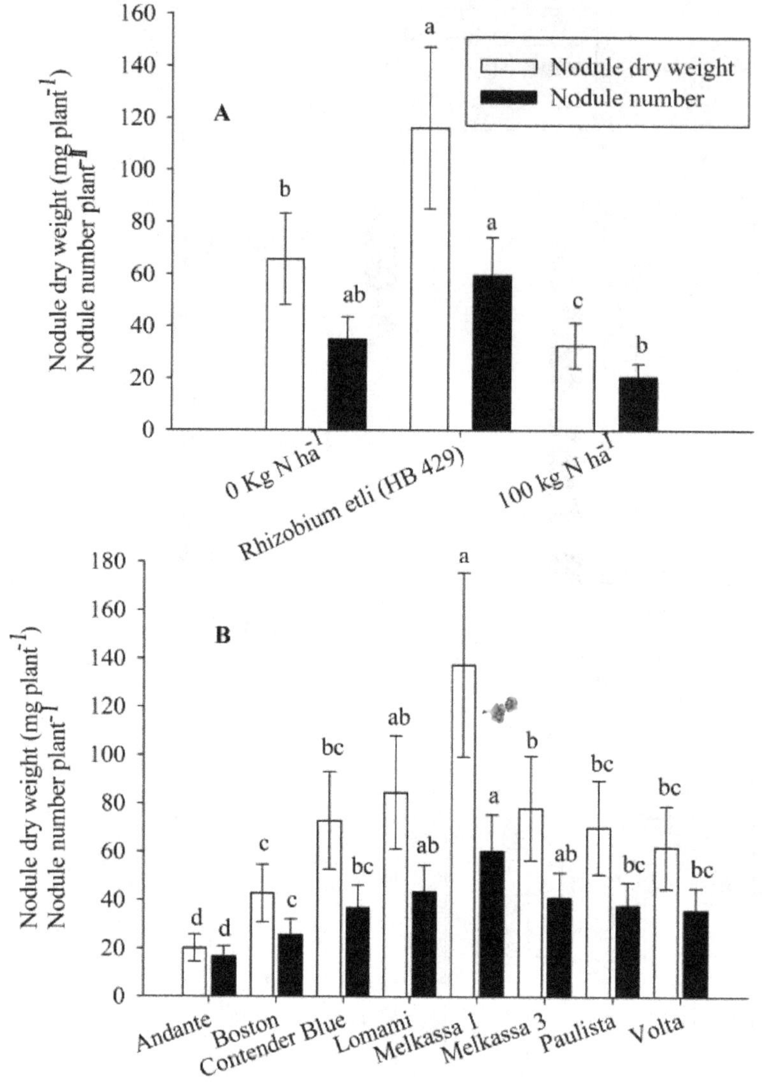

Figure 3. Nodule number and nodule dry weight of snap bean influenced by nitrogen treatment (**A**) and cultivar (**B**) in 2011 and 2012. The same lower letter on the bars with the same legend within panels (**A**) and (**B**) indicates that data are not statistically significant according to LSD at $p < 0.05$

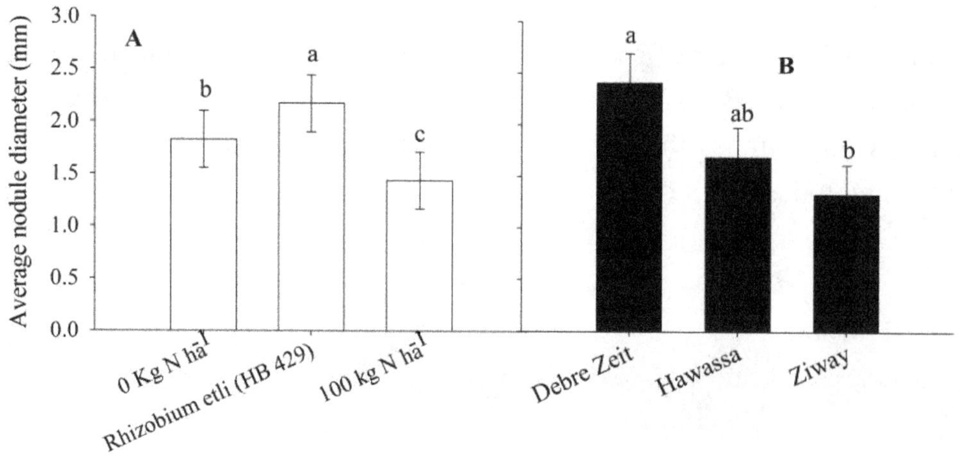

Figure 4. *Cont.*

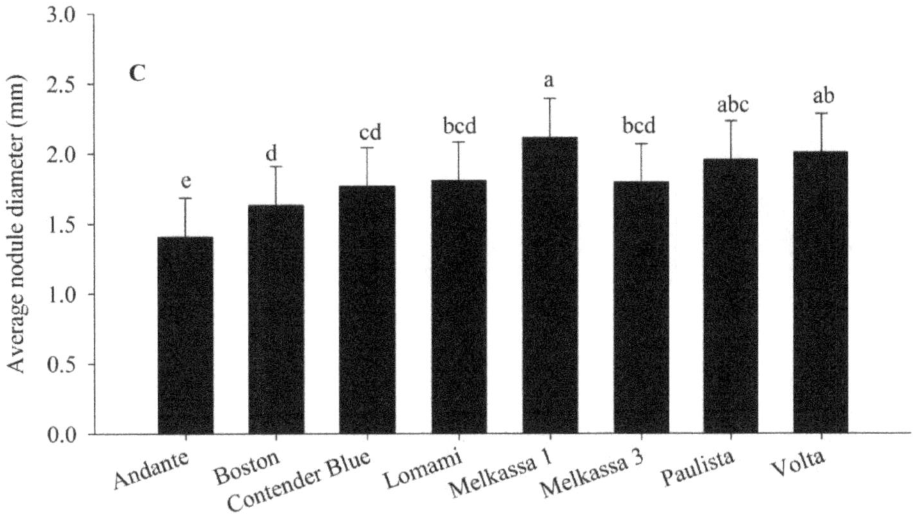

Figure 4. Average nodule diameter of snap bean influenced by nitrogen treatment (**A**), location (**B**) and cultivar (**C**) in 2011 and 2012. The same lower letter on the bars within panels (**A**), (**B**) and (**C**) indicates that data are not statistically significant according to LSD at $p < 0.05$.

There were significant ($p < 0.05$) differences among cultivars for %Ndfa which was determined by both ^{15}N dilution (Debre Zeit, 2011) and natural ^{15}N abundant methods (all the three sites, 2012) (Figure 5A). Results from both ^{15}N dilution and ^{15}N abundance methods indicated that Melkassa 1 had the highest %Ndfa. This cultivar was significantly different from Andante, Paulista and Volta in %Ndfa according to estimates using the natural ^{15}N abundance method and from all other cultivars according to the ^{15}N dilution method (Figure 5A). According to the ^{15}N dilution method Contender Blue was the lowest N fixer of all cultivars N (Figure 5A). The results of this experiment also indicated that %Ndfa was highest at Debre Zeit followed by Ziway, and the least at Hawassa (Figure 5B). The correlation analysis showed that there was a positive significant correlation (r = 0.91; p < 0.05) in 2011 and (r = 0.93; p < 0.05) in 2012, between %Ndfa and yield of snap bean cultivars at Debre Zeit. This correlation was weak and non-significant (r = 0.61; p > 0.05) both at Hawassa and Ziway in 2012.

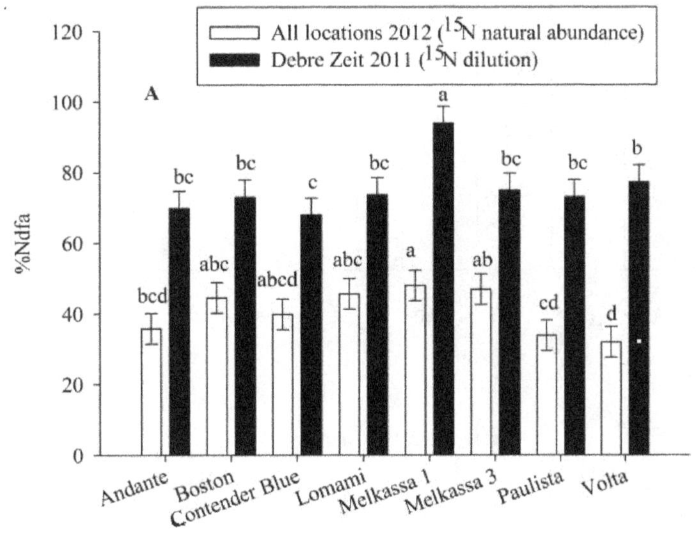

Figure 5. *Cont.*

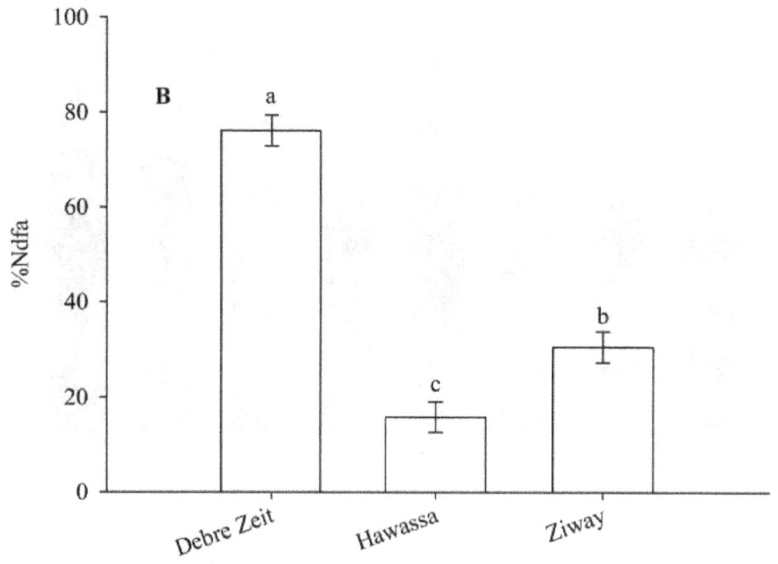

Figure 5. Biological nitrogen fixation by snap bean as influenced by cultivar (**A**) and location (**B**). The same lower letter on bars with the same legend within panels (**A**) and (**B**) indicates that the data are not statistically significant according to LSD at $p < 0.05$.

Total fixed N ha^{-1} was significantly affected by cultivar and location ($p < 0.05$). Melkassa 1 fixed the highest N ha^{-1} (Figure 6A). Andante fixed the lowest N followed by Paulista (Figure 6A). Melkassa 1 cultivar fixed N almost twice of Andante (Figure 6A). These two cultivars are contrasted also for plant size and fresh pod yield. The greatest fixed N ha^{-1} was from Debre Zeit (Figure 6B). Total fixed N was low at Hawassa and Ziway (Figure 6B). Total fixed N followed similar pattern with %Ndfa with the exceptions of cultivars Contender Blue and Volta (Figures 5A and 6A).

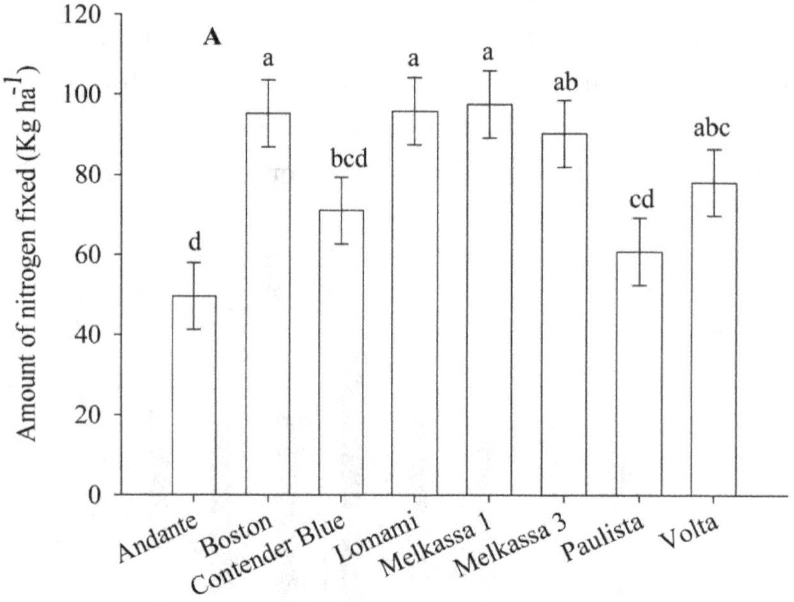

Figure 6. *Cont.*

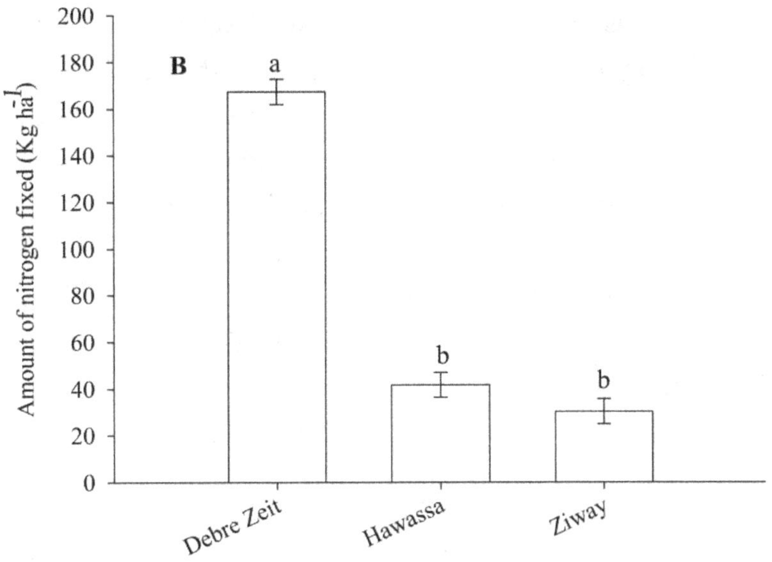

Figure 6. Total nitrogen fixed by inoculated snap bean as influenced by cultivar (**A**) and location (**B**) 2012. The same lower letter on the bars within panels (**A**) or (**B**) indicates that data are not statistically significant according to LSD at $p < 0.05$.

4. Discussion

This study demonstrated that N application consistently increased total yield, pod dry weight and pod number per plant of snap bean cultivars. Rhizobia inoculation increased the total yield and yield components of snap bean relative to an unfertilized and uninoculated treatments. Increased yield (42%) due to applied inorganic N fertilizer was far greater than yield increases achieved by rhizobia inoculation (18%), suggesting that not all N requirements were met via biological N_2 fixation, or alternatively, that N_2 fixation taxed the energy requirements of the plant. Legumes need extra energy to fix N_2 from the air into useable forms by themselves [28,29]. Moreover, common bean has greater mineral N uptake efficiency as compared to when it is fixing N_2 [14]. This may indicate that snap bean's early N demand could have been satisfied from the applied N source than N from fixation so early N supply would have contributed to the rapid growth of leaves and accumulation of dry matter. Additionally, N_2 fixation by legumes is affected by a number of edaphic, climatic and biotic factors [30] that may reduce its effectiveness when compared to applied N.

According to our preliminary survey on snap production in the Rift Valley regions of Ethiopia, the productivity of commercial snap bean ranges between 10–20 $t \cdot ha^{-1}$. The commercial producers applied on average of 100 $kg \cdot N \cdot ha^{-1}$ for snap bean production under irrigated conditions. The current experiment showed that similar range of productivity levels can be achieved by rhizobial inoculation. However, applied N still can do better than the rhiziobial inoculation. Given the high cost of N fertilizer, small-scale resource-limited farmers can be benefited from using rhizobial inoculation.

The greater pod number was obtained by 100 $kg \cdot N \cdot ha^{-1}$ at Hawassa. Application of 100 $kg \cdot N \cdot ha^{-1}$ and rhizobia inoculation at Debre Zeit resulted in similar pod numbers with that of rhizobia inoculation and the control at Hawassa. Within each location, rhizobium inoculation was significantly different from the control only at Debre Zeit. This result clearly indicated that rhizobia inoculation was most effective at Debre Zeit. Hawassa was the most suitable area for increasing pod number of snap bean. A greater number of pods lead to higher yield as these two traits have direct association [31]. Although applied N resulted in the highest pod yield and yield components of snap bean, rhizobial inoculation also improved the vegetable yield of snap bean (green pod yield) as compared to the control treatment. This result suggests that pod yield increases can be achieved by a N_2 fixation system sustained until commercial maturity of snap bean (green pod), and viable snap bean production can be realized using rhizobial inoculant under low input systems.

Inoculation with *R. etli* (HB 429) increased nodule number, nodule dry weight and diameter of nodules, whereas, N application suppressed these parameters. The current result was similar with previous reports [10,32–35] conducted on common bean and other grain legumes. Nodule numbers under 0 $kg \cdot N \cdot ha^{-1}$ were comparable to the inoculated treatment. This may be due to higher number of indigenous rhizobia in the research site which usually resulted in small ineffective nodules as reflected by nodule diameter and nodule dry weight.

Yield variability was observed among the current cultivars. Genetic variability affected the performance of common bean cultivars in terms of shoot biomass and yield [36]. Our result indicated that Melkassa 1 was the best cultivar to grow under rain fed conditions. This cultivar exhibited substantial N_2 fixation and greatest yield compared to other cultivars under rain fed conditions. Melkassa 1 also had the greatest pod dry weight and was among the top cultivars for number of pods produced under rain fed conditions.

This study demonstrated that the cultivar Melkassa 1 has the potential to produce large numbers of nodules with larger nodule sizes resulting in greater overall nodule dry weights. Moreover, Melkassa 1 achieved relatively high levels of %Ndfa which was determined by both ^{15}N dilution method from Debre Zeit site in 2011 and ^{15}N natural abundance method across all locations in 2012. Most importantly Melkassa 1 also fixed the highest $N \cdot ha^{-1}$. Previous investigations explained that effectiveness of N_2 fixation by rhizobia and further conversion into yield is affected by genotype in field bean [9,15,17]. The current result further indicated the occurrence of variations among snap bean cultivars for effective N_2 fixation. The ^{15}N dilution method in 2011 resulted in higher %Ndfa as compared to ^{15}N natural abundance in 2012. In 2011, the ^{15}N dilution method was used to determine %Ndfa only from single location (Debre Zeit). The nodulation and N_2 fixation data showed that higher N_2 fixation occurred at Debre Zeit as compared to other two locations. However, %Ndfa was determined by natural abundance method from all the three locations in 2012. Nodulation and N_2 fixation were lower at Hawassa and Ziway. The lower %Ndfa from Hawassa and Ziway may lower the average %Ndfa in 2012. This may be the main reason for the discrepancy between ^{15}N dilution and natural abundance methods on %Ndfa in the current experiment.

The greater N_2 fixation response in this experiment compared to previous reports on common bean [10,37] may be due to the effectiveness of the strain we used, the response of the cultivars, appropriate site selection or suitable growing season during experiments. Many other factors including improved crop management may also contribute to a favorable outcome. Moreover, several reports demonstrated that significant improvement of yield and yield components in common bean was due to N_2 fixation [9,32,38–40].

Our study also demonstrated that %Ndfa and total fixed $N \cdot ha^{-1}$ were highest at Debre Zeit which was also reflected in the number of pods per plant from interactions between N treatment and location. The correlation analysis also showed that fresh pod yield had a significant positive correlation with %Ndfa at Debre Zeit both in 2011 and 2012. In addition to other factors [18,19], higher %Ndfa and total fixed $N \cdot ha^{-1}$ at Debre Zeit may be due to greater copper concentration in the soil that contributed to the effectiveness of N_2 fixation [41,42]. Nodule size was greater at Hawassa than at Ziway, but the reverse was observed for %Ndfa for these two locations. This may indicate that the presence of nodules and their average size may not necessarily guarantee effective N_2 fixation. However, large biomass at Hawassa may contribute to high total N leading to high fixed N compared to at Ziway.

5. Conclusions

Snap bean pod yield improvement can be achieved by N_2 fixation system sustained until the commercial maturity of snap bean, and viable snap bean production can be realized using rhizobial inoculant in low input systems. Rhizobia inoculation was not as effective as high rates of inorganic N fertilizer, but inoculation still remains a viable and potentially less expensive alternative for improving the pod yield of snap bean.

The demonstrations that snap bean can be successfully produced without irrigation is an important finding for farmers with no irrigation infrastructure. Melkassa 1 was the highest yielding cultivar and most suitable for a reduced input production system due to its successful nodulation character, highest N_2 fixation and consistently high performance across locations under rain fed conditions. Conditions at Debre Zeit were the most conducive for supporting biological N_2 fixation for snap bean production. This shows potentially the best possible use of rhizobial inoculant as the main N source in areas at other parts of the world with similar agro-ecological characteristics of Debre Zeit. Ziway, which was characterized by a semi-arid environment with unpredictable rain fall, was a less suitable area for snap bean production.

Acknowledgments

Financial support for the research was provided by the *Canadian International Food Security Research Fund* (CIFSRF) of the *Department of Foreign Affairs, Trade and Development* and the International Development Research Center (IDRC). We thank Melkassa Agriculture Research Center, Ethioflora and SolAgro PLc (all in Ethiopia) for the supply of snap bean cultivars and for allowing use of their research sites. Our thanks extend to Myles Stocki, University of Saskatchewan and Dr. Sheleme Beyene, Hawassa University, for their technical expertise and support during the research period.

Author Contributions

Conceived, designed and managed the experiments: Hussien Mohammed Beshir, Fran L. Walley, Rosalind Bueckert and Bunyamin Tar'an. Performed the field and laboratory experiments: Hussien Mohammed Beshir. Wrote and edited the manuscript: Hussien Mohammed Beshir, Bunyamin Tar'an, Rosalind Bueckert and Fran L. Walley. All authors read and approved the final manuscript.

Conflicts of Interest

The authors declare no conflicts of interest.

References

1. Stephen, F. *Cornucopia II: A Source Book of Edible Plants*; Kampong Publications: Vista, CA, USA, 1998; p. 713.
2. FAOSTAT. Food and Agriculture Organizations of the United Nations: Statistics Division, 2014. Available online: http://faostat.fao.org/site/567/default.aspx (accessed on 9 September 2014).
3. Chianu, J.N.; Nkonya, E.M.; Mairura, F.S.; Chianu, J.N.; Aknnifesi, F.K. Biological nitrogen fixation and socioeconomic factors for legume production in sub-Saharan Africa: A review. *Agron. Sustain. Dev.* **2011**, *31*, 139–154.
4. Safapour, M.A.; Rdakani, M.; Khaghani, S.; Rejali, F.; Zargari, K.; Changizi, M.; Teimuri, M. Response of yield and yield components of three red bean (*Phaseolus vulgaris* L.) genotypes to co-inoculation with *Glomus intraradices* and *Rhizobium phaseoli*. *Am. Eurasian J. Agric. Environ. Sci.* **2011**, *11*, 398–405.
5. Zahran, H.H. Rhizobium-legume symbiosis and nitrogen fixation under severe conditions and in an arid climate. *Microbiol. Mol. Biol. Rev.* **1999**, *63*, 968–989.
6. Bliss, F.A. Utilizing the potential for increased nitrogen fixation in common bean. *Plant Soil* **1993**, *152*, 157–160.
7. Bhuiyan, M.A.H.; Khanam, D.; Hossain, M.F.; Ahmed, M.S. Effect of Rhizobium inoculation on nodulation and yield of chickpea in calcareous soil. *Ban. J. Agric. Res.* **2008**, *33*, 549–554.
8. Salvagiotti, F.; Cassman, K.G.; Specht, J.E.; Walters, D.T.; Weiss, A.; Dobermann, A. Nitrogen uptake, fixation and response to fertilizer N in soybeans: A review. *Field Crops Res.* **2008**, *108*, 1–13.
9. Bildirici, N.; Yilmaz, N. The effect of different nitrogen and phosphorus doses and bacteria inoculation (*Rhizobium phaseoli*) on the yield and yield components of field bean (*Phaseolus vulgaris* L.). *J. Agron.* **2005**, *4*, 207–215.
10. Otieno, P.E.; Muthomi, J.W.; Chemining'wa, G.N.; Nderitu, J.H. Effect of Rhizobia inoculation, farmyard manure and nitrogen fertilizer on growth, nodulation and yield of selected food grain legumes. *J. Biol. Sci.* **2009**, *9*, 326–332.
11. Maingi, J.M.; Shisanya, C.A.; Gitonga, N.M.; Hornetz, B. Nitrogen fixation by common bean (*Phaseolus vulgaris* L.) in pure and mixed stands in semi-arid south-east Kenya. *Euro. J. Agron.* **2001**, *14*, 1–12.
12. Manrique, A.; Manrique, K.; Nakahodo, J. Yield and biological nitrogen fixation of common bean (*Phaseolus vulgaris* L.) in Peru. *Plant Soil* **1993**, *152*, 87–91.
13. Tsai, S.M.; Da Silva, P.M.; Cabezas, W.L.; Bonetti, R. Variability in nitrogen fixation of common bean (*Phaseolus vulgaris* L.) intercropped with maize. *Plant Soil* **1993**, *152*, 93–101.
14. George, T.; Singleton, P.W. Nitrogen assimilation traits and dinitrogen fixation in soybean, and common bean. *Agron. J.* **1992**, *84*, 1020-1028.
15. Bliss, F.A. Breeding common bean for improved biological nitrogen fixation. *Plant Soil* **1993**, *152*, 71–79.
16. Pena-Cabriales, J.J.; Grageda-Cabrera, O.A.; Kola, V.; Hardarson, G. Time course of N_2 fixation in common bean (*Phaseolus vulgaris* L.). *Plant Soil* **1993**, *152*, 115–121.

17. Diouf, A.; Diop, T.A.; Gueye, M. Nodulation *in situ* of common bean (*Phaseolus vulgaris* L.) and field outcomes of an elite symbiotic association in Senegal. *Res. J. Agric. Biol. Sci.* **2008**, *4*, 810–818.
18. Bordeleau, L.M.; Prevost, D. Nodulation and nitrogen fixation in extreme environments. *Plant Soil* **1994**, *161*, 115–125.
19. Toro, N. Nodulation competitiveness in the rhizobium-legume symbiosis. *World J. Microbiol. Biotechnol.* **1996**, *12*, 157–162.
20. Purcell, L.C.; De Silva, M.; King, C.A.; Kim, W.H. Biomass accumulation and allocation in soybean associated with genotypic differences in tolerance of nitrogen fixation to water deficits. *Plant Soil* **1997**, *196*, 101–113.
21. Kigel, J. Culinary and nutritional quality of *Phaseolus vulgaris* seeds as affected by environmental factors. *Biotechnol. Agron. Soc. Environ.* **1999**, *3*, 205–209.
22. Ministry of Agriculture. *Agro-Ecological Zones of Ethiopia*; Natural Resources Management and Regulatory Department: Addis Ababa, Ethiopia, 2000.
23. Danso, S.K.A.; Hardarson, G.; Zapata, F. Misconceptions and practical problems in the use of ^{15}N soil enrichment techniques for estimating N_2 fixation. *Plant Soil* **1993**, *152*, 25–52.
24. Unkovich, M.; Herridge, D.; Peoples, M.; Cadisch, G.; Boddey, B.; Giller, K.; Alves, B.; Chalk, P. Measuring plant-associated nitrogen fixation in agricultural systems. The Aust. Centre for Int. Agric. Res. (ACIAR): Canberra, Australia, 2008; Volume 1–5, p. 258.
25. Hardarson, G.; Danso, S.K.A. Methods for measuring biological nitrogen fixation in grain legumes. *Plant Soil* **1993**, *152*, 19–23.
26. Mariotti, A. Atmospheric nitrogen is a reliable standard for 15N natural abundance measurements. *Nature* **1983**, *303*, 685–687.
27. SAS Institute, Inc. *The SAS System for Windows. Release 9.3*; SAS Institute, Inc.: Cary, NC, USA, 2012.
28. Pate, J.S.; Layzell, D.B.; Atkins, C.A. Economy of carbon and nitrogen in a nodulated and nonnodulated (NO_3-grown) legume. *Plant Physiol.* **1979**, *64*, 1083–1088.
29. Kaschuk, G.; Kuyper, T.W.; Leffelaar, P.A.; Hungria, M.; Giller, K.E. Are the rates of photosynthesis stimulated by the carbon sink strength of rhizobial and arbuscular mycorrhizal symbioses? *Soil Biol. Biochem.* **2009**, *41*, 1233–1244.
30. Mulongoy, K. Technical paper 2: Biological nitrogen fixation. In *The AFNETA Alley Farming Training Manual: Source Book for Alley Farming Research*; Tripathi, B.R., Psychas, P.J., Eds.; International Institute of Tropical Agriculture: Ibadan, Nigeria, 1992; Volume 2, pp. BNF1–14.
31. Araújo, L.C.; Gravina, G.A.; Marinho, C.D.; Almeida, S.N. C.; Daher, R.F.; Amaral, J.A.T. Contribution of components of production on snap bean yield. *Crop Breed. Appl. Biotechnol.* **2012**, *12*, 206–210.
32. Da Silva, P.M.; Tsai, S.M.; Bonetti, R. Response to inoculation and N fertilization for increased yield and biological nitrogen fixation of common bean (*Phaseolus vulgaris* L.). *Plant Soil* **1993**, *152*, 123–130.
33. Fan, S.; Lifang, H.; Jin, H.; Li, Z. Improvement of root nodule nitrogen fixation and soil fertility by balanced fertilization of broad bean. *Better Crops Int.* **1997**, *11*, 22–23.

34. Giller, K.E.; Amijee, F.; Brodrick, S.J.; Edje, O.T. Environmental constraints to nodulation and nitrogen fixation of *Phaseolus vulgaris* Tansania. II. Response to N and P fertilizers and inoculation with *Rhizobium*. *Afr. Crop Sci. J.* **1998**, *6*, 171–178.
35. Amba, A.A.; Agbo, E.B.; Garba, A. Effect of nitrogen and phosphorus fertilizers on nodulation of some selected grain legumes at Bauchi, Northern Guinea Savana of Nigeria. *Int. J. Biosci.* **2013**, *3*, 1–7.
36. Mourice, S.K.; Tryphone, G.M. Evaluation of common bean (*Phaseolus vulgaris* L.) genotypes for adaptation to low phosphorus. *ISRN Agron.* **2012**, doi:10.5402/309614.
37. Musandu, A.A.O.; Joshua, O.O. Response of common bean to Rhizobium inoculation and fertilizers. *J. Food Technol. Afr.* **2001**, *6*, 121–125.
38. Daba, S.; Haile, M. Effects of rhizobial inoculant and nitrogen fertilizer on yield and nodulation of common bean under intercropped conditions. *J. Plant Nutr.* **2002**, *25*, 1443–1455.
39. Cardoso, E.J.B.N.; Nogueira, M.A.; Ferraz, S.M.G. Biological N_2 fixation and mineral N in common bean–maize intercropping or sole cropping in southeastern Brazil. *Exp. Agric.* **2007**, *43*, 319–330.
40. Kellman, A.W. Rhizobium inoculation, cultivar and management effects on the growth, development and yield of common bean (*Phaseolus vulgaris*). Ph.D. Thesis, Lincoln University, Canterbury, New Zealand, 2008.
41. Snowball, K.; Robson, A.D.; Loneragan, J.F. The effect of copper on nitrogen fixation in subterranean clover (*Trifolium subterraneum*). *New Physiol.* **1980**, *85*, 63–72.
42. Seliga, H. Nitrogen fixation in several grain legume species with contrasting sensitive to copper nutrition. *Acta Physiol. Pantarum* **1998**, *20*, 263–267.

10

Tools for Optimizing Management of a Spatially Variable Organic Field

Thomas Panagopoulos [1,*], Jorge de Jesus [2] and Jiftah Ben-Asher [3]

[1] Research Center of Spatial and Organizational Dynamics (CIEO), University of Algarve, Campus Gambelas, Faro 8005-139, Portugal
[2] Ben-Gurion University of the Negev, Beer Sheva 84105, Israel; E-Mail: jorge.jesus@gmail.com
[3] Katif research center for coastal deserts development, Ministry of Science Sedot Negev Academic Campus, Sedot 86200, Israel; E-Mail: benasher@bgu.ac.il

* Author to whom correspondence should be addressed; E-Mail: tpanago@ualg.pt

Academic Editor: Ole Wendroth

Abstract: Geostatistical tools were used to estimate spatial relations between wheat yield and soil parameters under organic farming field conditions. Thematic maps of each factor were created as raster images in R software using kriging. The Geographic Resources Analysis Support System (GRASS) calculated the principal component analysis raster images for soil parameters and yield. The correlation between the raster arising from the PC1 of soil and yield parameters showed high linear correlation ($r = 0.75$) and explained 48.50% of the data variance. The data show that durum wheat yield is strongly affected by soil parameter variability, and thus, the average production can be substantially lower than its potential. Soil water content was the limiting factor to grain yield and not nitrate as in other similar studies. The use of precision agriculture tools helped reduce the level of complexity between the measured parameters by the grouping of several parameters and demonstrating that precision agriculture tools can be applied in small organic fields, reducing costs and increasing wheat yield. Consequently, site-specific applications could be expected to improve the yield without increasing excessively the cost for farmers and enhance environmental and economic benefits.

Keywords: GRASS; raster images; principal component analysis; organic farming; precision agriculture; geostatistics

1. Introduction

Wheat (*Triticum turgidum* var. durum) is cultivated over more than 13 million hectares worldwide [1]. In recent years, the management regime of those crops has undergone a series of changes as a result of an increase in average field size. New tools are consequently required to enable a global view of these larger-sized fields and to determine the heterogeneous zones that often appear within them. The use of yield prediction maps is an important tool for the delineation of within-field management zones.

Yield prediction maps are of great importance to ensure that yields are maximized with fewer inputs, less waste and consequently less environmental impact. Accurate estimation of yield can be used for zonal management of the most productive areas, to plan the best time for harvesting and its transport for industrial processing, and to locate any water and nutritional deficiencies in the field [2]. Yield monitoring and mapping have given producers a direct method for measuring spatial variability in yield [1]. Along with yield mapping, producers have expressed increased interest in characterizing soil variability.

Wheat yield is spatially variable because of inherent spatial variability of factors affecting the yield at field scale [3]. Precision agriculture is an emerging management strategy that combine geographic information systems (GIS), global positioning systems (GPS), computer modeling, remote sensing, expert systems, and advanced information processing with the goal of optimizing returns on inputs while preserving resources [4].

Precision agriculture can provide a knowledge-based management of agricultural production to reduce environmental impact and increase profit margins [5]. According to Virgilio *et al.* [6], the core of precision farming theory is to understand field spatial variation and the relations with crop response, resulting in a substantial increase in the input effectiveness and in the average biomass yield and obtaining economic and environmental benefits. Precision agriculture in conventional agriculture is already recognized, although it is uncommon to see similar research in organic agriculture because of the high heterogeneity of the plant production factors and yield [7].

The yield variability is not an independent phenomenon, and in theory, it should be influenced by the soil variability and by the scarcest resource that controls the growth and not by the sum of all resources available [8]. However, in the field, many production factors conjointly act, and some positive effects may be hidden by the negative ones, such as low water content [9]. GIS is a powerful tool for analyzing spatial data and establishing a process for decision support [10]. Geostatistics offers the possibility to represent the spatial dependence of soil and yield variable distribution [11].

The approach of precision agriculture generates an intensive stream of data, ranging from soil parameters to yield factors that need to be submitted to data mining methodologies so that concrete relations of the several factors influencing the yield can be shifted from the raw data. Raster maps with detailed information on soil properties and yield components have some degree of correlation; thus, principal component analysis can group them in a new and reduced set of images that can be more easily analyzed and understood [12].

Therefore, the present research addresses the following: (i) to estimate the spatial variation of durum wheat yield under organic agriculture field conditions, (ii) to assess the spatial variation of soil properties, (iii) to produce thematic maps of yield and soil parameters using geostatistical kriging approaches to find possible relations between soil parameters and yield, and (iv) to reduce the level of complexity between the measured parameters using principal component analysis.

2. Materials and Methods

2.1. Study Area

The study area was located in the organic farm of Kibbutz Nirim (34°35′ N/31°20′ E) in the North of the Negev Desert in Israel. Average temperature varies between 16 °C in January and 27 °C in August. The average annual rain is 250 mm, and the annual evaporation from class A pan is 1550 mm (30 years of meteorological data). The soil is Loess (Calcic haploxeralf) with an average bulk density of 1320 kg/m^3 and pore volume of 0.5 m^3/m^3; clay, silt, and sand are 150, 300, and 550 g/kg, respectively. The texture is sandy loam, and the cation exchange capacity is 18 cmol/kg.

The research field was in cultivation during data collection and was sowed with wheat (*Triticum turgidum* var. *durum* Desf.) on January, which is usual sowing time in Negev Desert, and harvested on 10 June 2004. The grid used to gather the samples followed the characteristics of the yield and was designed in a triangular way, with the objective of maximizing the covered area with the minimum number of samples, fitting the field with 73 samples, an amount in concordance with other studies and authors [13]. In precision agriculture, the rule is to use sampling intervals equal to half of the semivariogram range [14]. To know the optimal grid size, "cross validation" was used to compare the prediction performances of the semivariograms [15]. This grid fitted an area of 1.9 ha, which is enough for a precision agriculture study according to previous studies [16]. Each grid point was determined in the field by the use of GPS.

Soil samples were collected during the emergence at February 17, at 15 March and at 6 June 2004. Those dates represent the initial, middle, and final stages of crop growth prior to harvesting. Soil specific area (SSA), nitrate, soil water content (SWC), and carbon flow (CF) were determined in the upper 0.15 m. SSA was measured with the Ethylene Glycol Monoethyl Ether method [17], nitrate with Ion Specific Probe [18] and SWC with gravimetric method as was described by Gardener [19] using 100 gr of soil measured for their initial and final mass after being dried for 24 h in a 110 °C oven. Measurement of the soil carbon flux was performed using the LI-6400 Soil CO_2 Flux system, which is an InfraRed Gas Analyzer (IRGA) working in a closed system chamber [20]. Measurements were performed several days after harvest. A PVC ring, buried at a depth of about 1 cm, was placed at each sampling point. The LI-6400 gas chamber was then placed on each ring and the soil carbon flux was measured.

Before harvesting of wheat field, yields at the 73 predetermined georeferenced points were measured using the total number of grains and plants, the weight of 1000 kernels as well as the total weight of grain, average weight of one kernel, stems, plants, and grain/plant were determined in a sampling area of 0.2 m^2 per point [21]. The leaf area index (LAI) measurements were determined only at the two initial dates and were not measured on the last date because the crop was completely dry. LAI was measured with the ΔT—SunScan Canopy Analysis System as described by Welles and Norman [22].

2.2. Data Transformation and Trend Removal

The gathered data were checked for skewness, and data transformation functions of log(x), sqrt(x), inv(x), and sq(x) were applied to bring the skewness closer to zero [23]. After the data were checked for trend using the mean-polish trend removal method [24,25]. This method divides the grid system of the sample into columns and rows and tests for an increase or decrease of data values. The $Z(x)-m-r_k-c_l$ (residual) and $(\mathrm{k}-k)(\mathrm{l}-l)$ were calculated. The m is the overall mean; r and c are the row and column mean, respectively; and k and l are the mean of number of rows and columns. The values of $Z(x)-m-r_k-c_l$ and $(\mathrm{k}-k)(\mathrm{l}-l)$ were plotted, and trend line was fitted.

For each parameter, the trend line was tested for slope using an analysis of variance (ANOVA) with F distribution table [26]. A slope equal to zero indicated that there was no trend, whereas a value different from zero indicated a trend in the data. If the ANOVA proved that there was a parameter with trend, then the residual values were used instead of the measured values for the interpolation because residuals are correlated even if the observations are not.

2.3. Interpolation Method

The interpolation method used for raster creation was ordinary kriging, which is a common method for data interpolation [27]. This method is based on the creation of a semivariogram graphic and from the information contained in the best-fitted model [28,29].

The models used were gathered from common geostatistics publications [30,31]. The semivariogram is a plot between the distances of ordered data and their value of semi variance; this plot explains the spatial relation between the samples and is given by Equation (1).

$$\gamma(h) = \frac{1}{2\mathrm{N(h)}} \sum [Z_i - Z_{i+h}]^2 \tag{1}$$

The most related samples have lower values of semi variance ($\gamma(h)$). N(h) is the number of samples that can be grouped using vector h, Z_i represents the value of the sample, and Z_{i+h} is the value of another sample located at a distance h from the initial sample Z_i. The semivariogram is a point graphic with points plotted at specific distance intervals. Because there is a need to know the semi-variance value at distances not defined in the plot, a model was fitted using the lowest possible root square error.

The fitted model provided two important parameters to determine if the samples are spatially correlated, which were the Nugget (N) and Sill (S). The "nugget/sill ratio" (N/S) was introduced by Cambardella *et al.* [32] as a measure to quantify the level of spatial structure. A low N/S ratio (<0.25) indicates that the samples are spatially correlated, while a high N/S ratio (>0.75) means that the samples have a very low spatial correlation.

According to Issaks and Srivastava [31], kriging tries to have a mean residual error equal to zero with the lowest possible value of the standard deviation of the error and, at the same time, estimates the weighted linear combinations (wi) of the available data for the interpolation result (Equation (2)).

$$z(x_0) = \sum_{i=1}^{n} w_i \,.\, z(x_i) \quad \wedge \quad \sum_{i=1}^{n} w_i = 1 \tag{2}$$

The linear weight necessary for the interpolation is obtained by the ordinary kriging system (Equation (3))

$$C \quad . \quad w \quad = D$$

$$\begin{bmatrix} C_{11} & \cdots & C_{1n} & 1 \\ \vdots & \ddots & \vdots & \cdots \\ C_{nl} & \cdots & C_{nn} & 1 \\ 1 & \cdots & 1 & 0 \end{bmatrix} \cdot \begin{bmatrix} w_1 \\ \vdots \\ w_n \\ \mu \end{bmatrix} = \begin{bmatrix} C_{10} \\ \vdots \\ C_{n0} \\ 1 \end{bmatrix} \tag{3}$$

$$(n+1)*(n+1) \qquad (n+1)*1 \quad (n+1)*1$$

The matrix C contains the covariance from all the samples surrounding the sample to be interpolated. The matrix w contains the weights as well as the parameter called Lagrange parameter. The matrix D contains the covariance from the sample to be determined and the surrounding ones. The final objective is to determine matrix w.

2.4. Principal Component Analysis

Principal component analysis (PCA) is a linear transformation of a set of numerical variables (or images), which creates a new variable set (or images), called principal components (PCs). In this framework, the new variables are uncorrelated and ordered in terms of the amount of variance explained from the original data [33]. The advances made in the last years concerning PCA and image analysis can be useful when there is sufficient spatial information in agricultural fields, mainly, raster images that can be generated by sampling and interpolation using geostatistics [34,35].

Each PC is a combination of the original images with coefficients equal to the eigenvector of the covariance matrix [36]. Aside from the eigenvector, the eigenvalues are also obtained and can be used to determine how well the PC can explain the variability of the original values. It can therefore be useful in determining the number of PCs necessary for in-depth analysis.

To begin the transformation, a covariance matrix $\vec{C}$ of the original data has to be found. Using the covariance matrix, the eigenvalues λ_i and the eigenvectors $\vec{e}_i$ are obtained from the following equation (Equation (4)).

$$\left|\vec{C} - \lambda_i \vec{I}\right| = 0 \quad \equiv \quad \left(\vec{C} - \lambda_i \vec{I}\right) \vec{e}_i = 0 \tag{4}$$

where $i \in [1,2,3....n]$, n is the total number of images, and $\vec{I}$ is the identity matrix. The PC images are then given by Equation (5):

$$\vec{P}C = \vec{T} \circ DN \tag{5}$$

where $\vec{D}N$ is the digital number matrix of the original images and T is the transformation matrix given by Equation (6):

$$\vec{T} \equiv \begin{bmatrix} e_{11} & \cdots & e_{1n} \\ \vdots & \ddots & \vdots \\ e_{nl} & \cdots & e_{nn} \end{bmatrix} \tag{6}$$

where e_{nm} is the value of the values of the eigenvectors.

The generated PC images are uncorrelated. The transformed data points are linear combinations of their original data values weighed by the eigenvectors. The percentage in each of the components is given by the next equation (Equation (7)).

$$Variance_i = \frac{\lambda_i * 100}{\sum_{i=1}^{n} \lambda_i} \tag{7}$$

where λ_i is the calculated eigenvalues obtained in Equation (4).

By computing the correlation of each original band with each PC, it is possible to determine how the images (and therefore, the parameters that they represent) are associated with each principal component. This enables determining which of the actual variables are more important, as well as the relation between them. One of the contributions of this study was the conversion of PCs into actual variables.

The variables required for the calculation of the correlation between the map and a PC are as allows: the eigenvalue (λ_i), the variance of the image (Var_i), and the eingenvector value for image i and component p (e_{ip}) (Equation (8)).

$$R_{ip} = e_{ip} \frac{\sqrt{\lambda_i}}{\sqrt{Var_i}} \tag{8}$$

GRASS (Geographic Resources Analysis Support System) and R software were used as tools for the present study. For the spatial analysis and semivariogram creation it was used the "sgeostat" software (Iowa State University). The initial data were imported from the spreadsheet (after it was processed for trend and skewness), using a comma-separated value (CSV) file format. For the semivariogram estimation, a lag of 10 and a maximal range of 100 were used. The lag and range values were obtained after several simulations of estimated semivariogram creation. For interpolation and export to GRASS, it was necessary that R run from inside GRASS so that the spatial information of the study area can be imported to R (projection, grid resolution, mask, *etc.*) and then be exported and integrated on the GRASS database [37].

Figure 1 presents the flowchart describing the PCA calculation methodology using the raster maps produced from geostatistics. The raster images created with geostatistics were imported from R software and rescaled from their original values to an 8-bit scale using the module r.rescale. At the rescale, the minimal interpolated value represented by 0 digital number (DN) and the maximal interpolated value was represent by 255 DN. This rescale was necessary to confirm all the rasters and to ease their analysis [38]. For the PCA calculation, the i.pca module was used; this module processes n input raster map layers and produces n output raster map layers containing the principal component in decreasing order of variance and also output of the eigenvector matrix.

Afterward, the raster images were separated into the two groups of plant production factors and yield factors. For each group, a PCA was conducted, obtaining the PC rasters and the eigenvector. To calculate the eigenvalues, the covariance array was passed to R software and using the eigenfunctions, the eigenvalues were obtained. From Equation (8), it was determined which factor contributed the most for each of the principal components because the most important factors will have the highest correlation with the principal component to where they belong.

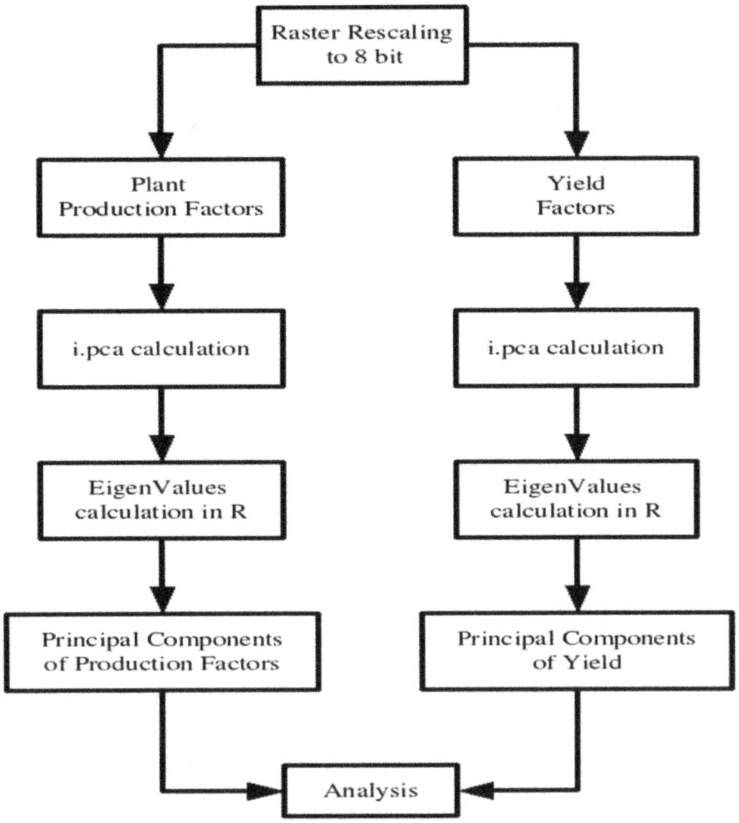

Figure 1. Flowchart describing the PCA calculation methodology using the raster maps produced from geostatistics.

3. Results

The trend analysis revealed that only the nitrate recorded on February 17 had a trend for a 10% confidence interval. This parameter was detrended, and the residuals were used for the semivariogram creation and interpolation. The majority of the parameters had a moderate-strong or moderate-weak spatial correlation (Table 1). The number of plants had a null spatial relation between samples, which was caused by the random sowing procedure.

Table 1. Basic statistics of the parameters, results of the semivariograms, and spatial dependence.

Parameter	Min.	Max.	Mean	CV	SK	Nugget	Sill	Range (m)	Model	N/S	Spatial Dep.
Soil specific area (m^2/g)	67.5	104.7	82.17	11.12	0.51	7.94×10^{-7}	1.86×10^{-6}	50	spherical	0.43	Moderate
Nitrate in February (ppm)	2.78	22.83	5.94	52.9	2.82	1.22×10^{-3}	4.83×10^{-3}	15.85	exponential	0.25	Mod.-strong
Nitrate in March (ppm)	2.92	16.67	6.01	39.96	1.89	2.42×10^{-3}	3.60×10^{-3}	64.25	exponential	0.67	Mod.-weak
Nitrate in June (ppm)	2.39	20.16	5.25	58.31	2.17	0.01	0.02	796.2	exponential	0.48	Moderate
SWC in. February (%)	6.92	11.99	10.21	9.47	−0.55	131.19	388.02	11.88	exponential	0.34	Mod.-strong
SWC in March (%)	6.99	11.99	10.16	9.48	−0.39	239.74	394.7	39.33	exponential	0.61	Mod.-weak
SWC in June (%)	1.91	6.97	4.03	24.52	0.78	0.02	0.06	10.79	exponential	0.35	Mod.-strong
Carbon flux ($\mu mol\ m^{-2} \cdot s^{-2}$)	0.16	2.35	0.96	47.06	1.08	0.19	0.22	48.59	exponential	0.88	weak
LAI in February	1.3	4.2	2.56	26.14	0.41	0.02	0.05	43.14	exponential	0.3	Mod.-strong
LAI in March	2	4.6	3.55	20.42	−0.62	7.06	24.45	13.76	exponential	0.29	Mod.-strong
Number of grains	430	1247	713.2	21.44	0.27	4.91	9.41	59.75	spherical	0.52	Moderate
Number of plants	7	36	20.4	33.17	0.33	0.55	—	—	nugget	—	None
Weight of 1000 grains (g)	31.67	72.49	53.42	12.73	−0.6	1.00×10^{-6}	494634	6.58	exponential	0.00	Strong
Total weight of grain (g)	22.22	56.87	37.68	20.07	−0.06	48.56	69.72	158.64	spherical	0.69	Mod.-weak
Weight of kernel (g)	8.85	42.18	19.63	40.46	1.08	0.05	0.2	48.89	exponential	0.25	Mod.-strong
Weight of stems (g)	6.74	58.52	24.95	48.78	0.8	0.66	1.39	12.01	exponential	0.48	Moderate
Weight of plants (g)	22.22	56.84	37.68	42.43	0.92	0.08	0.18	19.08	exponential	0.42	Mod.-strong
Weight grain/plant (g)	0.31	2.4	0.99	43.34	0.68	0.08	0.22	60.69	spherical	0.35	Mod.-strong

Figure 2 shows the semivariograms of the main parameters studied. The parameter weight of stems is moderate-strong spatial correlated with semi-variogram showing a spatial range below 20 meters. The parameter weight of plants is moderate-strong spatial correlated, but despite the slow increase of the sill of the model, the majority of the spatial correlation is on the first three bins, which is a range a bit over 20 meters. The semivariogram for Nitrate on February 17, which is the only parameter that has shown a non-stationary behavior, shows a very good model fitting.

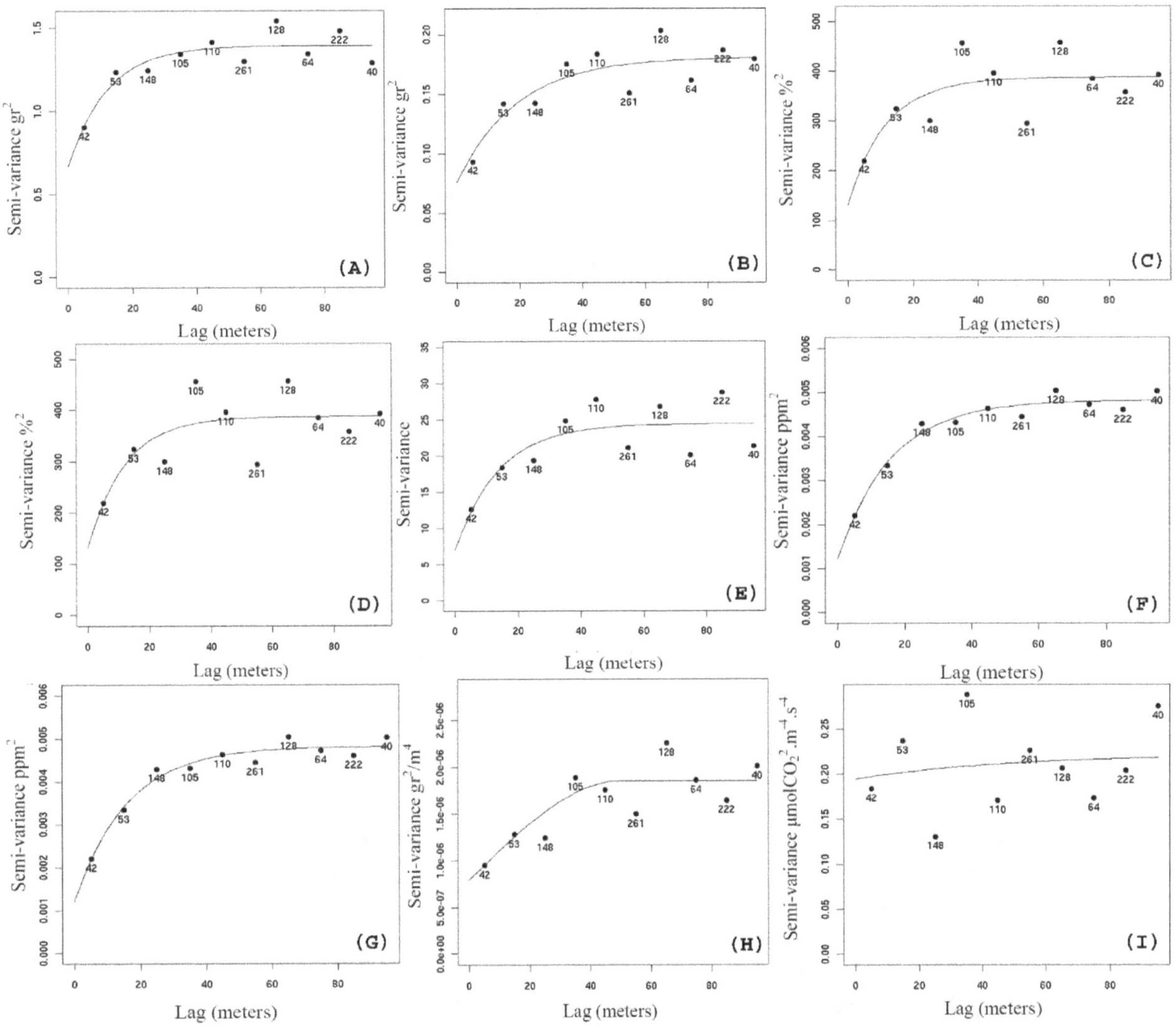

Figure 2. Results of semivariogram and model fitting for the parameter weight of stems (**A**); weight of plants (**B**); the soil water content recorded in February 17 (**C**); the soil water content recorded in June 6 (**D**); leaf area index recorded March 15 (**E**); Nitrate in February 17 (**F**); SSA (**G**); total weight of grain (**H**); Carbon flux (**I**).

The spatial distribution of the interpolated maps is caused by the behavior of the semivariogram and the interpolation method. The SSA ranged from 73.20 to 95.17 m^2/g (Figure 3a) using the average value of 81.29 ± 4.91 m^2/g and the general equations from the work of Banin and Amiel [39] for Israeli soils; it was determined that the soil had a 16.65% ± 3.45% clay content, and this corresponds to a bulk density of 1.46 g/cm^3 (considering a 55% sand content). The spatial behavior of the nitrate was somehow irregular; this is caused by the ability of this nutrient to move because it is not well adsorbed by the soil particles. Nevertheless, the average nitrate values decreased as the crop developed from 8.24 ± 4.30 ppm in February and 5.42 ± 0.72 ppm in March (Figure 3c) to 4.21 ± 0.21 ppm.

The SWC maps had a similar spatial distribution (Figure 4a–c), with increasing water contents from east to west. The average SWC values decreased from 11.46% ± 0.40% in February (Figure 4a), to 11.20% ± 0.40% in March (Figure 4b) to 3.91% ± 0.37% in June (Figure 4c). The carbon flux had the average measured value of 0.98 ± 0.34 μmol CO_2 $m^{-2} \cdot s^{-2}$, and the distribution of the carbon flux indicates that the upper right part of the field had more organic matter.

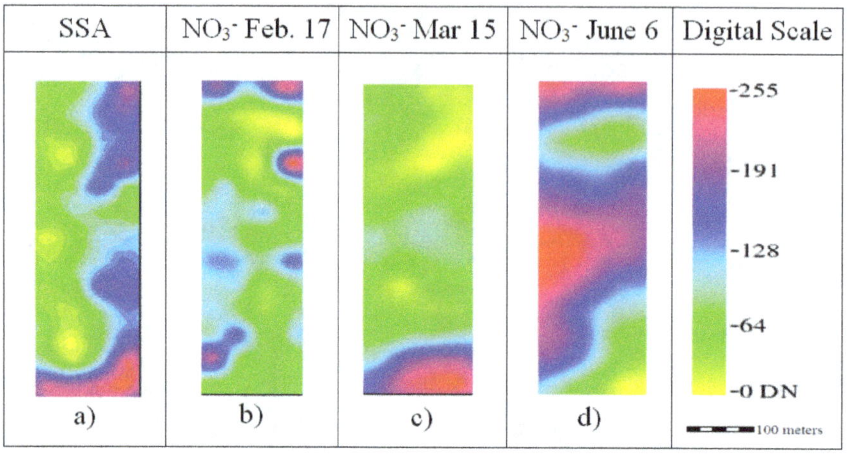

Figure 3. Ordinary kriging raster images rescaled to digital number (DN) range for the production factors Soil Specific Area (SSA) and Nitrate.

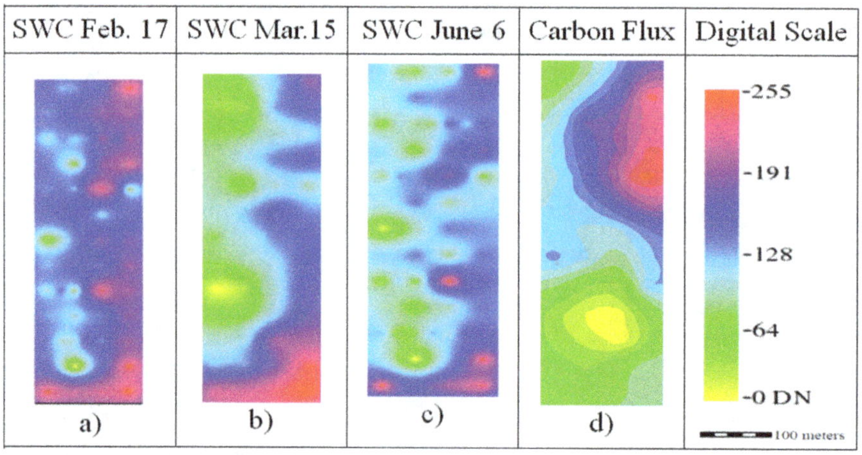

Figure 4. Ordinary kriging raster images rescaled to digital number (DN) range for the production factors Soil Water Content (SWC) and Carbon flux.

The LAI for February and March (Figure 5a,b) had similarities to SSA and SWC. In fact, the LAI increased from 2.51 ± 0.41 in February (Figure 5a) to 3.56 ± 0.35 in March (Figure 5b) and from east to west. The average number of grains was 3547.3 ± 399.25 n/m^2 (Figure 5c), and the average number of plants was 503.25 plants/m^2 (Figure 5d). These two yield parameters seem to be inversely related, with areas of high number of grain density related to low plant density areas.

The average weight of 1000 kernels was 37.6 ± 2.75 g (Figure 6a). As indicated by the nugget/sill ratio (Table 1), this trait showed a strong spatial correlation, and the range of the semivariogram is small;

the map that was obtained had some "bull eyes", a feature that is more typical of IDW. The total weight of grain (Figure 6b) is the most important yield parameter because it represents the actual economic production; the average value was 1.88 ± 0.11 t/ha.

The weight of stems (Figure 6c) and kernel (Figure 6d) shared the same behavior because they represent the lower and upper biomass of the plant; the average weight of the stem was 0.93 ± 0.21 t/ha, and the average weight of the kernel was 1.16 ± 0.21 t/ha.

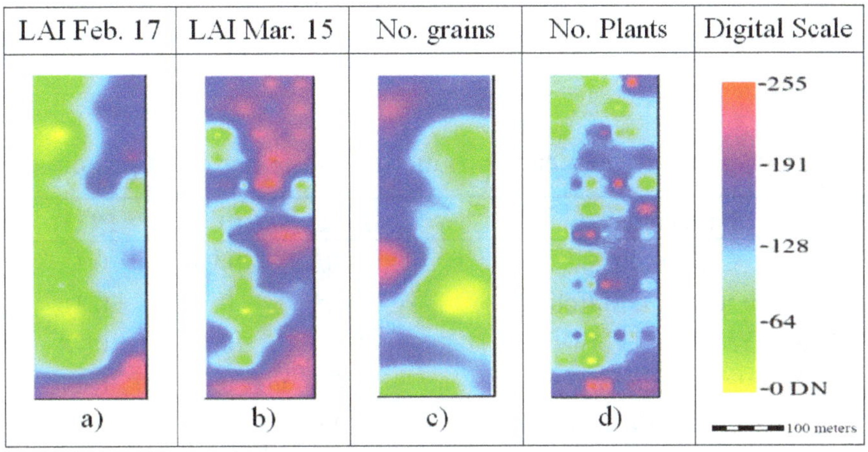

Figure 5. Ordinary kriging raster images rescaled to digital number (DN) range for the yield components Leaf Area Index (LAI), Number of grains and Number of plants.

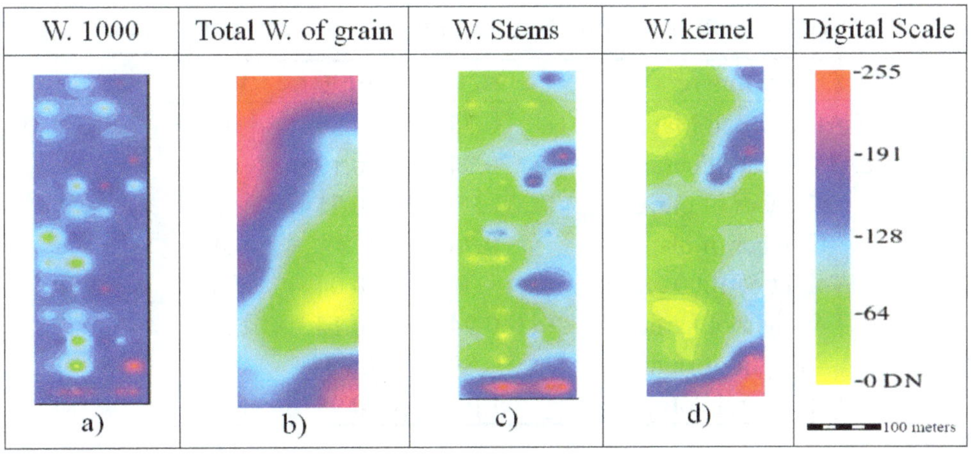

Figure 6. Ordinary kriging raster images rescaled to digital number (DN) range for the yield components weight of 1000 grains, weight of stems and weight of kernel.

Principal Component Analysis Results

The data show that durum wheat yield is strongly affected by variability in soil parameters (soil water content recorded in 17 February 2004, soil water content recorded in June 6, Nitrate in February 17, CF and, SSA), and thus, the average production can be substantially lower than its potential. However, the objective of this research is to give a mechanistic explanation of the yield variation, something that depends on the conjoint effects of several contrasting or additive factors. PCA results could adequately investigate which main production factors affect yield spatial variation. In this research, it was shown that areas with the lowest biomass production were also those characterized by low carbon flux and high

SSA and SWC, parameters that were all significantly related to the yield. Therefore, the use of an appropriate site-specific practice may be expected to substantially increase the average yield. Surprisingly, the effect of nitrate on the crop yield was low because the SWC was a limiting factor.

Correlation coefficients based on kriged maps are presented on Tables 2 and 3. The correlation coefficient of the PCA from the production factors is given in Table 2. The cumulative results of PC1, PC2, and PC3 account for 85.52% of the data variance. The PC1 explained 48.50% of the data variance, whereas PC2 and PC3 explained 23.53% and 13.48%, respectively. This high cumulative result indicates that these three maps should contain sufficient information to explain internal relations between different producing factors.

Table 2. Correlation coefficient between each production factor and each principal component.

Parameters	PC1	PC2	PC3
Carbon flux	−0.02	0.94	−0.11
Nitrate recorded March 15	−0.65	−0.57	−0.16
Nitrate recorded February 17	0.3	−0.19	−0.74
Nitrate recorded June 6	0.78	−0.07	−0.57
Soil specific area	−0.89	0.23	−0.26
Soil water content recorded March 15	−0.89	−0.13	−0.19
Soil water content recorded February 17	−0.82	0.27	−0.28
Soil water content recorded June 6	−0.81	0.2	−0.28

Table 3. Correlation coefficient between each yield factor and each principal component.

Parameters	PC1	PC2	PC3
LAI recorded March 15	−0.78	−0.39	−0.36
LAI recorder February 17	−0.94	−0.17	−0.16
Number of grains	0.55	−0.75	−0.01
Number of plants	−0.78	0.09	0.47
Weight of 1000 grains	−0.8	−0.04	0.34
Weight of kernel	−0.94	−0.14	0.02
Weight of stems	−0.89	0.03	0.38
Total weight of grain	0.10	−0.98	0.11

The correlation coefficient of PCA for the yield is given in Table 3. The performance of the PCA of the yield was slightly better than the PCA of the production factor because PC1, PC2, and PC3 explained 92.04% of the data's variability. The PC1 explained 51.52% of the data variance and showed a specific grouping of biological parameters: LAI; number of plants; weight of kernel, stem, and weight of 1000 grains. The major economic yield parameter (total weight of grain) was in PC2 together with the number of grains. The PCA results and the accumulative percentage of explained variance from both production factor and yield were acceptable, and the three PCs from the two groups were sufficient for data analysis of the rasters and their relations.

Factor Loading

Analysis of correlation between the PC and the factors responsible for its creation was used to determine the most important rasters for the creation of each PC. This analysis is called factor loading. By determining which input maps were important for each PC, it was possible to determine the relation between several factors.

SWC and SSA were the parameters that exerted the most influence among all the production factors in PC1. Nitrate had some influence in PC1, but the result was unclear because of different signs and a high correlation of one of the nitrate dates in PC3. The most influencing production factor in PC2 was the carbon flux. All the results for PC3 have a negative value of correlation coefficient; normally, each PC has factors that influence the cloud of points in one direction or another, but in this case, all the factors had a negative correlation, meaning that all push the cloud only in one direction [40].

The majority proved that PCA can be used to study the relation between soil properties and that all the different yield factors can be obtained by a correlation between the different PCs of each group. Figure 7 present the correlation between the two rasters, one belonging to the production factor group and the other from yield parameters. It showed a very high coefficient of determination ($r = 0.75$) with an almost linear behavior ($y = 0.9247 + 0.833x$). Despite the relation between SSA-SWC with biological yield, this is less important than the relation of economic yield with production factors; therefore there is the need to see how yield parameters are related to SSA-SWC.

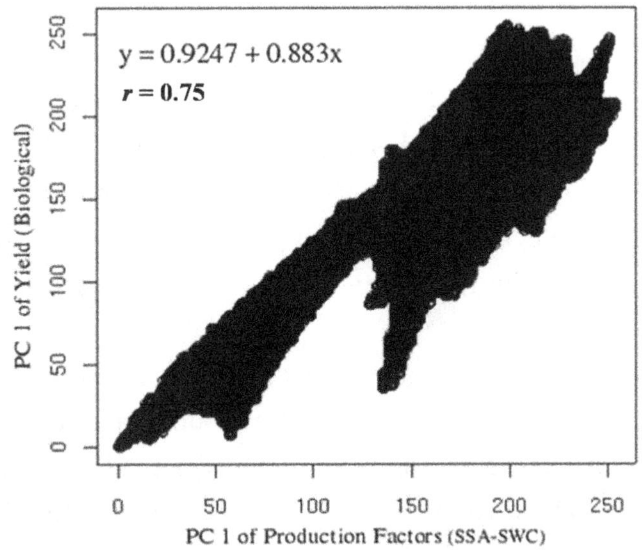

Figure 7. Correlation between PC1 of production factors (Soil Specific Area and Soil Water Content) and PC1 of yield components (Leaf Area Index, weight of kernel and stems). Data rescaled to digital number (DN) range of 0–255.

Figure 8 presents the correlation between the PC1 of production factors SSA and SWC and PC2 of yield (total weight and number of grains). This correlation was weak $r = 0.17$ ($y = 83.51 + 0.36x$). In the same correlation, it was observed that until the value of 125 DN range, the correlation was high ($r = 0.91$) and almost linear, but the correlation became null after this value. This change of behavior around the value of 125 DN range also exists on the biological yield correlation, where one part of the cloud starts to divert from the almost perfect linearity. According to Webster and Oliver [41], this behavior may indicate biological saturation response, where above certain threshold, the effect of one or several of the parameters decreases or stop having an impact.

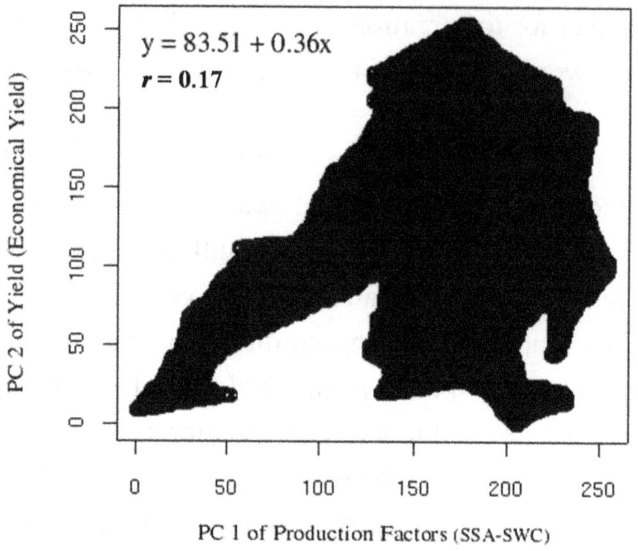

Figure 8. Correlation between PC1 of production factors (Soil Specific Area and Soil Water Content) and PC2 of yield components (total weight and number of grains). Data rescaled to digital number (DN) range of 0–255.

4. Discussion

The use of precision agriculture tools and PCA in plant production factors and yield of an organic field helped reduce the level of complexity between the measured parameters, eliminating data redundancy and resulting in feasible relations. According to Lopez-Granados *et al.* [42], knowledge on spatial dependence helps to calculate the sampling interval and develop an accurate site-specific application scheme. Therefore, the feasibility of precision farming applications may increase with the degree of spatial dependence.

Data analysis showed a wide variability within the organic field, resulting in inefficient use of resources. Therefore, the use of an appropriate site-specific practice may be expected to substantially increase the average yield [43]. In fact, some parameters change gradually across the field, whereas others show a patchy distribution. The carbon flux, SWC, and nitrate recorded in March were weakly spatially dependent; thus, additional samples at smaller lag-distances may be needed for those parameters. Nonetheless, higher sampling density could be uneconomic.

Instead of having several data rasters, the PCA "compressed" the information in a reliable way, facilitating further analysis. The PCA results showed the relation between multiple parameters as well as which plant production factors were explanatory of the yield results. The carbon flux does not relate to any other parameter; therefore, it had a specific PC. Nitrate had two parameter associated to PC3, but the behavior of nitrate is not clear because it also integrates in part PC1. The most important plant production factor was SWC followed by SSA.

For yield factor, the weight of 1000 grains was grouped in PC1, concluding that when the biomass of the crop increases, the weight per grain also increases. The major economic yield parameter (total weight of grains) in PC2 indicated the distinctive behavior between the biomass and the total grain production.

Also, an analysis to the different signs showed that the higher the biomass, the lower was the number of grains and total weight of grain. Therefore, the increase of weight per grain caused by an increase of biomass is not sufficient to increase total production.

The correlation (r) between the biological yield parameters and the PC of SSA-SWC was high, further proving the proper arrangement of plant production factors and yield according to the most important PCs, the influence of SWC-SSA as the major influencing factor for biomass.

The PC1 also showed a different cloud orientation for the group of biological yield, such as total weight and number of grains. This indicates that when LAI and weight of kernel, stem, and weight of 1000 grains are higher, there should be a decrease on the number of grains and total weight of grain (economic yield) and *vice versa*. This indicates that higher biological biomass will produce less grain, but heavier (because the weight of 1000 grains will be higher), and a lower economic yield. Similar results were found by Dong *et al.* [44] when wheat cultivars were grown in a dry year and without irrigation.

5. Conclusions

The data show that durum wheat yield is strongly affected by soil parameter variability, and thus, the average production can be substantially lower than its potential. From the Principal component analysis raster images for soil parameters and yield, PC1 explained 48.50% of the data variance. The correlation between the rasters arising from the PC1 of soil and yield parameters showed high linear correlation ($r = 0.75$). The objective of reducing the number of raster necessary to analyze the reasons behind the specific yield was achieved and still preserving a high percentage of useful information. Soil water content was the limiting factor to grain yield and not nitrate as in other similar studies.

The use of precision agriculture tools and PCA helped reduce the level of complexity between the measured parameters by the grouping of several parameters in the different PC, creating a sort of "compression", eliminating data redundancy, and resulting in feasible relations. The present research shows that precision agriculture tools can be applied in small organic fields, reducing costs and increasing average wheat yield. Consequently, some site-specific applications could be expected to improve the yield without increasing excessively the cost for farmers and, at the same time, enhance environmental and economic benefits [45]. For example the carbon flux distribution indicates that the upper right part of the field had more organic matter and with this knowledge it can be saved money when applying manure. Also, it can be provided more water in the areas of the field that have lowest soil water content using precision irrigation and by using less manure at the southern part of the field that presented high nitrate in March (Figure 3c).

Acknowledgments

This research was supported by the Foundation for Science and Technology (Fundação para a Ciência e a Tecnologia), Portugal, Ph.D. grant number SFRH/BD/8303/2002, and the Research Center of Spatial and Organizational Dynamics (CIEO). Part of this research was supported by a grant from the Ministery of Science, Culture and Sport, Israel and the Bundesmenisterium fuer Bildung and Forschung (BMBF).

Author Contributions

Jorge de Jesus, Jiftah Ben-Asher and Thomas Panagopoulos design the research and performed the analysis; Jorge de Jesus and Thomas Panagopoulos wrote the paper. Thomas Panagopoulos made the final revision of the paper and answer to the reviews questions.

Conflicts of Interest

The authors declare no conflict of interest.

References

1. Panagopoulos, T.; Jesus, J.; Blumberg, D.; Ben-Asher, J. Spatial variability of wheat yield as related to soil parameters in an organic field. *Commun. Soil Sci. Plant Anal.* **2014**, *45*, 2018–2031.
2. Aggelopoulou, K.D.; Wulfsohn, D.; Fountas, S.; Gemtos, T.A.; Nanos, G.D.; Blackmore, S. Spatial variation in yield and quality in a small apple orchard. *Precis. Agric.* **2008**, *11*, 538–556.
3. Basso, B.; Cammarano, D.; Chen, D.; Cafiero, G.; Amato, M.; Bitella, G.; Rossi, R.; Basso, F. Landscape position and precipitation effects on spatial variability of wheat yield and grain protein in Southern Italy. *J. Agron. Crop Sci.* **2009**, *195*, 301–312.
4. McBratney, A.; Whelan, B.; Ancev, T. Future directions of precision agriculture. *Precis. Agric.* **2005**, *6*, 7–23.
5. Booltink, H.W.G.; van Alphen, B.J.; Batchelor, W.D.; Paz, J.O.; Stoorvogel, J.J.; Vargas, R. Tools for optimizing management of spatially variable fields. *Agric. Syst.* **2001**, *70*, 445–476.
6. Virgilio, N.; Monti, A.; Venturi, G. Spatial variability of switchgrass (*Panicum virgatum* L.) yield as related to soil parameters in a small field. *Field Crops Res.* **2007**, *101*, 232–239.
7. Panagopoulos, T.; Jesus, J.; Antunes, M.D.C.; Beltrão, J. Analysis of spatial interpolation for optimising management of a salinized field cultivated with lettuce. *Eur. J. Agron.* **2006**, *24*, 1–10.
8. Mendas, A.; Delali, A. Integration of multicriteria decision analysis in GIS to develop land suitability for agriculture: Application to durum wheat cultivation in the region of Mleta in Algeria. *Comput. Electron. Agric.* **2012**, *83*, 117–126.
9. Chatterjee, A.; Lal, R. On farm assessment of tillage impact on soil carbon and associated soil quality parameters. *Soil Till. Res.* **2009**, *104*, 270–277.
10. Ferreira, V.; Panagopoulos, T. Seasonality of soil erosion under Mediterranean conditions at the Alqueva dam watershed. *Environ. Manag.* **2014**, *54*, 67–83.
11. Bojaca, C.R.; Gil, R.; Cooman, A. Use of geostatistical and crop growth modelling to assess the variability of greenhouse tomato yield caused by spatial temperature variations. *Comput. Electron. Agric.* **2009**, *65*, 219–227.
12. Carr, J. A visual basic program for principal components transformation of digital images. *Comput. Geosci.* **1998**, *24*, 209–218.
13. Carter, M. *Soil Sampling and Methods of Analysis*; Canadian Society of Soil Science, Lewis Publishers: Charlottetown, PE, Canada, 1993.
14. Kerry, R.; Oliver, M. Variograms of ancillary data to aid sampling for soil surveys. *Precis. Agric.* **2003**, *4*, 261–278.

15. Castrignano, A.; Wong, M.T.F.; Stelluti, M.; de Benedetto, D.; Sollitto, D. Use of EMI, gamma-ray emission and GPS height as multi-sensor data for soil characterisation. *Geoderma* **2012**, *175–176*, 78–89.
16. Van Meirvenne, M. Is the soil variability within the small fields of flanders structured enough to allow precision agriculture? *Precis. Agric.* **2003**, *4*, 193–201.
17. Eltanawy, I.; Arnold, P. Reappraisal of Ethylene Glycol Mono-Ethyl Ether (EGME) method for surface area estimations of clays. *J. Soil Sci.* **1973**, *24*, 232–238.
18. Morf, W. *The Principles of Ion-Selective Electrodes and of Membrane Transport*; Elsevier: Amsterdam, The Netherlands, 1981.
19. Welles, J.; Demetriades-Shah, T.; McDermitt, D. Considerations for measuring ground CO_2 effluxes with chambers. *Chem. Geol.* **2001**, *177*, 3–13.
20. Gardner, W. Water Content. In *SSA Book Series: 5—Methods of Soil Analysis, Part 1—Physical and Mineralogical Methods*; Klute, A., Ed.; Soil Science Society of America: Madison, WI, USA, 1986.
21. Leilah, A.; Al-Khateeb, S. Statistical analysis of wheat yield under drought conditions. *J. Arid Environ.* **2005**, *61*, 483–496.
22. Welles, J.; Norman, J. Instrument for indirect measurement of canopy architecture. *J. Agron.* **1991**, *83*, 818–825.
23. Crawley, M. *Statistical Computing—An Introduction to Data Analysis using S-Plus*; John Wiley & Sons Ltd., The Atrium, Southern Gate: Chichester, West Sussex, UK, 2002.
24. Cressie, N. *Statistics for Spatial Data*; John Wiley & Sons: New York, NY, USA, 1993.
25. Berke, O. Modified median polish kriging and its application. *Environmetrics* **2001**, *12*, 731–748.
26. Dalgaard, P. *Introductory Statistics with R*; Springer-Verlag: New York, NY, USA, 2002.
27. Ferreira, V.; Panagopoulos, T.; Cakula, A. Prediction of seasonal soil erosion risk at the Alqueva dam watershed, Portugal. *Fresenius Environ. Bull.* **2013**, *22*, 1997–2005.
28. Clark, I. *Practical Geostatistics*; Applied Science Publishers Ltd.: London, UK, 1979.
29. Chabala, L.M.; Mulolwa, A.; Lungu, O. Mapping the spatial variability of soil acidity in Zambia. *Agronomy* **2014**, *4*, 452–461.
30. Wackernagel, H. *Multivariate Geostatistics: An Introduction with Applications*; Springer: Berlin, Germany, 1995.
31. Isaaks, E.; Srivastava, R. *An Introduction to Applied Geostatistics*; Oxford University Press: New York, NY, USA, 1989.
32. Cambardella, C.; Moorman, T.; Novak, J.; Parkin, T.; Karlen, D.; Turco, R.; Konopka, A. Field-scale variability of soil properties in Central Iowa soils. *Soil Sci. Soc. Am. J.* **1994**, *58*, 1501–1511.
33. Eastman, J.; Fulk, M. Long sequence time series evaluation using standardized principal components. *Photogramm. Eng. Remote Sens.* **1993**, *59*, 991–996.
34. Goovaerts, P.; Jacquez, G.; Marcus, A. Geostatistical and local cluster analysis of high resolution hyperspectral imagery for detection of anomalies. *Remote Sens. Environ.* **2005**, *95*, 351–367.
35. Yalouris, K.; Kollias, V.; Lorentzos, N.; Kalivas, D.; Sideridis, A. An integrated expert geographical information system for soil suitability and soil evaluation, *J. Geogr. Inf. Decis. Anal.* **1997**, *1*, 90–100.
36. Ricotta, C.; Avena, C. The influence of principal component analysis on the spatial structure of a multispectral dataset. *Int. J. Remote Sens.* **1999**, *20*, 3367–3376.

37. Grunsky, E. R: A data analysis and statistical programming environment—An emerging tool for geosciences. *Comput. Geosci.* **2002**, *28*, 1219–1222.
38. Bivand, R. Using the R statistical data analysis language on GRASS 5.0 GIS data base files. *Comput. Geosci.* **2000**, *26*, 1043–1052.
39. Banin, A.; Amiel, A. A correlative study of the chemical and physical properties of a group of natural soils of Israel. *Geoderma* **1969**, *3*, 185–198.
40. Richards, J.; Jia, X. *Remote Sensing Digital Image Analysis: An Introduction*, 3rd ed.; Springer: Berlin, Germany, 1999.
41. Webster, R.; Oliver, M. Statistical Methods in Soil and Land Resource Survey. Oxford University Press: New York, NY, USA, 1990.
42. Lopez-Granados, F.; Jurado-Exposito, M.; Atenciano, S.; Garcia-Ferrer, A.; Sanchez de la Orden, M.; Garcia-Torres, L. Spatial variability of agricultural soil parameters in southern Spain. *Plant Soil* **2002**, *246*, 97–105.
43. Panagopoulos, T.; Rodrigues, S.; Neves, N.; Cruz, S.; Antunes, D. Decision support tools for optimising kiwifruit production and quality. *Acta Hortic.* **2007**, *753*, 407–414.
44. Dong, B.; Shi, L.; Shi, C.; Qiao, Y.; Liu, M.; Zhang, Z. Grain yield and water use efficiency of two types of winter wheat cultivars under different water regimes. *Agric. Water Manag.* **2011**, *99*, 103–110.
45. Panagopoulos, T.; Antunes, M.D.C. Integrating geostatistics and GIS for assessment of erosion risk on low density *Quercus suber* woodlands of South Portugal. *Arid Land Res. Manag.* **2008**, *22*, 159–177.

11

Effect of Foliar Boron Fertilization of Fine Textured Soils on Corn Yields

Gurpreet Kaur [1,*] and Kelly A. Nelson [2]

[1] Department of Soil, Environmental and Atmospheric Sciences, University of Missouri, Columbia, MO 65211, USA

[2] Division of Plant Sciences, University of Missouri, Novelty, MO 63460, USA; E-Mail: NelsonKe@missouri.edu

* Author to whom correspondence should be addressed; E-Mail: GK478@mail.missouri.edu

Academic Editor: Roel Merckx

Abstract: Boron (B) is an essential micronutrient needed for normal plant growth and development. To evaluate the response of corn to foliar B applications at V4–V6 (4–6 leaves with visible collars) and VT (tasseling) growth stages on fine textured soils, a field experiment was conducted at four sites from 2008 to 2010 in Northeast Missouri. The treatments included a non-treated control; V4–V6 applied B at 0.56, 1.12 and 2.24 $kg \cdot ha^{-1}$; and VT applied B at 0.28, 0.56 and 1.12 $kg \cdot ha^{-1}$. Foliar B, applied at V4–V6 at 2.24 $kg \cdot ha^{-1}$, resulted in higher yields than VT applications. No significant differences in yield were found for B applications at different timings for concentrations of 0.56 and 1.12 $kg \cdot ha^{-1}$. Boron applied at V4–V6 and 2.24 $kg \cdot ha^{-1}$ increased yield 0.29 $Mg \cdot ha^{-1}$ compared to the non-treated control. The B applications at VT increased ear leaf tissue B concentration compared to V4–V6 applications and non-treated control, but it had no significant effect on corn yields. No significant difference between B treatments was observed for grain oil, protein, starch or extractable starch concentration; severity of anthracnose stalk rot or common rust; and ear tip fill. The B application of 2.24 $kg \cdot ha^{-1}$ at V4–V6 decreased the severity of gray leaf spot, but increased the severity of northern leaf blight compared to the non-treated control. Boron applied at V4–V6 at 2.24 $kg \cdot ha^{-1}$ was the most beneficial timing and concentration evaluated in these fine textured soils.

Keywords: boron; corn; foliar fertilizer; V4–V6; VT

1. Introduction

Boron (B) is an essential micronutrient needed for normal plant growth and development. It is involved in many plant processes such as sugar transport, cell wall synthesis, lignification, meristematic tissue cell division, petal and leaf bud formation, cell wall structure integrity, sugar and hydrocarbon metabolism and their transport, ribose nucleic acid (RNA) metabolism, respiration, indole acetic acid (IAA) metabolism, cytokinin production and transfer, phenol metabolism, nitrogen fixation, pollen germination, pollen tube formation and seed formation [1–4]. Intensive cropping systems and the use of high yielding hybrids has resulted in depletion of soil micronutrients [5]. Globally, B deficiency has been recognized as the second most important micronutrient constraint in crops after zinc (Zn) [1]. In the USA, B deficiency was the most widespread among micronutrients [6,7]. In soil, concentrations of total B typically ranged from 20 to 200 $mg \cdot B \cdot kg^{-1}$, but generally the availability to plants is less than 5%–10% [8]. Boron requirements may vary by plant type, but the range for soil solution B concentration between deficiency and toxicity is smaller than other nutrient elements [9]. Corn has a low requirement for B, but can be very sensitive to excess B. Several factors affect B uptake by plants including soil type (texture, pH, organic matter content), B concentration, moisture, and plant species. The interaction of B with other nutrients (N, P, K, Ca, Mg, Al, and Zn) can be synergistic or antagonistic which can influence B availability to plants [2]. The bioavailability of B reduces as soil dries due to low rainfall or limited irrigation because of decreased B mobility in soil by mass flow to plant roots [1,9]. Boron is absorbed by plants as boric acid, which is easily leached in soils [10]. Boron is relatively immobile in a corn plant and its availability is essential at all growth stages, particularly during fruit and seed development [2]. Boron deficiency in corn was first observed during the 1960s in the United States [11] and B applications showed more than a 10% increase in yield on coarse textured soils [12]. The B sufficiency range for corn was from 4 to 25 ppm in the ear leaf [13]. In corn, B deficiency caused barren ears and blank stalks at concentrations below 0.05 ppm which resulted in lower yields [13]. Woodruff *et al.* reported that B interacted with N, K and lime while B fertilization at 2.24 $kg \cdot ha^{-1}$ was necessary for preventing a reduction in corn yields when higher K fertilizer rates were applied in South Carolina [12]. In B deficient soils, a B application increased plant B concentration which helped to improve the quality of corn fodder for animals without causing any significant increase in dry matter yield [14]. On a calcareous soil, B interacted with zinc (Zn) and antagonistically affected nutrient concentration and synergistically affected growth [15]. Minimum amounts of B accumulated in corn during initial growth stages and maximum accumulation was observed after 100 days of seedling emergence in two corn hybrids in Brazil while the total amount of B required to produce one ton of corn was 0.9 g [16]. However, B decreased P uptake and dry weight of corn genotypes [17], while a Ca application antagonized shoot B concentrations of four corn hybrids [18]. Boron applied with high Zn levels resulted in higher NPK concentrations in corn grains [19].

Foliar nutrient sprays may be an effective way to correct micronutrient deficiencies, which sometimes results in higher yields and crop quality [20,21]. Advantages of a foliar application compared to a soil application included rapid plant response, increased convenience and effective placement [22]. The reported negative effects of foliar applications include leaf necrosis due to the direct effects of the foliar salts, which reduce effective leaf area and photosynthate production [23]. Boron can be soil applied (broadcast or banded) or foliar applied [5]. An in-row application of B has showed higher plant uptake compared to foliar application while both of these application methods were more effective than a broadcast soil application [24]. In some southeastern states, foliar spray applications increased tissue B concentration compared to a broadcast soil application [25]. In contrast, foliar applied B at high rates have caused severe toxicity to corn compared to soil applied rates [26]. In *Vigna radiata*, soil applied B had a greater impact on dry matter yield whereas a foliar application increased grain yield [5]. In Missouri, foliar applied B at 1.12 $kg{\cdot}ha^{-1}$ increased the number of soybean (*Glycine max*) branches per plant and the formation of pods on branches [27]. A pre-plant soil application of B, at 3 $kg{\cdot}ha^{-1}$, had a greater effect on corn growth and average dry matter accumulation, but had lower yields compare to the non-treated control [28]. In a greenhouse experiment, corn plants showed injury to soil solution B concentrations of 20 $mg{\cdot}B{\cdot}L^{-1}$ [29].

Boron has also helped to reduce disease severity in some crops because of the effect that B has on plant metabolism, cell membranes and cell wall structure [30–33]. Boron reduced the infection of pathogens by improving cell wall and membrane strength with cross-linked polymers and by strengthening the plant's vascular bundles [34]. Plant disease development and management were affected by the environment especially when nutrient deficiencies and toxicity occurred [33]. A B application of 0.5 $kg{\cdot}ha^{-1}$ along with a combination of nitrogen (100 $kg{\cdot}ha^{-1}$) and Zn (0 and 1.0 $kg{\cdot}ha^{-1}$) resulted in reduced fungal mycotoxin production, which were responsible for rotting of corn ears [35]. On fine textured soils in Northeastern Missouri, B applied at VT (tasseling) with pyraclostrobin or pyraclostrobin alone reduced the disease severity most consistently [36]. There has been limited research on a suitable B application timing (early *vs.* late application) and concentration of foliar B on fine textured claypan soils in the Midwestern US. Therefore, the objective of this research was to evaluate the effects of foliar-applied B at different application concentrations and timings on corn yield, tissue B concentration, severity of diseases, grain oil, protein, starch, and extractable starch concentration.

2. Materials and Methods

Field experiments were conducted from 2008 to 2010 at four sites in Knox and Shelby counties in Northeastern Missouri. This research was conducted simultaneous to fungicide pyraclostrobin research at VT with foliar B applications at a single concentration [36]. The two sites in Knox County were at the University of Missouri's Greenley Research Center near Novelty (40°01′ N, 92°11′ W) and a cooperator's farm nearby Bee Ridge (40°04′ N, 92°04′ W). The locations in Shelby County were at the University of Missouri's Ross Jones Farm near Bethel (39°56′ N, 92°03′ W) and a cooperator's farm nearby Leonard (39°54′ N, 92°16′ W). Different sites were selected each year for each experiment at individual locations. All locations had a silt loam soil texture and were planted with corn (Table 1). The experiment was randomized complete block design with five replications and had seven B treatments including a non-treated control. Treatments included foliar applied B (Solubor, US Borax Inc., Valencia, CA, USA) at different timings and concentrations (Table 2). Boron was applied at V4–V6 (4–6 leaves with visible collars) [37] growth stage at 0.56, 1.12 and 2.24 $kg{\cdot}B{\cdot}ha^{-1}$ as well as VT at 0.28, 0.56 and

1.12 kg·B·ha^{-1}. Foliar B was applied using a CO_2 propelled backpack sprayer with 8002 flat fan nozzles (TeeJet Technologies, Wheaton, IL, USA) at 140 L·ha^{-1} and no surfactant was used. The B concentrations used in this study were in the range recommended for foliar applications in corn [38,39]. Plants were 18 to 36 cm tall at V4–V6 and 183 to 305 cm tall at VT. The individual plot size was 3 by 15.2 m. All plots were planted in 0.76 m wide rows. Field information about the four sites, selected management practices and details about foliar B application are shown in Tables 1 and 2 [36]. These sites had different tillage operations (conventional *vs.* no-till), previous crops, hybrids and seeding rates, which varied from 70,400 to 86,500 seeds ha^{-1} (Table 1). Fertilizer was applied based on recommendations from the University of Missouri soil test lab at Novelty and Bethel and by private labs at Bee Ridge and Leonard. Supplemental irrigation was scheduled using the Woodruff chart at the Novelty site [40].

Initial soil samples before planting and fertilizer application were collected from each replication from 15-cm deep soil cores (20 cores per replication). The soil samples were dried at approximately 50 °C, ground, and analyzed using standard methods by Clemson University Agricultural Service Laboratory (Clemson, SC, USA). Soil test information at all 12 site-years are presented in Table 3.

Before the V4–V6 B application and 10 days after application (DAA), 10 plants per plot from treatments with only foliar applied B at V4–V6 and the non-treated control were harvested, oven dried at 60 °C for 48 h, ground, weighed and tissue analyzed for B concentration. Ear leaf tissue was collected at VT and again 10 days after the VT B application, oven dried at 60 °C, weighed and analyzed for tissue B concentration. Plant stand counts were determined in order to calculate the plant population hectare^{-1}, while ear tip fill and barren stand counts were also recorded prior to harvest. A digital caliper (Performance Tool, Tukwila, WA, USA) was used for measuring the ear tip that remained with no kernels from 20 plants row^{-1} at each site-year. Foliar injury due to B applications was visually rated on a scale of 0 (no crop injury) to 100% (complete plant death). Corn plants were rated for diseases severity including gray leaf spot (*Cercospora zea-maydis*) and common rust (*Puccinia sorghi)* at all site-years. There was no apparent disease in 2008 (data not presented). The severity of diseased plants in each plot was assessed 42 days after VT based on a percentage of the plants (0%–100%) showing symptoms of gray leaf spot, common rust, northern corn leaf blight (*Exserohilum turcicum*), or anthracnose stalk rot (*Colletotrichum graminicola*). The percentage of leaf area with lesions over the entire canopy (0 = no lesions to 100 = complete plant coverage) was used for rating gray leaf spot, common rust, and northern corn leaf blight. The grey leaf spot measurements were taken at all sites in 2009 and at Bethel and Bee Ridge in 2010. The northern corn leaf blight measurements were taken only at Leonard in 2010. The percentage of the stalk with lesions was rated for anthracnose stalk rot (0 = no lesions to 100 = complete stalk coverage). The centermost two rows in each plot were harvested using a small plot combine (Wintersteiger Delta, Salt Lake City, UT, USA) and weighed for grain yields and test weights were determined. The seed moisture was measured at harvest and adjusted to 150 g·kg^{-1} before analyzing data. Ten subsamples from collected grain samples were analyzed for oil, protein, starch, and extractable starch (Foss Infratec 1241, Eden Prairie, MN, USA).

Table 1. Field information and selected management practices at 12 site-years [36].

Field	Novelty			Bethel			Leonard			Bee Ridge		
Information †	**2008**	**2009**	**2010**	**2008**	**2009**	**2010**	**2008**	**2009**	**2010**	**2008**	**2009**	**2010**
Previous crop	Soybean	Corn	Corn	Corn	Soybean	Soybean	Soybean	Soybean	Soybean	Soybean	Soybean	Soybean
Tillage	Conv.	Conv.	Conv.	NT	NT	NT	NT	NT	NT	Conv.	Conv.	Conv.
Planting date	19 May	11 May	26 May	14 June	19 May	28 May	21 May	23 May	29 May	21 May	22 May	28 May
Hybrid	DKC63–42VT3	DKC63–42VT3	DKC63–42VT3	Burrus 795t	DKC63–42 VT3	DKC63–42 VT3	Crow's 4835	Mycogen 2D653	Mycogen 2K594	Pioneer 33D13	Pioneer 33D13	Pioneer 33T57
Seeding rate, (seeds ha^{-1})	86,500	79,000	84,000	74,000	74,000	73,400	74,000	74,000	74,000	70,400	75,300	76,600
Fertilizer applications ($N–P_2O_5–K_2O$, $kg·ha^{-1}$)												
Fall date		15 December 2008				21 November 2009				21 November 2007		
Rates		34–90–135				66–170–252				34–90–180		
Source(s)		DAP				DAP				DAP		
Preplant date	19 May	4 May	12 April	14 June	8 April	20 May	25 April	20 May		19 May	25 April	15 April
Rates	224–0–0	202–0–0	280–90–170	202–100–135	202–0–0	56–0–0	44–112–135	56–0–0		163–0–0	40–103–0	44–112–112
Source(s)	AN	AN	AN + DAP	AN + DAP	AA	UAN	DAP	UAN		AA	DAP	DAP
Preemergence date		4 May					1 June				21 May	15 April
Rates		90–67–22					34–0–0				190–0–0	179–0–0
source(s)		AN + DAP					UAN				AA	AA
Top-dress date		17 June	11 June			14 June	20 June	23 & 28 June				
Rates		90–0–0	202–0–0			112–0–0	112–0–0	135–0–0 & 370–0–0				
source(s)		AA	UAN			UAN	UAN	UAN				

† Abbreviations: AA, anhydrous ammonia; AN, ammonium nitrate; Conv., Conventional; DAP, diammonium phosphate; NT, no-till; and UAN, 32% urea ammonium nitrate.

Table 2. Foliar B application information at V4–V6 and VT at the 12 site-years [36].

B Application	Novelty			Bethel			Leonard		Bee Ridge		
Information	2008	2009	2010	2008	2009	2010	2008	2009 2010	2008	2009	2010
					V4–V6 †						
Application date	16 June	12 June	17 June	16 July	15 June	19 June	15 June	17 June	20 June	15 June	10 June
Air temperature (°C)	21	22	23	33	24	27	26	33	38	19	27
Relative humidity, %	61	90	92	56	82	65	83	71	45	95	69
Height, cm	25	25	30	36	28	36	27	25	36	33	18
					VT						
Application date	23 July	20 July	23 July	15 August	27 July	28 July	27 July	23 July	28 July	27 July	23 July
Air temperature (°C)	25	23	28	27	29	34	31	37	34	29	33
Relative humidity, %	52	77	33	75	60	60	61	33	60	61	33
Height, cm	183	183	244	305	183	183	183	213	183	183	213
Harvest date	5 November	2 November	27 September	13 November	24 November	28 October	13 November	4 October	19 November	7 December	20 October
Soil type		Kilwinning			Putnam			Arbela	Wabash	Chariton	Wabash

† As described by Abendroth *et al.* [37].

Table 3. Initial soil test information at Novelty, Bethel, Leonard, and Bee Ridge in 2008, 2009, and 2010 [36].

Soil Test Information	Novelty	Bethel	Leonard	Bee Ridge
	2008			
pH (1:1 water)	5.8 ± 0.3 †	6.1 ± 0.1	6.2 ± 0.1	6.3 ± 0.2
Mehlich-1				
P, $kg \cdot ha^{-1}$	30 ± 8	76 ± 21	62 ± 9	89 ± 4
K, $kg \cdot ha^{-1}$	220 ± 43	173 ± 13	176 ± 12	220 ± 4
Ca, $kg \cdot ha^{-1}$	3660 ± 310	3920 ± 140	4110 ± 200	3720 ± 110
Mg, $kg \cdot ha^{-1}$	509 ± 48	336 ± 15	298 ± 24	246 ± 8
B, $kg \cdot ha^{-1}$	0.67 ± 0.08	0.58 ± 0.04	0.47 ± 0.04	0.96 ± 0.09
CEC ‡, $cmol_c kg^{-1}$	14.2 ± 0.9	12.8 ± 0.3	13.4 ± 0.6	11.8 ± 0.4
	2009			
pH (1:1 water)	6.3 ± 0.1	6.3 ± 0.2	7.2 ± 0	7.0 ± 0.2
Mehlich-1				
P, $kg \cdot ha^{-1}$	71 ± 10	37 ± 1	75 ± 8	100 ± 16
K, $kg \cdot ha^{-1}$	387 ± 74	181 ± 10	136 ± 4	163 ± 13
Ca, $kg \cdot ha^{-1}$	4260 ± 140	3610 ± 360	4800 ± 420	4790 ± 620
Mg, $kg \cdot ha^{-1}$	538 ± 24	360 ± 36	286 ± 47	361 ± 6
B, $kg \cdot ha^{-1}$	1.05 ± 0.06	1.03 ± 0.04	1.01 ± 0.13	0.81 ± 0.09
CEC, $cmol_c kg^{-1}$	16.0 ± 0.5	12.9 ± 0.9	14.1 ± 1.4	14.3 ± 1.2
	2010			
pH (1:1 water)	5.7 ± 0.3	7.3 ± 0.2	6.4 ± 0.1	6.7 ± 0.2
Mehlich-1				
P, $kg \cdot ha^{-1}$	28 ± 4	34 ± 2	48 ± 4	92 ± 27
K, $kg \cdot ha^{-1}$	166 ± 16	101 ± 9	137 ± 27	186 ± 47
Ca, $kg \cdot ha^{-1}$	3010 ± 380	4870 ± 340	4870 ± 210	3990 ± 490
Mg, $kg \cdot ha^{-1}$	377 ± 34	367 ± 24	343 ± 27	341 ± 50
B, $kg \cdot ha^{-1}$	0.47 ± 0.27	0.76 ± 0.27	0.43 ± 0.27	0.54 ± 0.27
CEC, $cmol_c kg^{-1}$	12.3 ± 0.6	14.4 ± 0.8	15.4 ± 0.6	12.8 ± 1.1

† Standard deviation of the five replications; ‡ CEC, cation exchange capacity.

All the collected data were analyzed using analysis of variance (ANOVA) in PROC GLM with the SAS statistical computer program [41]. If the overall F was significant, Fisher's Protected Least Significant Difference (LSD) at $p = 0.1$ was used for mean separation. In absence of a significant interaction between B treatments and site-years, data were averaged over the 12 site-years. Pearson correlation analysis (PROC CORR procedure of SAS) was used to determine the relationship between corn yield and other plant measurements.

3. Results and Discussion

The initial soil B concentration ranged from 0.43 $kg \cdot ha^{-1}$ at Leonard in 2010 to 1.05 $kg \cdot ha^{-1}$ at Novelty in 2009 (Table 3). All sites had acidic to neutral pH soils. The soils at all site-years were high in potassium concentration (101 to 387 $kg \cdot ha^{-1}$). The monthly precipitation at Novelty and Bee Ridge in Knox County was higher than the 10-year average precipitation for northeast Missouri except in the month of August in 2010. In 2010, there was 62% and 63% lower monthly precipitation in August at all sites compare to the 10-year average precipitation (Table 4) which made it drier compare to other years. In Shelby County during 2008 and 2010, the average monthly precipitation was 19% and 33% lower in May compared to the 10-year average precipitation. Since rainfall early in the season and during pollination was generally high, the overall availability of B to corn plants should have been readily available during critical portions of plant development. In general, overall corn yields were high for the period this research was conducted.

Table 4. Monthly precipitation average (10-year) for Northeast Missouri during the growing season and at Novelty, Bee Ridge, Bethel, and Leonard in 2008, 2009, and 2010. Novelty and Bee Ridge are in Knox County, and Bethel and Leonard are in Shelby County [36].

Month	Northeast Missouri 10-Year Average †	Knox County (Novelty and Bee Ridge) (mm)			Shelby County (Bethel and Leonard) (mm)		
		2008	2009	2010	2008	2009	2010
April	100	116	121	146	116	120	135
May	113	112	170	160	92	170	76
June	124	257	145	163	133	148	168
July	93	272	108	326	195	79	93
August	122	108	167	45	206	141	46
September	85	201	86	242	315	96	250
Total	637	1066	797	1082	1057	754	768

† Averaged from 2000 to 2009.

The mean corn yield averaged over the 12 site-years and B treatments was 11.7 Mg·ha^{-1}. A significant effect of B treatments, site-years and B treatment × site-year interaction was found for corn yields (Table 5). Boron applied at V4–V6 at 2.24 kg·ha^{-1} was the highest yielding treatment in five of the 12 site-years, and had the highest average yield of 11.96 Mg·ha^{-1} compared to all other treatments (Figure 1). The V4–V6 B application at 2.24 kg·ha^{-1} increased yield 0.29 Mg·ha^{-1} compared to the non-treated control. None of the other B treatments affected yield when averaged over the 12 site-years in this research. The V4–V6 B application at 0.56 and 1.12 kg·ha^{-1} had higher yields than VT B application at the same concentrations, but they were not significantly different. Corn yields from B applications at VT ranged from 11.56 to 11.58 Mg·ha^{-1}. Foliar B applications at VT resulted in 0.24 Mg·ha^{-1} lower yields than V4–V6 B applications. Higher yields due to an early application of B compared to late B application were reported for rice in Missouri [42]. Woodruff *et al.* [12] also reported an increase in corn yields due to B at 2.24 kg·ha^{-1} for soils receiving higher K fertilization (131–317 kg·ha^{-1}). A beneficial effect of early applications of B for grain yields were also reported for wheat, rice and cotton [43]. In contrast to this, no differences in soybean yields were observed due to early or late B applications in Northeast Arkansas [44]. There were no differences in yield due to B application in 2010 at all sites except Bee Ridge (Table 5). At Leonard, greater corn yields were obtained with a V4–V6 B application at 0.56 kg·ha^{-1}, which was significantly different from the lowest yields obtained with a VT application at 0.28 and 0.56 kg·ha^{-1} in 2009 only. The higher total rainfall received in Knox County compared to Shelby County may have resulted in greater B solubility and availability to corn plants [9]. The maximum corn yield obtained was 15.32 Mg·ha^{-1} at Novelty in 2009, which had the highest initial B concentration, of 1.05 kg·ha^{-1}, among all site-years and had sufficient rainfall along with supplemental irrigation (Table 5). However, the lowest yield obtained was 6.46 Mg·ha^{-1} in 2008 at Leonard, which had only 0.47 kg·ha^{-1} of initial B in the soil.

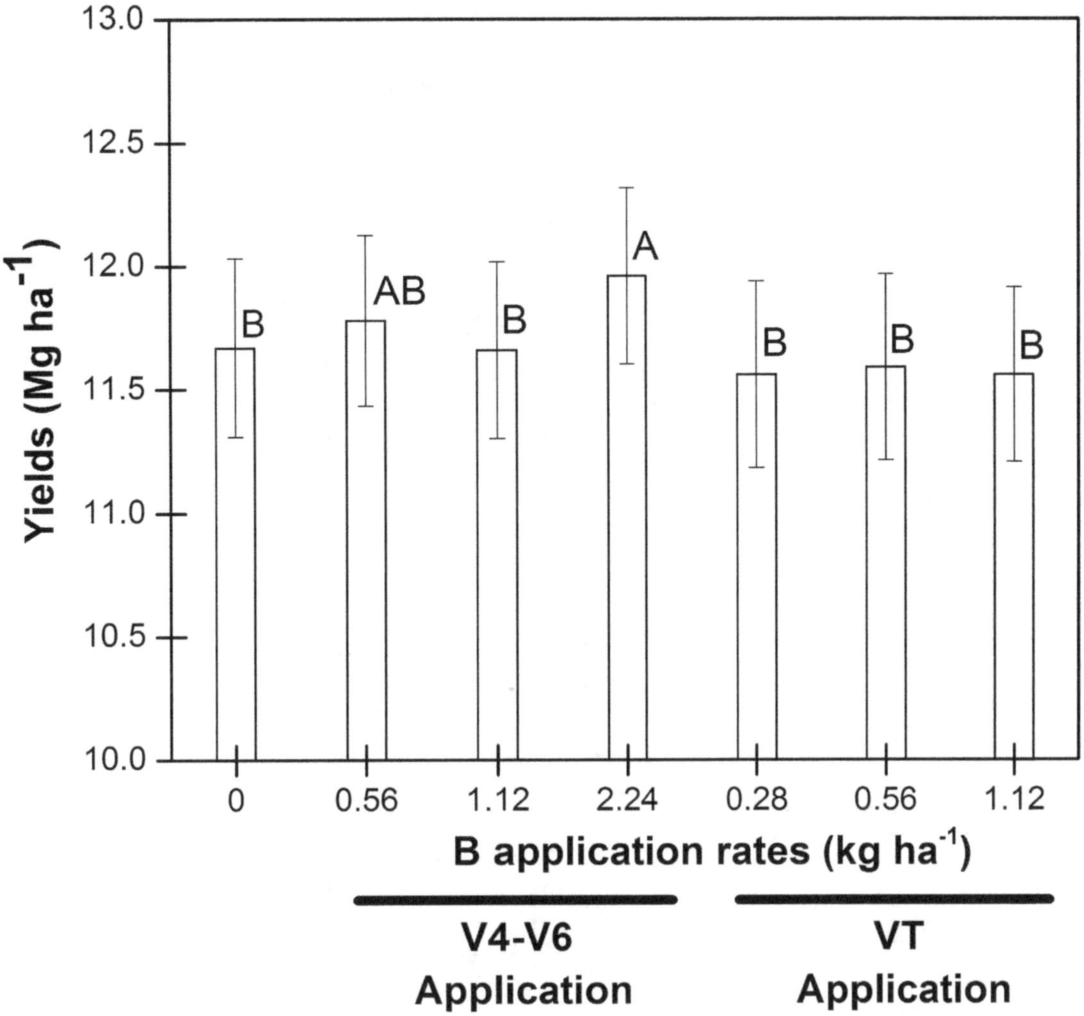

Figure 1. Corn grain yield in response to V4–V6(4–6 leaves with visible collars) and VT (tasseling) foliar B applications at different concentrations. Data were averaged over 12 site-years and the LSD (p = 0.1) was 0.24 Mg·ha^{-1}. Similar letters on bars indicate no significant differences between treatments. Error bars represents standard error.

Table 5. Corn grain yields at 12 site-years for V4–V6 and VT foliar B applications at different concentrations.

Application Timing	Application Concentration ($kg \cdot ha^{-1}$)	Novelty ($mg \cdot ha^{-1}$)			Bethel ($mg \cdot ha^{-1}$)			Bee Ridge ($mg \cdot ha^{-1}$)			Leonard ($mg \cdot ha^{-1}$)		
		2008	2009	2010	2008	2009	2010	2008	2009	2010	2008	2009	2010
Non-treated	0	8.6	14.9	9.3	11.6	10.8	14.1	14.3	14.3	9.4	6.5	12.3	14.0
V4–V6	0.56	8.9	14.6	8.6	11.4	11.3	14.1	13.7	14.3	10.8	6.7	13.0	14.1
V4–V6	1.12	8.8	15.0	8.3	10.6	11.3	14.2	14.0	14.4	9.7	6.9	12.6	14.1
V4–V6	2.24	9.2	15.3	9.7	11.9	11.1	14.2	13.9	14.5	9.7	6.8	12.7	14.4
VT	0.28	8.9	14.9	9.6	11.6	10.6	14.4	14.0	14.0	7.7	6.5	12.1	14.3
VT	0.56	8.7	14.7	7.8	11.7	11.3	14.4	13.7	14.4	9.1	6.9	11.9	14.4
VT	1.12	9.0	14.0	8.2	11.2	10.8	14.6	13.6	14.3	9.1	7.0	12.6	14.4
LSD ($p = 0.1$)		0.5	0.9	ns †	1.2	ns	ns	ns	0.5	1.8	ns	0.9	ns

† ns: not significant.

Table 6. Whole plant tissue B concentrations 10 days after V4–V6 B application at different concentrations.

Application Rate ($kg \cdot ha^{-1}$)	Tissue B Concentration ‡	Novelty ($mg \cdot kg^{-1}$)			Bethel ($mg \cdot kg^{-1}$)			Bee Ridge ($mg \cdot kg^{-1}$)			Leonard ($mg \cdot kg^{-1}$)		
		2008	2009	2010	2008	2009	2010	2008	2009	2010	2008	2009	2010
Non treated	5.0	5.4	5.2	5.2	6.0	4.8	4.8	6.0	4.0	5.2	4.6	5.0	4.0
0.56	5.5	5.4	5.4	5.4	8.2	5.0	5.0	5.6	4.2	6.8	5.2	5.4	4.2
1.12	5.8	6.2	5.2	5.5	9.6	5.8	4.8	5.8	4.2	6.4	5.8	5.4	5.2
2.24	7.3	6.6	6.2	5.6	14.4	7.4	5.4	10.2	4.4	7.6	8.4	5.6	5.2
LSD ($p = 0.1$)	0.4	0.4	0.8	ns †	2.0	2.2	ns	4.1	ns	1.8	1.2	ns	0.7

† ns: not significant; ‡ Tissue B concentrations were averaged over 12 site-years.

The initial B concentration in plant tissue ranged from 3.8 to 6.8 mg·kg^{-1}. There were no significant differences in tissue B concentration between B treatments from plant samples taken before the V4–V6 application of B among all site-years except at Bee Ridge in 2008 and Leonard in 2009 (data not presented). This might be due to higher initial soil B concentration at Bee Ridge in 2008 and Leonard in 2009, which was 0.96 and 1.01 kg·ha^{-1}, respectively. At Bee Ridge in 2008, treatments containing B applied at V4–V6 at 0.56 kg·ha^{-1} had 0.8 mg·kg^{-1} greater B concentration than treatments having B applied at 2.24 kg·ha^{-1}. In 2009 at Leonard, both of these treatments showed no differences in tissue B concentration, but a V4–V6 B application at 0.56 kg·ha^{-1} had 0.8 mg·kg^{-1} lower tissue B concentration than B application at 1.12 kg·ha^{-1} (data not presented). Tissue B concentration from samples taken 10 days after B application at V4–V6 showed differences among treatments and an interaction between treatments and site-years was observed (Table 6). Foliar B applied at V4–V6 at 2.24 kg·ha^{-1} had 2.3 mg·kg^{-1} higher tissue B concentration when averaged over the 12 site-years compared to the non-treated control. A V4–V6 B application at 0.56 and 1.12 kg·ha^{-1} had 0.5 to 0.8 mg·kg^{-1} greater tissue B concentration than the non-treated control, but 1.5 to 1.8 mg·kg^{-1} lower tissue B concentration than the treatment having a V4–V6 B application at 2.24 kg·ha^{-1}. No differences were observed between treatments at Novelty and Bethel in 2010 and at Bee Ridge and Leonard in 2009. In the remaining site-years, a V4–V6 B application at 2.24 kg·ha^{-1} had higher B concentrations compared to the non-treated control. A V4–V6 application of B increased B concentration in the plant tissue among all site-years as the rate of B increased.

Boron concentration in plant tissue samples taken at VT before VT B application did not show any differences at the 12 site-years (data not presented). The ear leaf B concentration averaged over site-years ranged from 3.71 to 8.42 mg·kg^{-1} among treatments from plant samples taken 10 days after a foliar VT application (Figure 2). The treatments having a V4–V6 application at 0.56 and 1.12 kg·ha^{-1} had similar ear leaf B concentrations as the non-treated control plots; while V4–V6 applied B at 2.24 kg·ha^{-1} had a 0.37 mg·kg^{-1} higher ear leaf B concentration compared to other B application concentrations at V4–V6. All VT B application treatments resulted in significantly higher ear leaf B concentration (5.2 to 8.41 mg·kg^{-1}) compared to non-treated (3.7 mg·kg^{-1}) and V4–V6 treatments. Ear leaf B concentration increased with increasing concentrations of foliar B application from 0.28 kg·ha^{-1} to 1.12 kg·ha^{-1} at VT. A foliar VT B application at 1.12 kg·ha^{-1} had 4.7 mg·kg^{-1} greater ear leaf B concentration compared to the non-treated control. A significant interaction was observed between ear leaf B concentration and site-years (Table 7). Among all site-years, a VT B application at 1.12 kg·ha^{-1} had higher ear leaf B concentrations compared to the other B treatments including the non-treated control. An increase in leaf B concentration with increasing B application concentrations were also reported in soybean in Northeast Arkansas [45].

The plant population varied from 69,500 to 71,500 plants·ha^{-1} over the 12 site-years. There was no effect of B treatments on plant population, but there were differences in plant population among site-years (Table 8). Similar results were obtained by Nelson *et al.* [36] for B applications with fungicides in corn. The highest plant population of 86,064 plants·ha^{-1} was found at Leonard in 2010 and lowest population of 61,132 plants·ha^{-1} was found at Novelty in 2009 (Table 8). Woodruff *et al.* [12] reported increased corn yields with B with plant populations more than 50,000 plants·ha^{-1}. Oil, protein, starch and extractable starch content of grains were not affected by the B treatments and no significant interaction of B treatments and site-years were found. Grain protein, oil, starch and extractable starch concentrations varied by site-years (Table 8). The differences between site-years might be because of different environmental conditions, management systems, soil properties and hybrid differences. No differences in protein and oil concentration in corn grains due to B at 0.28, 0.56 and 1.12 kg·ha^{-1} were also observed in Georgia [46]; however, B at 0.45 kg·ha^{-1} increased seed oil and protein concentration in soybean [47]. Boron alone increased corn grain protein and starch concentration by 0.7% and 3.7%, respectively, in wheat [48]. Corn yields were negatively correlated with ear leaf B concentration, plant population, grain starch and extractable starch concentration (Table 9). There was positive correlation of corn yields with grain protein concentration. However, a highly positive correlation was reported for seed yield and leaf B content of *Vigna radiata* [5].

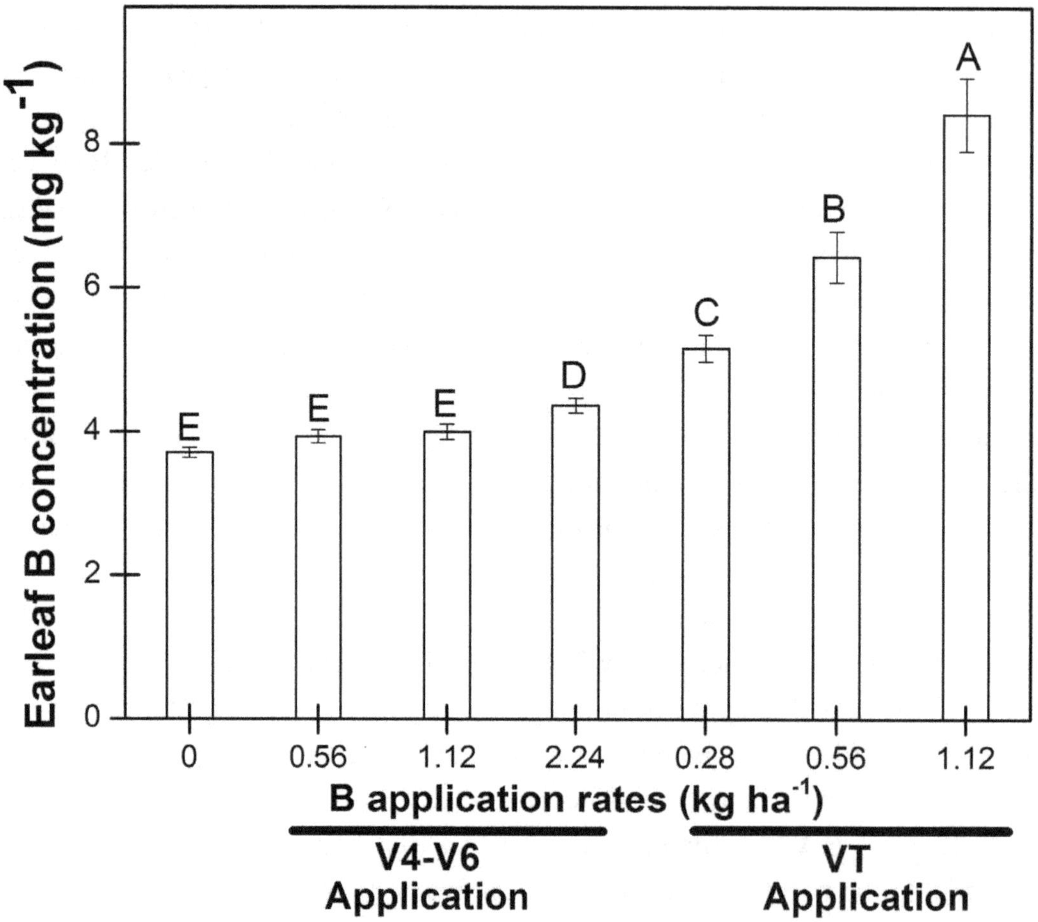

Figure 2. Ear leaf B concentration in response to V4–V6 (4–6 leaves with visible collars) and VT (tasseling) foliar B application 10 days after VT application. Data were averaged over 12 site-years and the LSD ($p = 0.1$) was 0.3 mg·kg^{-1}. Similar letters on bars indicate no significant differences between treatments. Error bars represents standard error.

Table 7. Ear leaf B concentration ten days after VT at the 12 site-years for V4–V6 and VT foliar B applications.

Application Timing	Application Concentration (kg·ha^{-1})	Novelty (mg·kg^{-1})			Bethel (mg·kg^{-1})			Bee Ridge (mg·kg^{-1})			Leonard (mg·kg^{-1})		
		2008	2009	2010	2008	2009	2010	2008	2009	2010	2008	2009	2010
Non-treated	0	3.8	3.8	4.2	3.6	4.0	4.4	3.8	3.2	3.0	3.0	3.9	3.8
V4–V6	0.56	3.8	4.2	4.4	4.0	4.4	4.6	4.0	3.8	3.0	3.6	3.6	3.8
V4–V6	1.12	4.0	3.8	4.0	4.2	4.6	5.0	3.8	3.8	3.0	3.8	3.8	4.0
V4–V6	2.24	4.2	4.2	4.8	4.4	5.4	5.0	4.2	3.8	3.2	4.4	4.6	4.4
VT	0.28	4.8	4.8	6.2	4.4	6.2	4.6	4.2	5.0	3.4	6.2	8.0	4.2
VT	0.56	5.8	5.8	5.2	5.8	7.8	4.8	5.4	6.8	3.2	12.2	10.4	4.0
VT	1.12	8.2	7.2	6.0	8.2	10.4	5.6	8.0	7.4	4.0	14.0	16.4	5.6
LSD ($p = 0.1$)		1.3	0.9	1.1	1.6	1.5	0.9	1.4	0.9	0.5	1.9	2.1	0.7

Table 8. Oil, protein, starch and extractable starch concentration in corn grains, ear tip fill, and common rust severity at 12 site-years. Data were averaged over treatments at each site-year.

Grain Components	Novelty			Bethel			Bee Ridge			Leonard			LSD ($p = 0.1$)
	2008	2009	2010	2008	2009	2010	2008	2009	2010	2008	2009	2010	
Oil (g·kg^{-1})	46	42	44	43	43	42	42	37	34	42	55	45	3
Protein (g·kg^{-1})	66	85	67	71	90	75	85	84	74	62	84	89	1
Starch (g·kg^{-1})	739	720	739	714	719	723	740	734	734	745	702	716	2
Extractable starch (g·kg^{-1})	702	675	679	693	649	690	688	687	703	709	653	639	2
Ear tip fill (mm)	10	4	29	2	2	28	1	7	25	7	24	18	2
Common rust	- ‡	3	10	-	12	11	-	6	33	-	5	27	1
Plant population (plants·ha^{-1})	76,041	61,132	64,958	69,679	65,758	65,745	66,708	72,221	69,654	69,782	77,822	86,063	2169

‡ Common rust measurements were not taken at all sites in 2008.

Table 9. Correlation analysis between corn grain yields and corn response measurements.

Measurement	Yield
Plant population	−0.13
$p > \|r\|$	0.0059
Grain protein	0.74
$p > \|r\|$	<0.0001
Grain starch content	−0.47
$p > \|r\|$	<0.0001
Grain extractable starch	−0.44
$p > \|r\|$	<0.0001
TipFill	−0.13
$p > \|r\|$	0.0075
Common Rust	−0.21394
$p > \|r\|$	0.0003

No toxicity symptoms and barren stalks were observed due to foliar B treatments at all 12 site-years (data not presented). Production of barren ears in corn was reported for B levels less than 0.05 ppm [49]. Ear tip fill was used to evaluate the effect of B treatments on ear filling. A higher number for ear tip fill indicated a greater amount of ear was barren. Although ear tip fill did not show differences between B treatments, it varied by site-years (Table 8). The ear tip fill was greater in 2010 than in 2008 and 2009.

The severity of disease varied by sites and years, and the impact of B applications were generally subtle. The occurrence of common rust was not significantly different between B treatments, but it varied by site-year. Common rust severity was 22% higher at Bee Ridge in 2010 compared to Leonard. Ear tip fill and common rust were negatively correlated with the corn yields (Table 9). In 2010, corn was rated for the severity of anthracnose stalk rot at Bethel and Leonard. Severity of anthracnose stalk rot at Bethel and Leonard in 2010 was not affected by B treatments (data not presented). At Leonard in 2010, severity of northern corn leaf blight was affected by the timing of foliar B application, but not by the amount of B applied (Table 10). A reduction in the severity of common rust, gray leaf spot, and northern corn leaf blight in corn due to pyraclostrobin, not B was found by some other researchers [36,50]. The V4–V6 B application at 2.24 $kg \cdot ha^{-1}$ and all VT applications increased the severity of northern corn leaf blight compared to the non-treated control. The severity of gray leaf spot was 6% and 12% lower with V4–V6 applied B at 1.12 and 2.24 $kg \cdot ha^{-1}$ and VT applied B at 0.56 $kg \cdot ha^{-1}$, respectively, compared to other B treatments including non-treated control (Table 10). Boron applied at V4–V6 at 2.24 $kg \cdot ha^{-1}$ decreased gray leaf spot severity, but the severity of northern leaf blight increased compared to non-treated control. Reductions in disease severity due to B have been reported by other researchers [35,36], but may be inconsistent depending on the disease, environmental conditions, and hybrids. Frequent rains along with cool and cloudy weather has favored fungal disease development, in addition to nutrient deficiencies and toxicities [32].

Table 10. Effect of B application timing and concentrations on severity of grey leaf spot and northern corn leaf blight. Data were averaged over 6 site-years for grey leaf spot. Severity of northern corn leaf spot was measured in 2010 at Leonard.

Application Timing	Application Concentration ($kg \cdot ha^{-1}$)	Grey Leaf Spot (%)	Northern Corn Leaf Blight (%)
Non-treated	0	16	7
V4–V6	0.56	16	8
V4–V6	1.12	15	8
V4–V6	2.24	15	9
VT	0.28	16	11
VT	0.56	14	9
VT	1.12	16	9
LSD ($p = 0.1$)		1	2

4. Conclusions

A B application along with recommended NPK fertilizers affected corn grain yields and severity of diseases. Boron applied at V4–V6 and 2.24 $kg \cdot ha^{-1}$ was more beneficial than VT applications at 0.28, 0.56 and 1.12 $kg \cdot ha^{-1}$ for high yield production systems even though VT B applications resulted in higher tissue B concentrations in the ear leaf. The ear leaf B concentrations were not positively correlated with corn yield. No significant differences in yield were found for B applications at different timings for same concentration of 0.56 and 1.12 $kg \cdot ha^{-1}$. A V4–V6 application of B at 2.24 $kg \cdot ha^{-1}$ had the greatest average corn yields. The V4–V6 foliar B application concentrations up to 2.24 $kg \cdot ha^{-1}$ were not toxic to corn plants and increased corn yields. A higher concentration of B for VT application was not included in this study. Since B is needed by corn plants throughout the growing period, foliar application of B at earlier growth stages (V4–V6) was more beneficial for high yields. Boron showed no significant effect on plant populations, grain oil, protein, starch, or extractable starch concentration. Boron application concentrations up to 2.24 $kg \cdot ha^{-1}$ did not cause any visual injury to crop plants. The B applications at V4–V6 had a slight decrease severity of diseases including gray leaf spot and leaf blight compared to foliar VT applications. The V4–V6 B application at 2.24 $kg \cdot ha^{-1}$ was the best option for B fertilization in corn on fine-textured soils with low soil test B.

Acknowledgments

The authors would like to thank Clinton Meinhardt and Randall Smoot for their technical assistance with this research.

Author Contributions

Gurpreet Kaur was responsible for the interpretation of results and manuscript preparation. Kelly Nelson was responsible for planning, design, site selection, coordination of the field research, data collection, interpretation of results, and editing of the manuscript.

Conflicts of Interest

The authors declare no conflict of interest.

References

1. Ahmad, W.; Zia, M.H.; Malhi, S.S.; Niaz, A.; Saifullah. Boron Deficiency in Soils and Crops: A Review. Available online: http://www.intechopen.com/books/crop-plant/boron-deficiency-in-soils-and-crops-a-review (accessed on 4 June 2014).
2. Gupta, U.C. *Boron and Its Role in Crop Production*; CRC Press: Boca Raton, FL, USA, 1993.
3. Marschner, H.; Rimmington, G. *Mineral Nutrition of Higher Plants*; Academic Press: London, UK, 1996.
4. Pilbeam, D.; Kirkby, E. The physiological role of boron in plants. *J. Plant Nutr.* **1983**, *6*, 563–582.
5. Padbhushan, R.; Kumar, D. Influence of soil and foliar applied boron on green gram in calcareous soils. *Int. J. Agric. Environ. Biotechnol.* **2014**, *7*, 129–136.
6. Berger, K.C. Micronutrient shortages, micronutrient deficiencies in the United States. *J. Agric. Food Chem.* **1962**, *10*, 178–181.
7. Sparr, M. Micronutrient needs—which, where, on what—in the United States. *Commun. Soil Sci. Plant Anal.* **1970**, *1*, 241–262.
8. Mengel, K.; Kosegarten, H.; Kirkby, E.A.; Appel, T. *Principles of Plant Nutrition*; Springer Science and Business Media: Dordrecht, The Netherlands, 2001.
9. Goldberg, S. Reactions of boron with soils. *Plant Soil* **1997**, *193*, 35–48.
10. Nable, R.O.; Bañuelos, G.S.; Paull, J.G. Boron toxicity. *Plant Soil* **1997**, *193*, 181–198.
11. Shorrocks, V.; Blaza, A. Boron nutrition of maize. *Field Crop* **1973**, *25*, 25–27.
12. Woodruff, J.; Moore, F.; Musen, H. Potassium, boron, nitrogen, and lime effects on corn yield and earleaf nutrient concentrations. *Agron. J.* **1987**, *79*, 520–524.
13. Vitosh, M.; Johnson, J.; Mengel, D. Tri-state fertilizer recommendations for corn, soybeans, wheat and alfalfa. Avaiable online: https://www.google.com/url?sa=t&rct=j&q=&esrc=s&source=web&cd=1&cad=rja&uact=8&ved=0CCAQFjAA&url=https%3A%2F%2Fwww.extension.purdue.edu%2Fextmedia%2FAY%2FAY-9-32.pdf&ei=sEaYVJfPD8b4yQTOmoGgAw&usg=AFQjCNHDHk60GAqRaXSbx_Ikjzrwunz7wA&sig2=NFd1S-kC6iGiABAL4_zQKw (accessed on 5 July 2014).
14. Jahiruddin, M.; Harada, H.; Hatanaka, T.; Sunaga, Y. Adding boron and zinc to soil for improvement of fodder value of soybean and corn. *Commun. Soil Sci. Plant Anal.* **2001**, *32*, 2943–2951.
15. Hosseini, S.; Maftoun, M.; Karimian, N.; Ronaghi, A.; Emam, Y. Effect of zinc × boron interaction on plant growth and tissue nutrient concentration of corn. *J. Plant Nutr.* **2007**, *30*, 773–781.
16. Borges, I.D.; von Pinho, R.G.; de Andrade Pereira, J.L. Micronutrients accumulation at different maize development stages. *Ciênc. Agrotecnol.* **2009**, *33*, 1018–1025.
17. Günes, A.; Alpaslan, M. Boron uptake and toxicity in maize genotypes in relation to boron and phosphorus supply. *J. Plant Nutr.* **2000**, *23*, 541–550.

18. Kanwal, S.; Rahmatullah; Aziz, T.; Maqsood, M.A.; Abbas, N. Critical ratio of calcium and boron in maize shoot for optimum growth. *J. Plant Nutr.* **2008**, *31*, 1535–1542.
19. Aref, F. The effect of boron and zinc application on concentration and uptake of nitrogen, phosphorous and potassium in corn grain. *Indian J. Sci. Technol.* **2011**, *4*, 785–791.
20. Asad, A.; Blamey, F.; Edwards, D. Effects of boron foliar applications on vegetative and reproductive growth of sunflower. *Ann. Bot.* **2003**, *92*, 565–570.
21. Perveen, S. Effect of foliar application of zinc, manganese and boron in combination with urea on the yield of sweet orange. *Pak. J. Agric. Res.* **2000**, *16*, 135–141.
22. Rimar, J.; Balla, P.; Princik, L. The comparison of application effectiveness of liquid fertilizers with those in solid state in conditions of the east slovak lowland region. *Rostl. Vyroba* **1996**, *42*, 127–132.
23. Fernández, V.; Sotiropoulos, T.; Brown, P.H. *Foliar Fertilisation: Principles and Practices*; International Fertiliser Industry Association (IFA): Paris, France, 2013; p. 140.
24. Peterson, J.; MacGregor, J. Boron fertilization of corn in minnesota. *Agron. J.* **1966**, *58*, 141–142.
25. Touchton, J.; Boswell, F. Boron application for corn grown on selected southeastern soils. *Agron. J.* **1975**, *67*, 197–200.
26. Ben-Gal, A. The contribution of foliar exposure to boron toxicity. *J. Plant Nutr.* **2007**, *30*, 1705–1716.
27. Schon, M.K.; Blevins, D.G. Foliar boron applications increase the final number of branches and pods on branches of field-grown soybeans. *Plant Physiol.* **1990**, *92*, 602–607.
28. Palta, C.; Karadavut, U. Shoot growth curve analysis of maize cultvars under boron deficiency. *J. Anim. Plant Sci.* **2011**, *21*, 696–699.
29. Bingham, F.T.; Garber, M. Zonal salinization of the root system with nacl and boron in relation to growth and water uptake of corn plants. *Soil Sci. Soc. Am. J.* **1970**, *34*, 122–126.
30. Donald, C.; Porter, I. Integrated control of clubroot. *J. Plant Growth Regul.* **2009**, *28*, 289–303.
31. Dordas, C. Role of nutrients in controlling plant diseases in sustainable agriculture: A review. *Agron. Sustainable Dev.* **2008**, *28*, 33–46.
32. Simoglou, K.B.; Dordas, C. Effect of foliar applied boron, manganese and zinc on tan spot in winter durum wheat. *Crop Prot.* **2006**, *25*, 657–663.
33. Thomidis, T.; Exadaktylou, E. Effect of boron on the development of brown rot (*Monilinia laxa*) on peaches. *Crop Prot.* **2010**, *29*, 572–576.
34. Liew, Y.; Husni, M.; Zainal, A.; Ashikin, N. Effects of foliar applied copper and boron on fungal diseases and rice yield on cultivar MR219. *Pertanika J. Trop. Agric. Sci.* **2012**, *35*, 339–349.
35. Hassegawa, R.H.; Fonseca, H.; Fancelli, A.L.; da Silva, V.N.; Schammass, E.A.; Reis, T.A.; Corrêa, B. Influence of macro-and micronutrient fertilization on fungal contamination and fumonisin production in corn grains. *Food Control* **2008**, *19*, 36–43.
36. Nelson, K.A.; Meinhardt, C. Foliar boron and pyraclostrobin effects on corn yield. *Agron. J.* **2011**, *103*, 1352–1358.
37. Abendroth, L.J.; Elmore, R.W.; Boyer, M.J.; Marlay, S.K. Corn Growth and Development. Available online: https://store.extension.iastate.edu/Product/Corn-Growth-and-Development (accessed on 20 June 2014).

38. Kelling, K.A. Soil and Applied Boron. Available online: http://www.google.com/url?sa=t&rct=j&q=&esrc=s&source=web&cd=1&cad=rja&uact=8&ved=0CCAQFjAA&url=http%3A%2F%2Fwww.soils.wisc.edu%2Fextension%2Fpubs%2FA2522.pdf&ei=Yj-YVLz8GMqLyATDkYGoAw&usg=AFQjCNH04QxJuklbBR2XOZXBq6lu1crNEA&sig2=QKXA61bkYxio5VikEW-ToQ (accessed on 20 June 2014)
39. Rehm, G.W.; Fenster, W.E.; Overdahl, C.J. Boron for Minnesota Soils. Available online: http://www.extension.umn.edu/distribution/cropsystems/DC0723.html (accessed on 29 December 2008).
40. Henggler, J. Woodruff Irrigation Charts. Available online: http://agebb.missouri.edu/irrigate/woodruff/ (accessed on 8 February 2008).
41. SAS Institute. *SAS User's Guide*; SAS Institute: Cary, NC, USA, 2010.
42. David, D.; Gene, S.; Andy, K. Boron fertilization of rice with soil and foliar applications. *Crop Manag.* **2005**, doi:10.1094/CM-2005-0210-01-RS.
43. Ahmad, R.; Irshad, M. Effect of boron application time on yield of wheat, rice and cotton crop in Pakistan. *Soil Environ.* **2011**, *30*, 50–57.
44. Slaton, N.; DeLong, R.; Thompson, R. Irrigated Soybean Yield Response to Boron Application Time and Rate. Available online: www.uark.edu/depts/agripub/publications (accessed on 20 June 2014).
45. Ross, J.R.; Slaton, N.A.; Brye, K.R.; DeLong, R.E. Boron fertilization influences on soybean yield and leaf and seed boron concentrations. *Agron. J.* **2006**, *98*, 198–205.
46. Jellum, M.D.; Boswell, F.; Young, C.T. Nitrogen and boron effects on protein and oil of corn grain. *Agron. J.* **1973**, *65*, 330–331.
47. Bellaloui, N. Effect of water stress and foliar boron application on seed protein, oil, fatty acids, and nitrogen metabolism in soybean. *Am. J. Plant Sci.* **2011**, *2*, 692–701.
48. Dani, H.; Saini, H.; Saini, S.; Sareen, K. Effect of boron on starch and protein contents of wheat grains. *Curr. Sci.* **1970**, *39*, 235–236.
49. Berger, K.; Heikkinen, T.; Zube, E. Boron deficiency, a cause of blank stalks and barren ears in corn. *Soil Sci. Soc. Am. J.* **1957**, *21*, 629–632.
50. Bradley, C.; Ames, K. Effect of foliar fungicides on corn with simulated hail damage. *Plant Dis.* **2010**, *94*, 83–86.

12

Screening for Sugarcane Brown Rust in the First Clonal Stage of the Canal Point Sugarcane Breeding Program

Duli Zhao [1,*], R. Wayne Davidson [2], Miguel Baltazar [2], Jack C. Comstock [1], Per McCord [1] and Sushma Sood [1]

[1] USDA-Agricultural Research Service, Sugarcane Field Station, Canal Point, FL 33438, USA;
E-Mails: jack.comstock@ars.usda.gov (J.C.C); per.mccord@ars.usda.gov (P.M.);
Sushma.sood@ars.usda.gov (S.S.)

[2] Florida Sugar Cane League, Inc., Clewiston, FL 33440, USA;
E-Mails: wayne.davidson@ars.usda.gov (R.W.D); miguel.baltazar@ars.usda.gov (M.B)

* Author to whom correspondence should be addressed; E-Mail: duli.zhao@ars.usda.gov

Academic Editor: Diego Rubiales

Abstract: Sugarcane (*Saccharum* spp.) brown rust (caused by *Puccinia melanocephala* Syd. & P. Syd.) was first reported in the United States in 1978 and is still one of the great challenges for sugarcane production. A better understanding of sugarcane genotypic variation in response to brown rust will help optimize breeding and selection strategies for disease resistance. Brown rust ratings were scaled from non-infection (0) to severe infection (4) with intervals of 0.5 and routinely recorded for genotypes in the first clonal selection stage of the Canal Point sugarcane breeding program in Florida. Data were collected from 14,272 and 12,661 genotypes and replicated check cultivars in 2012 and 2013, respectively. Mean rust rating, % infection, and severity in each family and progeny of female parent were determined, and their coefficients of variation (CV) within and among families (females) were estimated. Considerable variation exists in rust ratings among families or females. The families and female parents with high susceptibility or resistance to brown rust were identified and ranked. The findings of this study can help scientists to evaluate sugarcane crosses and parents for brown rust disease, to use desirable parents for crossing, and to improve genetic resistance to brown rust in breeding programs.

Keywords: sugarcane; Canal Point (CP) sugarcane breeding program; brown rust; screening for rust resistance

1. Introduction

Sugarcane (a complex hybrid of *Saccharum* spp.) is an important cash crop in Florida, USA with an annual economic impact of more than $677 million [1]. Consistent and continuous development of high-yielding sugarcane cultivars with resistance or tolerance to biotic and abiotic stresses is critical for commercial sugarcane production in South Florida [2]. The USDA-ARS Sugarcane Field Station at Canal Point (26.52° N; 80.36° W), Florida was initially established at its present site in 1920 to make sugarcane crosses and produce true seed for the Louisiana sugarcane industry. Since the 1960s, the Canal Point (CP) station has been developing sugarcane cultivars with CP prefixes for Florida under a three-party cooperative agreement among the USDA-ARS, the University of Florida, and the Florida Sugar Cane League, Inc. The CP Sugarcane Field Station also makes crosses for the USDA-ARS in Houma, Louisiana. The CP cultivars now account for more than 90% of the hectarage in Florida. In 2014, the top six major sugarcane cultivars grown in Florida were "CP 96-1252" [3], "CP 00-1101" [4],"CP 01-1372" [5], "CL 88-4730" (a cultivar of the United States Sugar Corporation), "CP 88-1762" [6], and "CP 89-2143" [7], and their percentage hectares were 16.8, 13.2, 11.1, 8.8, 8.8, and 7.7%, respectively [8].

The CP sugarcane breeding and cultivar development program (CP program) consists of six stages, namely Crossing, Seedlings, and Stages I, II, III, and IV [9]. It takes at least eight years to release a cultivar from the time a cross is made [10]. Cane yield (TCH, Mg ha^{-1}), commercial recoverable sucrose (CRS, kg sucrose Mg^{-1} cane), and sucrose yield (TSH, Mg ha^{-1}) with disease resistance are the major agronomic traits considered in advancing sugarcane clones during the selection stages. Edmé *et al.* [11] reported that CRS, TCH, and TSH of the Florida commercial sugarcane cultivars linearly increased by 26.0, 15.5, and 47.0%, respectively, over a 33-year period from 1968 to 2000. Underscoring the critical need for cultivar development for the Florida sugarcane industry, about 69% of the sugar yield gain in Florida was from genetic improvement attributable to the CP program [11], indicating importance of the CP program in sustaining sugarcane production in Florida.

Brown rust (Figure 1), caused by *Puccinia melanocephala* Syd. & P. Syd., is an important disease of sugarcane in many production areas around the world [12]. Sugarcane brown rust was first recorded in the United States in 1978. It can cause substantial losses of sugarcane growth, yields, and profits [13–17] and has been responsible for the withdrawal of commercial cultivars in the United States and in most sugarcane growing regions of the world [18]. Although the negative effects of brown rust on sugarcane growth and yield have been reduced by breeding and management practices, the disease is still one of most important concerns in Florida sugarcane production (Figure 1).

Currently, leaf rust diseases, including brown rust and orange rust (caused by *P. kuehnii* (W. Krüger) E.J. Butler), are great challenges for sustainable sugarcane production in Florida [19]. Most dominant commercial cultivars in Florida are susceptible to one or both rusts (Figure 1). The typical symptoms of sugarcane brown rust and orange rust are brown and orange colors for their spores, respectively, as their names imply. Pustules of brown rust are easier to distinguish from those of orange rust on the younger, upper leaves than on the older, lower leaves [20]. Numerous coalescing pustules of rusts cause premature tissue death on leaves. Growers use fungicides to control the negative effects of rusts on yields, but the cost of fungicide applications considerably reduce the production profits [21]. Therefore, development of rust resistant cultivars is the first priority for sustainable sugarcane production. Scientists in the CP program and in the sugarcane industry in Florida are using multiple approaches to develop new cultivars with rust resistance/tolerance and high yields. Cultivars developed in the CP program are used not only in the United States, but also in many other countries in Central America [22] and Asia where sugarcane industries use them for either breeding or commercial production. Therefore, evaluation and screening genotypes for rust resistance in the CP program are important for sustainable sugarcane production in the USA and other countries. Although Sood *et al.* [18] developed a whorl inoculation method to more accurately and efficiently test sugarcane genotype resistance in brown rust, it is still difficult to use the artificial inoculation test in the first clonal stage (Stage I) of the CP program because of a large number (12,000–15,000) of genotypes in this stage and because of limited resources [9]. Therefore, natural infection has been the primary means of assessing rust resistance in the Stage-I clones and further artificial inoculation tests for rust resistance are usually used in later stages (Stages II and III) of the CP program [23].

Figure 1. (**A**) A photo taken on 18 March 2015 in commercial sugarcane fields with a brown rust susceptible cultivar (left) and a resistance cultivar (right) and (**B**) brown rust pustules on leaf abaxial surface of the susceptible sugarcane.

The selection of which sugarcane clones to be used as crossing parents is a critical decision for breeders. Knowledge and better understanding of variability in rust infection and severity among genotypes may provide useful information for sugarcane genotype advancement and for efficient use of parents in future crossing efforts for rust resistance. A study was conducted in the Stage-I fields of the CP program at the USDA-ARS Sugarcane Field Station, Canal Point, Florida. The objectives of this study were to determine variability in brown rust rating among crosses (families) based on data collected from the Stage-I clones of the CP program in 2012 and 2013 and to use the information in cross appraisal. Overall, the rust diseases (both brown and orange rusts) in the 2012 and 2013 sugarcane growing seasons in the area were the most severe in the last seven years due to favorable environment conditions for the rusts. The data collected from these two years under natural infection conditions could partly reflect genotype variation in brown rust resistance or susceptibility. In this paper, we mainly focused on sugarcane brown rust because findings of sugarcane orange rust have been reported in a current article [19].

2. Results and Discussion

2.1. Brown Rust of Check Cultivars

There were significant differences among the four check cultivars and between years in mean rating of brown rust, and the cultivar × year interaction was also highly significant ($p < 0.0001$) (Table 1). Except for CP 80-1743, the other three cultivars had higher mean ratings of brown rust in 2012 than in 2013. In 2012, CP 78-1628 had a significantly higher mean rating of brown rust than other cultivars. In 2013, CP 78-1628 and CP 80-1743 had higher mean brown rust ratings than CP 88-1762 and CP 89-2143. The differences in % infection of brown rust among cultivars or between years were similar to those in the mean rust rating. On the other hand, the differences in rust severity were relatively less among cultivars or between years compared with % infection (Table 1). These results are consistent to those of orange rust in sugarcane [19]. Formal statistical tests of brown rust severity were not available because CP 89-2143 and CP 88-1762 had either no plots or only one to two lots were infected.

2.2. Distribution of Brown Rust Rating

Total numbers of clones used for rust rating in Stage I of the CP program were 14,272 in 2012 and 12,661 in 2013 (Figure 2). These did not include 158 and 166 replicated check plots in 2012 and 2013, respectively. The % infections of brown rust were 34.8% in 2012 and 38.5% in 2013. Distributions of brown rust ratings based on the number of clones at each rating level are given in Figure 2. Although the peak frequency of the rust rating distribution was 2 and the overall severity value was also approximately 2 in both years, the peak value of brown rust in 2012 was greater than that in 2013. The coefficients of variation (CVs) of brown rust ratings across the infected clones in 2012 (4,969) and 2013 (4,880) were 21.2% and 28.7%, respectively. Brown rust % infection and mean rating were similar between years. The differences in severity between years were small (Figure 2).

Table 1. Mean rating, % infection, and severity of brown rust for four check cultivars tested with sugarcane clones in the Stage-I field of the Canal Point (CP) sugarcane breeding and cultivar development program in 2012 and 2013.

Year	Cultivar	# of plots	Mean rating	% infection	Severity
	CP 78-1628	40	1.55 a[†]	77.5 a	2.00
	CP 80-1743	40	0.05 b	2.5 b	2.00
2012	CP 88-1762	40	0.14 b	7.5 b	1.83
	CP 89-2143	38	0.08 b	5.3 b	1.50
	Mean	40	0.46	23.2	1.83
	CP 78-1628	40	0.34 a	25.0 a	1.35
	CP 80-1743	40	0.43 a	20.0 a	2.13
2013	CP 88-1762	43	0.05 b	2.3 b	2.00
	CP 89-2143	43	0.00 b	0.0 b	0.00
	Mean	42	0.21	11.8	1.37

[†]Data following the same letter within a year are not significant at $p = 0.05$.

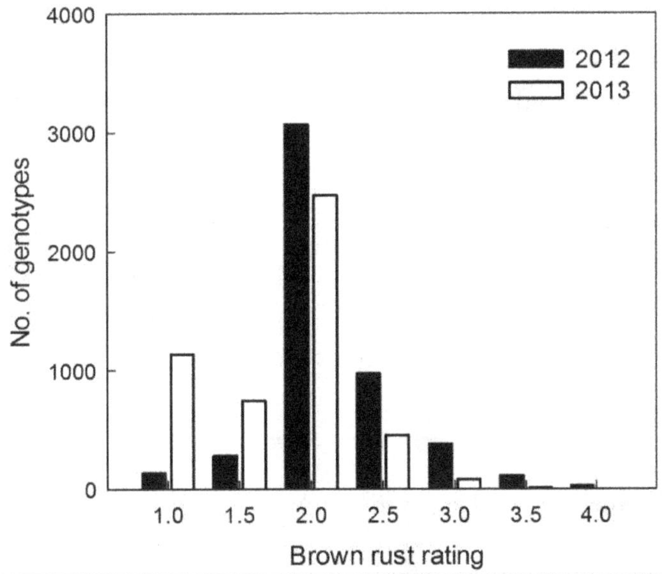

Parameter	2012	2013
Total clones planted	14,272	12,661
No. of clones no-Infected (rating = 0)	9,303	7,781
No. of clones Infected	4,969	4,880
% of infection	34.82	38.54
Severity (mean rating of infected clones)	2.16	1.76
Overall mean rating (including no-infected clones)	0.71	0.68

Figure 2. Distributions of brown rust based on the number of infected clones at different rust rating levels from 1 to 4 in Stage I of the Canal Point (CP) sugarcane breeding and cultivar development program in 2012 and 2013. Note: Brown rust ratings were assessed in July; the rust rating is equal to 0 or no rust infected clones; and other important parameters are also listed in the figure.

2.3. Variability in Clonal Numbers and Brown Rust Ratings among Families

Numbers of total crosses (families) advanced to Stage I from the seedling stage by individual selection in 2012 and 2013 were 576 and 455, respectively. Clone numbers among families ranged from 1 to 214 with a mean of 25 in 2012, and from 1 to 209 with a mean of 28 in 2013, and their CVs in two years were 113% and 100%, respectively (Table 2). This substantial variability in the number of clones among families is probably associated with differences in the number of viable seeds per cross in the crossing stage, variability in the number of seedlings planted per family in the seedling stage, and variable selection rates among families in the seedling stage [19]. On a family basis, mean brown rust ratings ranged from 0.0 to 2.5 in 2012 and from 0.0 to 2.2 in 2013. Among-family CV values of brown rust in 2012 and 2013 were 62.9% and 60.3%, respectively. Within-family CVs (136%–150%) were greater than their among-family CVs. Large variability in the number of clones and in the rust ratings (Table 2) among families in Stage I of the CP program suggested that Stage-I data can be used to identify useful parental combinations and individual parents for their progeny in resistance/tolerance to rusts in the test years. As variation in sugarcane orange rust discussed by Zhao *et al.* [19], the greater CV for rust rating within families than among families suggests that placing more emphasis on both individual clonal evaluation and family-based evaluation in Stage I of the CP program may help improve our knowledge and ability to select genotypes with potential for eliminating the negative effects of rusts on sugarcane growth and yields.

Table 2. Maximum, minimum, mean, standard deviation (SD), and coefficient of variation (CV) for clone number and mean brown rust ratings of 576 (2012) and 455 (2013) sugarcane families with a total of 14,272 and 12,661 clones, respectively, tested in the 2012 and 2013 Stage-I fields of the Canal Point (CP) sugarcane breeding and cultivar development program.

Parameter	2012		2013	
	Clone	Rust rating	Clone	Rust rating
	(No. family^{-1})		(No. family^{-1})	
Maximum	214	2.50	209	2.50
Minimum	1	0.00	1	0.00
Mean	25	0.46	28	0.96
SD	28	0.49	28	0.51
CV (%, among families)	113	104.9	100	52.7
CV (%, within families)	---	184.2	---	119.1

2.4. Correlations of Mean Brown Rust Ratings with % Infection and Severity

To properly evaluate sugarcane families, at least 10 to 20 clones per family are required [24]. Therefore, the families that had ≥15 clones were used in the present study to further investigate relationships between mean rust rating and % infection or rust severity. In 576 (2012) and 455 (2013) families planted in Stage I, 295 and 263 families, respectively, had ≥15 clones. From these families with ≥15 clones, the mean brown rust ratings, % infections, and severity values for each family were calculated and the relationships between these three parameters were further determined (Figure 3).

Overall, the mean ratings of brown rust among families ranged from 0.0 to 1.8; the values of % infection ranged from 0.0 to 87.0%; and the values of severity ranged from 0.0 to 3.3. Averaged across years, the CV values of the mean rating, % infection, and severity were 47.4, 44.4, and 13.6%, respectively, among families. The mean brown rust rating highly and linearly correlated with % infection (r^2 = 0.91 to 0.95****), but the relationship between the mean rating and severity value was poor (r^2 = 0.04 to 0.13) (Figure 3). The similar relationships between the three rust traits were also found in sugarcane orange rust [19]. Considering the variability (CV) and correlations between the three rust variables, rust severity may not be a good parameter to distinguish family differences in response to brown rust in the early selection stage of a sugarcane breeding and selection program such as the CP program.

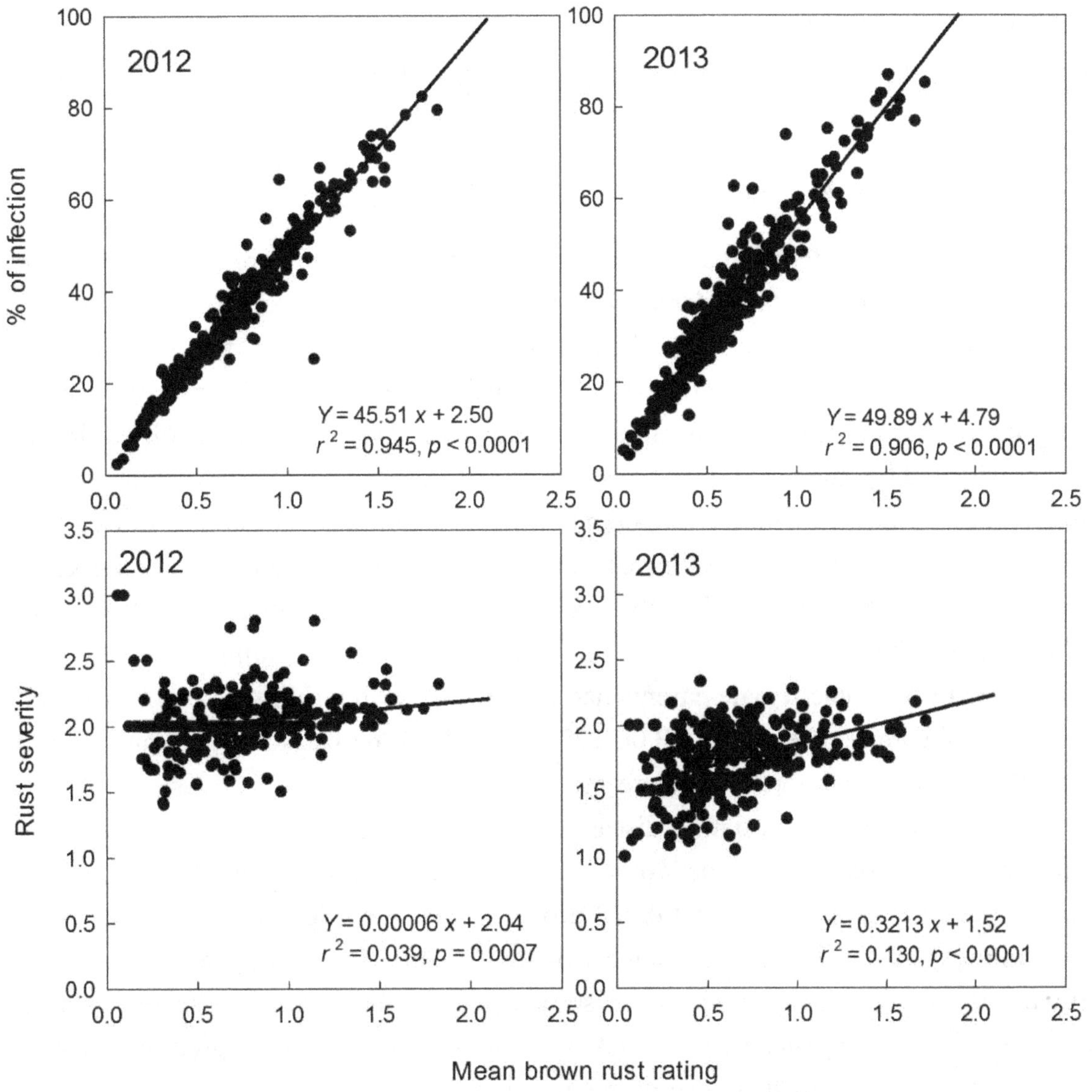

Figure 3. Relationships between mean rust ratings and % infection or severity of brown rust for 295 and 263 families with ≥15 clones, respectively, in the 2012 and 2013 Stage I of the Canal Point (CP) sugarcane breeding and cultivar development program.

2.5. Evaluation of Families Based on Mean Rust Rating

The families with ≥15 clones were also used to evaluate family tolerance to brown rust. The 295 and 263 families that had ≥15 clones planted in Stage I of the CP program in 2012 and 2013, respectively, were ranked based on their mean ratings of brown rust (Tables 3 and 4). The mean brown rust ratings across families were 0.8 (ranged from 0.1 to 2.0) in 2012 and 0.7 (ranged from 0.0 to 1.7) in 2013 with CV values of 47%–48%. The top 20 families with the highest (Table 3) and 20 families with the lowest (Table 4) mean ratings of brown rust in each year were identified. The % infection and severity of brown rust for these families and their parents (females and males) are also given in Tables 3 and 4. Overall, there was greater variability (CV) in the mean rating and in severity of brown rust within family than among families. Parents of these families could provide useful information for crossing combinations. For instance, reducing the use of parents listed in Table 3 (especially CL 90-4500, CP 95-1039, and CP 96-1252) and increasing the use of parents listed in Table 4 for crossing may improve brown rust tolerance of progeny in Stage I of the CP program.

2.6. Evaluation of Females Based on Mean Rust Ratings of Their Progeny

When data were analyzed by female parents regardless of males, there were 204 and 157 females, respectively, in 2012 and 2013. Of these females, 135 in 2012 and 113 females in 2013 had at least 15 progeny planted in the Stage-I fields. The progeny of these females were sorted by mean ratings of brown rust. The 20 females with their progeny having the greatest (Table 5) and 20 females with progeny having the lowest (Table 6) mean ratings of brown rust were further identified and ranked according to methods used by Zhao *et al.* [19].

There were 250 clones in 2012 with the female parent CL 90-4500 and 285 clones in 2013 with the female parent CP 92-1167. The mean brown rust ratings of these two female progeny ranked fourth and eighth, respectively (Table 5). Additionally, progeny of females CP 96-1252, CP 95-1039, and CPCL 02-8021 ranked in the top 20 in both years for their mean brown rust ratings. On the other hand, the progeny developed from some female parents had the lowest mean ratings of brown rust (Table 6). For instance, the clones with female parent CP 06-2274 or CP 88-1762 had low mean ratings of brown rust (*i.e.*, more tolerance to brown rust) in both years.

Brown rust is one of the most devastating diseases for sugarcane production in Florida, Louisiana, and many other production areas around the world [12,16–18]. Using proper parents with resistance to brown rust to make crosses is important in the CP program. There was great variation in brown rust ratings among families (Tables 3 and 4) in Stage I of the CP program. The great variation in mean ratings of brown rust among female parents was also detected (Tables 5 and 6). Clearly, using clones listed in Table 5 as female parents increases the risk of obtaining progeny with high brown rust ratings. These clones should be limited in their use as parents in the CP program in so far as brown rust resistance is concerned. Overall, the progeny developed from female parents listed in Table 6 had the lowest mean ratings of brown rust. Increased usage of these female clones for crossing may improve resistance to brown rust of progeny in Stage I of the CP program.

Table 3. Number of total sugarcane clones planted, mean brown rust (BR) rating, % infection, and severity and their parents for the 20 families with the highest mean BR ratings using 295 and 263 families with ≥15 clones in Stage I of the Canal Point (CP) sugarcane breeding and cultivar development program in 2012 and 2013, respectively.

2012							2013						
Family	No. clone	BR rating	% infect	Severity	Female†	Male†	Family	No. clone	BR rating	% infect	Severity	Female†	Male†
X09-1084	24	2.02	79.2	2.55	CL90-4500(-)	CP80-1743(+)	X10-1077	20	1.73	85.0	2.03	CP95-1039(-)	CP06-2170(-)
X09-0920	28	1.84	82.1	2.24	CL90-4500(-)	CPCL00-4027(-)	X05-0009	90	1.67	76.7	2.17	CP97-1989(-)	CP80-1743(+)
X09-0864	32	1.73	78.1	2.22	CP95-1039(-)	Poly09-23	X09-1235	32	1.58	81.3	1.94	CP01-2390(-)	CP04-1105(-)
X10-0369	24	1.71	66.7	2.56	TCP07-4820(-)	Poly10-09	X10-0909	38	1.57	78.9	1.98	CL84-3152(-)	Poly10-20
X10-0531	22	1.68	63.6	2.64	CP08-2409(-)	CP02-2281(+)	X09-0184	36	1.53	77.8	1.96	CPCL97-0393(-)	CP04-1105(-)
X09-1270	16	1.63	68.8	2.36	CPCL05-1777(-)	HoCP96-540(-)	X10-0286	30	1.52	86.7	1.75	CPCL02-8021(-)	L01-283(-)
X07-0943	23	1.61	73.9	2.18	CPCL06-8007(-)	CPCL01-0271(-)	X10-0771	23	1.48	82.6	1.79	CL84-3152(-)	CP06-2170(-)
X09-0181	21	1.60	71.4	2.23	CPCL01-6755(-)	CP04-1105(-)	X10-0816	21	1.45	81.0	1.79	CPCL01-6755(-)	CP06-2170(-)
X09-0186	21	1.60	71.4	2.23	CP96-1252(-)	TCP98-4447(-)	X10-1092	16	1.41	75.0	1.88	CP95-1039(-)	CPCL06-3470(?)
X09-1070	45	1.56	68.9	2.26	CL90-4500(-)	CP02-2281(+)	X10-0603	15	1.40	73.3	1.91	CPCL02-8021(-)	CPCL01-0271(-)
X10-0697	27	1.56	70.4	2.21	TCP00-4530(-)	Poly10-15	X10-0812	24	1.38	70.8	1.91	CPCL02-7386(-)	CP06-2170(-)
X09-0796	34	1.54	73.5	2.10	CL90-4500(-)	Poly09-21	X10-0645	34	1.35	73.5	1.84	CPCL02-7363(?)	CP06-2170(-)
X09-1232	17	1.53	70.6	2.17	CL88-4730(-)	HoCP96-540(-)	X09-0784	17	1.35	76.5	1.77	CP96-1252(-)	Poly09-21
X09-0949	44	1.52	63.6	2.39	CL90-4500(-)	CP06-2664(-)	X09-1146	27	1.28	66.7	1.92	CP80-1827(-)	MaleMix09G
X09-0893	33	1.50	66.7	2.25	CP97-1387(-)	Poly09-24	X09-0864	18	1.28	72.2	1.77	CP95-1039(-)	Poly09-23

Table 3. *Cont.*

2012							2013						
Family	No. clone	BR rating	% infect	Severity	Female[†]	Male[†]	Family	No. clone	BR rating	% infect	Severity	Female[†]	Male[†]
X09-0784	36	1.49	63.9	2.33	CP96-1252(-)	Poly09-21	X10-0569	29	1.26	58.6	2.15	CP89-2377(-)	Poly10-12
X09-1228	26	1.44	65.4	2.33	CL90-4500(-)	HoCP96-540(-)	X10-0479	23	1.24	60.9	2.04	CP08-1773(-)	CPCL06-3332(+)
X09-1234	24	1.44	62.5	2.30	CP95-1039(-)	CP04-1105(-)	X10-0691	24	1.23	66.7	1.84	CPCL96-2061(-)	Poly10-15
X10-0375	15	1.43	60.0	2.39	CP99-1893(-)	Poly10-09	X11-0497	16	1.22	68.8	1.77	CP97-1777(-)	TCP04-4709(-)
X07-1099	17	1.38	52.9	2.61	CPCL99-1371(-)	Mix07J	X11-0539	15	1.20	53.3	2.25	CP02-2281(+)	CR1009(?)
Mean[‡]	26 (42)	1.59 (0.78)	69 (36)	2.33 (2.19)			Mean[‡]	27 (43)	1.40 (0.67)	73 (38)	1.92 (1.75)		
Max	45 (214)	2.02 (2.02)	82 (82)	2.64 (3.00)			Max	90 (209)	1.73 (1.73)	87 (87)	2.25 (3.25)		
Min	15 (15)	1.38 (0.07)	53 (2.3)	2.10 (1.58)			Min	15 (15)	1.20 (0.00)	53 (0.0)	1.75 (0.00)		
CV (A)[¶]	33 (71)	10 (47)	10 (45)	7 (10)			CV(A)[¶]	60 (66)	11 (48)	12 (44)	8 (15)		
CV (W)[¶]		73 (158)		18 (17)			CV(W)[¶]		67 (151)		21 (26)		

[†] The "+", "-", and "?" in the parenthesis of each parent indicate presence, absence, and non-availability of the *Bru*1 gene, respectively; [‡] The first values of mean, maximum (Max), minimum (Min), and CV are calculated based on the top highest 20 families and the second values within parentheses are based on all 279 (2012) or 265 (2013) families with ≥15 clones planted; [¶] The CV(A) and CV(W) represent among- and within-family CVs, respectively.

Table 4. Number of total sugarcane clones planted, mean brown rust (BR) rating, % infection, and severity and their parents for the 20 families with the lowest mean BR ratings using 295 and 263 families with ≥15 clones in Stage I of the Canal Point (CP) sugarcane breeding and cultivar development program in 2012 and 2013, respectively.

2012							2013						
Family	No. clone	BR rating	% infect	Severity	Female†	Male†	Family	No. clone	BR rating	% infect	Severity	Female†	Male†
X09-0783	44	0.28	13.6	2.08	CP88-1762(+)	Poly09-21	X10-1194	28	0.25	14.3	1.75	CPCL99-2574(-)	Poly10-22
X09-0634	41	0.28	14.6	1.92	CPCL05-1108(?)	Poly09-19	X10-0347	18	0.25	16.7	1.50	CPCL99-2103(-)	CP06-2664(-)
X07-0933	30	0.27	13.3	2.00	CPCL99-1371(-)	CL89-5189(-)	X10-0709	16	0.25	18.8	1.33	CP06-2664(-)	CL95-5255(+)
X09-0978	17	0.26	11.8	2.25	CPCL02-8001(-)	Poly09-25	X10-0360	54	0.24	13.0	1.86	CPCL06-3458(?)	CP88-1762(+)
X09-0386	16	0.25	12.5	2.00	CP02-1143(+)	Poly09-13	X10-0136	37	0.23	18.9	1.21	CP06-2397(+)	Poly10-03
X09-1117	81	0.25	14.8	1.67	CPCL00-0458(-)	MaleMix09E	X09-1034	18	0.22	11.1	2.00	CPCL99-2103(+)	CL88-4730(-)
X09-0874	33	0.24	12.1	2.00	CPCL05-1108(?)	Poly09-23	X10-0536	70	0.22	15.7	1.41	CP05-1730(?)	Poly10-10
X09-1114	96	0.23	13.5	1.69	CP02-2454(+)	MaleMix09E	X10-0374	56	0.21	10.7	2.00	CP88-1762(+)	Poly10-09
X10-0351	22	0.23	9.1	2.50	CP06-2406(?)	CP06-2042(+)	X10-0467	26	0.21	15.4	1.38	CPCL02-7386(-)	CP06-2335(+)
X10-0378	47	0.22	10.6	2.10	CL89-5189(-)	Poly10-09	X09-0511	15	0.20	13.3	1.50	CP02-2065(+)	Poly09-16
X09-1133	52	0.22	9.6	2.30	CP89-1509(+)	MaleMix09F	X10-0543	25	0.18	12.0	1.50	CP06-3040(-)	Poly10-11
X07-1118	71	0.21	11.3	1.88	CP04-2166(+)	CPCL00-6131(-)	X10-1109	28	0.18	10.7	1.67	CPCL06-3458(?)	Mix10-06
X07-0837	33	0.20	9.1	2.17	CP03-1401(-)	Poly07-04	X09-1239	18	0.17	11.1	1.50	CL89-5189(-)	CP96-1252(-)
X10-0535	23	0.17	8.7	2.00	CP02-1554(-)	Poly10-10	X10-0693	22	0.16	9.1	1.75	CP05-1730(?)	Poly10-15
X07-1280	95	0.16	6.3	2.58	CP00-1630(+)	Poly07-16	X10-0318	21	0.14	9.5	1.50	CP07-1860(?)	CP97-1777(-)
X09-1150	25	0.16	8.0	2.00	CP05-1518(+)	CP88-1762(+)	X10-0895	28	0.13	10.7	1.17	CPCL96-4974(-)	Poly10-19
X09-1188	31	0.13	6.5	2.00	CP05-1518(+)	CP80-1743(+)	X10-0176	16	0.13	6.3	2.00	CP06-2897(-)	CL88-2747(-)
X10-0360	16	0.13	6.3	2.00	CPCL06-3458(?)	CP88-1762(+)	X10-0272	50	0.09	8.0	1.13	CP06-2274(-)	Poly10-05
X09-1269	30	0.10	3.3	3.00	CPCL05-1306(?)	HoCP96-540(-)	X10-0152	25	0.08	4.0	2.00	CP08-2003(?)	CP88-1762(+)
X09-1236	44	0.07	2.3	3.00	CP03-1160(-)	CP04-1105(-)	X10-0389	17	0.00	0.0	0.00	CPCL02-6848(+)	Poly10-09

Table 4. *Cont.*

2012							2013						
Family	No. clone	BR rating	% infect	Severity	Female[†]	Male[†]	Family	No. clone	BR rating	% infect	Severity	Female[†]	Male[†]
Mean[‡]	42(42)	0.20(0.78)	10(36)	2.16(2.17)			Mean[‡]	29(43)	0.18(0.67)	12(38)	1.59(1.75)		
Max	96(214)	0.28(2.02)	16(82)	3.00(3.00)			Max	70(209)	0.25(1.73)	19(87)	2.00(3.25)		
Min	16(15)	0.07(0.07)	2.3(2.3)	1.67(1.58)			Min	15(15)	0.00(0.00)	0.0(0.0)	0.00(0.00)		
CV(A)[¶]	59(71)	31(47)	37(45)	17(10)			CV(A)[¶]	55(66)	38(48)	41(44)	18(15)		
CV(W)[¶]		337(158)		14(17)			CV(W)[¶]		299(151)		28(26)		

[†] The "+", "-", and "?" in the parenthesis of each parent indicate presence, absence, and non-availability of the *Bru*1 gene, respectively; [‡] The first values of mean, maximum (Max), minimum (Min), and CV are calculated based on the lowest 20 families and the second values within parentheses are based on all 279 (2012) or 265 (2013) families with ≥15 clones planted; [¶]The CV(A) and CV(W) represent among- and within-family CVs, respectively.

Table 5. Number of total clones planted and mean brown rust (BR) rating for the 20 females in which their progeny had the highest mean BR ratings using 135 and 113 females with ≥15 progeny clones in Stage I of the Canal Point (CP) sugarcane breeding and cultivar development program in 2012 and 2013, respectively.

2012			2013		
Female†	**No. clone**	**Mean BR rating**	**Female**†	**No. clone**	**Mean BR rating**
US90-0018 (-)	19	2.05	CP97-1989 (-)	95	1.61
TCP07-4820 (?)	24	1.71	CPCL97-0393 (-)	38	1.55
CPCL05-1777 (-)	16	1.63	CP95-1039 (-)	95	1.26
CL90-4500 (-)	250	1.54	CPCL96-2061 (-)	24	1.23
CP97-1387 (-)	33	1.50	CPCL95-2287 (-)	17	1.15
CP99-1893 (-)	15	1.43	CP80-1827 (-)	48	1.08
TCP00-4530 (-)	53	1.39	CP08-2398 (-)	36	1.08
TCP97-4416 (-)	39	1.37	CP92-1167 (-)	285	1.03
CPCL01-6755 (-)	63	1.34	CL88-4730 (-)	91	1.00
CP04-1566 (-)	17	1.29	TCP04-4688 (-)	18	1.00
TCP07-4806 (?)	19	1.26	CPCL06-8004 (-)	113	0.97
CP96-1252 (-)	192	1.25	CL84-3152 (-)	630	0.97
CP06-3098 (-)	20	1.23	CP96-1252 (-)	243	0.93
CPCL06-8007 (-)	46	1.20	CPCL02-8021 (-)	158	0.92
CPCL99-1777 (-)	29	1.17	CP00-1301 (-)	28	0.91
CPCL02-8021 (-)	248	1.14	CPCL02-7363 (?)	90	0.90
CP95-1039 (-)	304	1.14	CP01-2390 (-)	204	0.88
CP06-2657 (?)	40	1.13	CP06-2214 (+)	252	0.86
CP08-2409 (-)	58	1.11	CL87-1630 (-)	61	0.85
CP78-1628 (-)	54	1.11	CP08-1965 (-)	128	0.85
Mean‡	77 (102)	1.35 (0.67)	Mean‡	133 (108)	1.05 (0.64)
Max	304 (639)	2.05 (2.05)	Max	630 (630)	1.61 (1.61)
Min	15 (15)	1.11 (0.14)	Min	17 (16)	0.85 (0.00)
CV (A)¶	118 (109)	18 (44)	CV (A)¶	108 (109)	21 (43)
CV (W)¶		87 (156)	CV (W)¶		95 (152)

† The "+", "-", and "?" in the parenthesis of each female parent indicate presence, absence, and non-availability of the *Bru*1 gene, respectively; ‡ The first values of mean, maximum (Max), minimum (Min), and coefficient of variation (CV) are calculated based on mean brown rust ratings of top 20 female parents and the second values within parentheses are based on all 136 (2012) or 113 (2013) female parents with ≥15 clones planted; ¶ The CV (A) and CV (W) represent among- and within-family CVs, respectively.

Table 6. Number of total clones planted and mean brown rust (BR) rating for the 20 females in which their progeny had the lowest mean BR ratings using 135 and 113 females with ≥15 progeny clones in Stage I of the Canal Point (CP) sugarcane breeding and cultivar development program in 2012 and 2013, respectively.

2012			2013		
Female†	**No. clone**	**Mean BR rating**	**Female**†	**No. clone**	**Mean BR rating**
CPCL00-1373 (-)	58	0.42	CP99-1896 (-)	48	0.40
CPCL02-7406 (+)	70	0.42	CL87-2882 (-)	36	0.38
CP02-2454 (+)	228	0.42	CPCL02-8001 (-)	35	0.37
CP04-1367 (+)	122	0.41	CP06-1730 (?)	50	0.37
CP89-1509 (+)	95	0.41	CP88-1762 (+)	96	0.33
CP05-1466 (+)	52	0.40	CL87-2282 (?)	20	0.33
CP06-2274 (-)	53	0.40	CPCL02-0843 (+)	20	0.33
CPCL05-1009 (-)	18	0.39	CPCL02-7610 (+)	49	0.31
HoCP04-856 (-)	16	0.38	CPCL02-7080 (-)	85	0.30
CL88-2747 (-)	103	0.36	CL94-0150 (-)	43	0.29
CP01-1178 (-)	38	0.36	CPCL06-3470 (?)	16	0.28
CP03-1912 (+)	64	0.35	CP05-1678 (-)	72	0.28
CP00-2164 (+)	51	0.34	CPCL99-2574 (-)	28	0.25
CPCL00-0458 (-)	138	0.31	CP06-2274 (-)	89	0.25
CPCL02-6334 (+)	151	0.29	CPCL99-2103 (-)	36	0.24
CP88-1762 (+)	44	0.28	CP05-1730 (?)	92	0.21
CP00-1630 (+)	138	0.21	CP06-3040 (-)	25	0.18
CP03-1160 (-)	72	0.18	CPCL96-4974 (-)	28	0.13
CP02-1554 (-)	23	0.17	CP08-2003 (?)	25	0.08
CP05-1518 (+)	56	0.14	CPCL02-6848 (+)	19	0.00
Mean‡	80 (102)	0.33 (0.67)	Mean‡	46 (108)	0.26 (0.64)
Max	228 (639)	0.42 (2.05)	Max	96 (630)	0.40 (1.61)
Min	16 (15)	0.14 (0.14)	Min	16 (16)	0.00 (0.00)
CV (A)¶	67 (109)	27 (44)	CV (A)¶	59 (109)	39 (43)
CV (W)¶		246 (156)	CV (W)¶		245 (152)

† The "+", "-", and "?" in the parenthesis of each female parent indicate presence, absence, and non-availability of the *Bru*1 gene, respectively; ‡ The first values of mean, maximum (Max), minimum (Min), and coefficients of variation (CV) are calculated based on mean brown rust ratings of the lowest 20 female parents and the second values within parentheses are based on all 136 (2012) or 113 (2013) female parents with ≥15 clones planted; ¶ The CV (A) and CV (W) represent among- and within-family CVs, respectively.

2.7. Bru1 in Parental Clones and Rust Rating in Their Progeny

*Bru*1 is a major gene for resistance to brown rust of sugarcane [25]. It is important for sugarcane breeders to develop a database of parental clones using agronomic and physiological traits and molecular markers for improving disease resistance and for sustaining yields and profits [19]. Phylogenetic analysis of sugarcane rusts based on rDNA sequences and phylogenetic relationships of sugarcane rust fungi have been reported [26,27]. The sugarcane brown rust resistance gene (*Bru*1) was recently found to be prevalent in 86% of brown rust-resistant clones in a sample of 380 modern cultivars and breeding materials covering worldwide diversity [28]. Therefore, the opportunity exists to utilize *Bru*1 in marker-assisted breeding and selection in order to improve brown rust resistance in sugarcane [29]. *Bru*1 has been used as a marker to identify if sugarcane genotypes are potentially resistant to brown rust [25,30,31]. Since 2009, this marker has also been utilized in the CP program to direct breeding strategies for brown rust resistance [29]. The information of *Bru*1 for parents (available on the CP database) used in this study is given in Tables 3, 4, 5, and 6. In general, dominant parents of the top 20 crosses that had the highest brown rust ratings (Table 3), or female parents that had high probability to produce progeny with high brown rust ratings (Table 5) showed absence of *Bru*1. However, parents of the crosses with the lowest brown rust ratings (Table 4) or female parents that had a high probability to produce progeny with the lowest brown rust ratings (Table 6) did not always show the presence of *Bru*1. For instance, the cross (family) X09-1236 in 2012 had the lowest mean brown rust rating among 295 families, but *Bru*1 was not detected in its parents CP03-1160 and CP04-1105 (Table 4). These results suggested that other genes may also be involved in the either durable or non-durable resistance of sugarcane plants to brown rust in addition to *Bru*1 [23] as reported by other studies [32–35]. Racedo *et al.* [34] recently found that the predominant source of resistance to brown rust in the sugarcane breeding program at Tucuman, Argentina would be a resistance source independent of the *Bru*1 gene. They suggested that it is necessary to characterize both genetic diversity of the pathogen and the alternative sources of resistance in order to improve brown rust disease management.

2.8. Relationships between Brown and Orange Rusts

Genotypic variation in sugarcane orange rust resistance/tolerance at family and female parent levels for the Stage I clones of the CP program has been reported [19]. However, it is unclear if there are any correlations between brown rust and orange rust ratings across genotypes. Relationships between mean ratings of brown rust and orange rust across a large number of families, females or males were determined in this study (Figure 4) based on mean values of rust ratings across families, females, and males that had ≥15 progeny clones in the 2012 and 2013 Stage I of the CP program. Overall, there were the trends of negative relationships between brown and orange rust ratings of sugarcane progeny clones across families, females, and males. Most of the linear relationships were significant even through r^2 values ranged 0.02 to 0.30 because of the large number of samples (Figure 4). These relationships were stronger in 2013 ($p < 0.0001$) than in 2012 ($p = 0.211$–0.0012) (Figure 4). The results suggested that sugarcane genotypes, which are resistant to brown rust disease, may be susceptible to orange rust. Therefore, caution must be taken for the reversible relationships between brown and orange rusts when using parents to make crosses for rust resistance. Some parents may produce progeny with resistance to one rust, but be highly susceptible to the other rust. For instance, the progeny of CP 88-1762 was resistant to brown rust (Table 6), but was highly susceptible to orange rust [19].

Multiple approaches have been and are being used for eliminating the negative effects of brown and orange rusts on yields, quality, and profits of sugarcane. These include standard breeding, molecular marker assisted selection, disease monitoring, adjustment of management practices, fungicide applications, and selection of rust resistant cultivars in the breed program by intensive rust screening [23]. Studies have shown that parents producing progeny with a high frequency of transgressive segregates for agronomic traits should provide the best opportunity for sugarcane breeders to select clones superior to their parents [36]. In addition to directly evaluating parental lines for brown rust and other diseases, therefore, parental evaluation based on their progeny performance for rust resistance and other agronomic traits in the early clonal stage of a sugarcane breeding program should help optimize parental selection and crossing combinations. Studies have suggested that family selection is effective in improving sugarcane populations in early selection stages [36–39], because it can identify those families that harbor the highest proportion of desirable clones and makes it possible to focus on selection for superior clones [36]. Availability of family data for rust diseases and agronomic traits helps sugarcane breeders improve crossing combinations for developing genotypes with resistance to rusts and high yields.

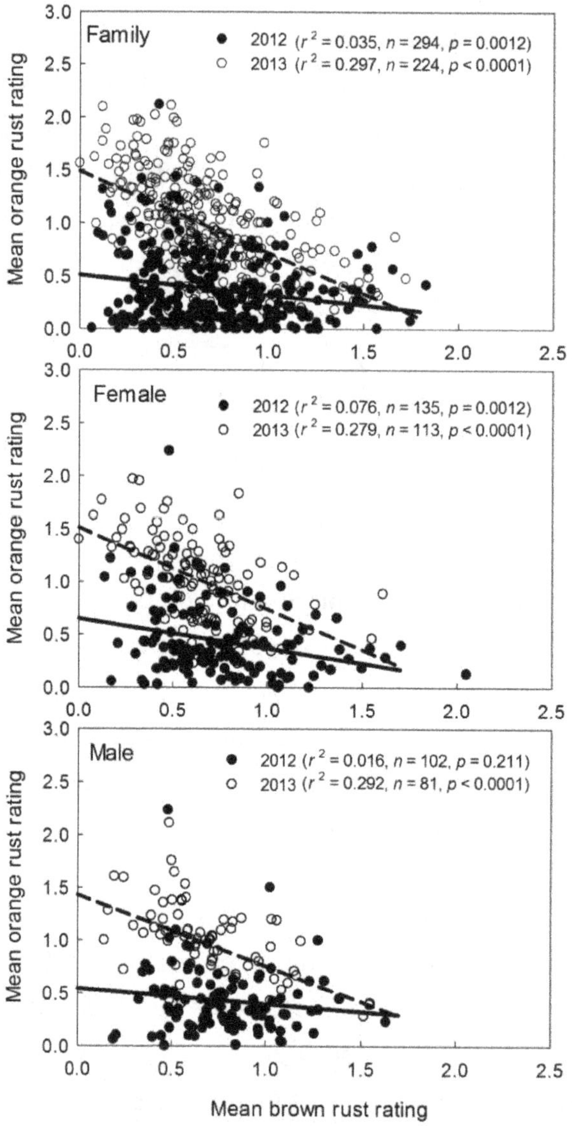

Figure 4. Relationships between mean brown rust ratings and mean orange rust ratings of clones based on the crossing families, females, and males those had ≥15 progeny clones in the 2012 and 2013 Stage I of the Canal Point (CP) sugarcane breeding and cultivar development program.

3. Materials and Methods

3.1. Plant Culture

A total of 14,272 genotypes (stools, each stool came from a true seedling plant) in 2012 and 12,661 genotypes in 2013 were visually selected at mature from fields of the 2011 and 2012 seedling stages. One stalk was cut in each of all these selected genotypes and planted in single-row plots in the Stage I fields in January of 2012 and 2013, respectively. To facilitate stalk transport and planting, two to five stalks (each stalk came from a true seed) in the seedling fields were bundled and labeled by family (*i.e.*, cross) prior to advancing them to Stage I [9]. These bundles were randomly distributed in the Stage-I fields. One stalk was placed in each plot and cut into two sections (each approximately 0.9 m long). The two sections were placed in the center of the plot as double lines of cane. The plot length was 2.4 m, with 1.5 m between-row spacing. There was a 1.5-m gap between adjacent clones within a row to allow scientists to recognize individual clones during disease rating, growth vigor evaluation, and selection. Four commercial cultivars, "CP 78-1628" [40], "CP 80-1743" [41], CP 88-1762, and CP 89-2143, were used as checks each year and randomly planted in a pattern of one check plot in approximately every 80 to 100 field plots. Approximately 40 replicated plots for each check cultivar were planted each year in the Stage-I fields. There was a 4.5 m alley every eight rows to facilitate field maintenance and genotype selection.

3.2. Data Collection of Brown Rust Disease

Although the peak season is in March to June for brown rust in most sugarcane production fields in Florida, the optimal plant age for brown rust development has been reported to be between 4 and 6 months [42,43]. The Stage-I planting dates of the CP program are routinely from late January to early February, therefore, brown rust and orange rust symptoms were evaluated in July to August in 2012 and 2013. Brown rust symptoms were differentiated from that of orange rust using the 20× pocket magnifying glass in the field, and positive identifications of inconclusive symptoms were made by observing spore morphology under a microscope in the laboratory. Rust diseases were recorded using a scale from 0 (no rust infection) to 4 (most severe rust infection) with intervals of 0.5. The 0 to 4 scale levels were defined as: 0 = no rust, 1 = one to a few pustules, 2 = patching presented, 3 = patching widespread up into the upper canopy with some lower leaf death, and 4 = massive amounts of rust pustules with heavy lower leaf death. In general, the plants with rust ratings of 0 to 1 were considered as resistant or tolerant, with rating 2 were considered as moderately susceptible, and with ratings of 3 and 4 were considered as susceptible and highly susceptible, respectively.

Additionally, all clones were visually evaluated for other diseases [leaf scald (*Xanthomonas albilineans*), smut (*Sporisorium scitamineum*), and mosaic] and plant vigor and agronomic traits [9] at the same time and in early September. A subjective plant vigor rating was determined for individual clones. All clones with high vigor rating or better than check cultivars in vigor rating, acceptable rust resistance (rust rating $\leq$1), and no other disease symptoms were further assessed for Brix value, an indicator of juice sucrose concentration, in early November. Approximately the 1,500 best clones with the largest vigor rate × Brix products were advanced to Stage II of the CP program. Recently, Zhao *et al.* [9] reported the details of vigor rating, Brix, and the Stage-I selection strategies as well as orange rust evaluation in the CP program [19]. Therefore, we mainly focused on brown rust on the basis of families and female parents in this paper. The information of sugarcane brown rust resistance gene (*Bru*1) for parents of the Stage-I clones was searched from a database of the CP program to investigate the relationships between *Bru*1 of parents and the rust resistance of their progeny. Three variables of mean rust rating, percentage of rust infection, and rust severity were used to determine variation of rust diseases among families (or female parents based on their progeny). These variables were defined and estimated using the following formulas [19]:

$$Mean\ rust\ rating = \frac{\sum Rust\ rating}{Total\ number\ of\ clones}\ \text{(Including clones with 0 rating)} \quad (1)$$

$$Percent\ of\ infection = \frac{Number\ of\ clones\ infected}{Total\ number\ of\ clones\ investigated} \times 100 \quad (2)$$

$$Severity = \frac{\sum Rust\ rating}{Number\ of\ clones\ infected}\ \text{(Not including clones with 0 rating)} \quad (3)$$

3.3. Data Analysis

The four check cultivars were randomly planted in the Stage-I field and each had approximately 40 replicated plots (Table 1) each year. Therefore, the MIXED procedure of SAS [44] was used to test the effects of check cultivar, year, and their interaction on the mean ratings of brown rust and on the % infection. If the hypothesis of equal means between the check cultivars was rejected by the *F* test, the trait means were separated with the least significant difference (LSD) at $p = 0.05$. The LSD values were calculated with the standard error (SE) values generated by the Diff option in the SAS MIXED procedure.

For the Stage-I clones, their parental combinations in the Crossing stage varied annually. Thus, the Stage-I data were analyzed separately for each year. Brown rust rating distributions were determined by pooling data across all clones within a year. Data of brown rust were analyzed for each family and female parent. Data of male parents were not analyzed because many of the progeny in the 2012 and 2013 Stage I were developed from poly crosses (*i.e.*, where a female tassel received pollen from several different male tassels). Thus, the specific males were unknown for all poly crosses. For families and female parents that had ≥15 progeny clones planted in Stage I, the mean ratings and coefficients of variation (CVs) were determined for brown rust to assess variability. The means and CVs were calculated using PROC MEANS of SAS [44]. Then, the CV values of among-families (females) were obtained based on their means and the CVs of within-families (females) were estimated by averaged CVs of individual families (females). Coefficients of variation for brown rust in each family were calculated from the individual clonal values of rust rating from all clones within a family or female according to Zhao *et al.* [9]. The within-family or within-female CV for each parameter was estimated by calculating the overall mean CV of all individual family (female) CVs for that trait. The among-family CVs were estimated using the mean (rather than individual clone) values of each family (female). For example, to calculate the among-family CV of 20 families for brown rust, we would have calculated the CV based on the standard deviation (SD) and overall mean from the 20 mean rust values of each of the 20 families (*i.e.*, CV = SD ÷ Mean × 100). The variability among- and within-families was described using respective CVs. The top 20 families that were most susceptible or most tolerant to brown rust in each year were further determined by ranking their mean rust ratings. The top 20 females in which their progeny were most susceptible or most tolerant to brown rust in each year were also determined based on mean ratings of brown rust.

4. Conclusions

Analyses of brown rust data collected from large numbers of individual clones in Stage I of the CP program in 2012 and 2013 revealed that there was great variation in rust ratings among genotypes and among families. Our data indicated that using brown rust rating data along with individual selection data on plant vigor and stalk juice Brix [9] and orange rust rating [19] could be useful to evaluate family performance in the first clonal stage of the CP program. The among- and within-family variability in these agronomic and disease resistance traits would improve our parental selection and optimize crosses among selected parents which should result in progeny with improved rust resistance and yield potential. Additional caution must be taken for the apparent inverse relationships between brown and orange rust resistance. These findings are useful not only for the local CP sugarcane breeding program, but also for others which import CP varieties around the world.

Acknowledgments

Authors thank Philip Aria for his valuable technical assistance and skilled data management. We also appreciate Tom Abbott and all other staff at the USDA-ARS Sugarcane Field Station for assistance on planting, field management, and harvest.

Author Contributions

Duli Zhao contributed to conception and design of the experiment, analyzed and interpreted data, wrote the manuscript and integrated all ideas and comments from other authors and peer reviewers. R. Wayne Davidson and Miguel Baltazar coordinated in rust data collection and interpretation. Jack C. Comstock made contributions to the original ideas of the experiment and disease data quality. Per McCord and Sushma Sood contributed to identification and interpretation of the brown rust resistance gene, *Bru*1.

Conflict of Interest

The authors declare no conflict of interest.

References

1. USDA-NASS. Available online: http://usda.mannlib.cornell.edu/usda/current/CropValuSu/ CropValuSu-02-14-2014.pdf. (accessed on 4 November 2014).
2. Zhao, D.; Glaz, B.; Comstock, J.C. Sugarcane response to water deficit stress during early growth on organic and sand soils. *Am. J. Agric. Biol. Sci.* **2010**, *5*, 403–414.
3. Edmé, S.J.; Tai, P.Y.P.; Glaz, B.; Gilbert, R.A.; Miller, J.D.; Davidson, J.O.; Dunckelman, J.W.; Comstock, J.C. Registration of 'CP 96-1252' sugarcane. *Crop Sci.* **2005**, *45*, 423.

4. Gilbert, R.A.; Comstock, J.C.; Glaz, B.; Edmé, S.J.; Davidson, R.W.; Glynn, N.C.; Miller, J.D.; Tai, P.Y.P. Registration of 'CP 00-1101' sugarcane. *J. Plant Reg.* **2008**, *2*, 95–101.
5. Edmé, S.J.; Davidson, R.W.; Gilbert, R.A.; Comstock, J.C.; Glynn, N.C.; Glaz, B.; del Blanco, I.A.; Miller, J.D.; Tai, P.Y.P. Registration of 'CP 01-1372' sugarcane. *J. Plant Reg.* **2009**, *3*, 150–157.
6. Tai, P.Y.P.; Shine, J.M., Jr.; Deren, C.W.; Glaz, B.; Miller, J.D.; Comstock, J.C. Registration of 'CP 88-1762' sugarcane. *Crop Sci.* **1997**, *37*, 1388.
7. Glaz, B.; Miller, J.D.; Deren, C.W.; Tai, P.Y.P.; Shine, J.M., Jr.; Comstock, J.C. Registration of 'CP 89-2143' sugarcane. *Crop Sci.* **2000**, *40*, 577.
8. Rice, R.; Baucum, L.; Davidson, W. Sugarcane variety census: Florida 2014. *Sugar J.* **2015**, *78*, 8–16.
9. Zhao, D.; Comstock, J.C.; Glaz, B.; Edmé, S.J.; Glynn, N.C.; Del Blanco, I.A.; Gilbert, R.A.; Davidson, R.W.; Chen, C.Y. Vigor rating and Brix for first clonal selection stage of the Canal Point sugarcane cultivar development program, *J. Crop Improv.* **2012**, *26*, 60–75.
10. Tai, P.Y.P.; Miller, J.D. Family performance at early stages of selection and frequency of superior clones from crosses among Canal Point cultivars of sugarcane. *J. Am. Soc. Sugar Cane Technol.* **1989**, *9*, 62–70.
11. Edmé, S.J.; Miller, J.D.; Glaz, B.; Tai, P.Y.P.; Comstock, J.C. Genetic contributions to yield gains in the Florida sugarcane industry across 33 years. *Crop Sci.* **2005**, *45*, 92–97.
12. Raid, R.N.; Comstock, J.C. Common rust. In *A Guide to Sugarcane Diseases*; Rott, P., Bailey, R.A., Comstock, J.C., Croft, B.J., Saumtally, A.S., Eds; Centre de Cooperation Internationale en Recherche Agronomique pour le Developpement (CIRAD) and International Society of Sugar Cane Technologists (ISSCT): Montpellier, France, 2000; pp. 85–89.
13. Comstock, J.C.; Raid, R.N. Sugarcane common rust. In *Current Trends in Sugarcane Pathology*; Bhargava, K.S., Rao, G.P., Gillaspie, A.G., Jr., Upadhyaya, P.P., Filino, A.B., Agnihotri, V.P. Chen, C.T., Eds.; International Books and Periodicals Supply Service: Delhi, India, 1994; pp. 1–10.
14. Comstock, J.C.; Shine, J.M., Jr.; Raid, R.N. Effect of rust on sugarcane growth and biomass. *Plant Dis.* **1992**, *76*, 175–177.
15. Comstock, J.C.; Shine, J.M., Jr.; Raid, R.N. Effect of early rust infection on subsequent sugarcane growth. *Sugar Cane* **1992**, *4*, 7–9.
16. Hoy, J. Brown rust is coming back—What can you do to prevent loss? *Sugar Bull.* **2012**, *90*, 19–21.
17. Hoy, J.W.; Hollier, C.A. Effect of brown rust on yield of sugarcane in Louisiana. *Plant Dis.* **2009**, *93*, 1171–1174.
18. Sood, S.G.; Comstock, J.C.; Glynn, N.C. Leaf whorl inoculation method for screening sugarcane rust resistance. *Plant Dis.* **2009**, *93*, 1335–1340.
19. Zhao, D.; Davidson, R.W.; Baltazar, M.; Comstock, J.C. Field evaluation of sugarcane orange rust for first clonal stage of the CP cultivar development program. *Am. J. Agric. Biol. Sci.* **2015**, *10*, 1–11.
20. Rott, P.; Sood, S.; Comstock, J.C.; Raid, R.N.; Glynn, N.C.; Gilbert, R.A.; Sandhu, H.S. Sugarcane orange rust. EDIS. Available online: http://edis.ifas.ufl.edu/sc099 (accessed on 20 July 2015).

21. Jiang, D.K. Chemical control of sugarcane rust *P. melanocephala*. *Rep. Taiwan Sugar Res. Inst.* **1985**, *108*, 25–34.
22. Machado, G.R., Jr. The spread of CP varieties, Canal Point, Florida, USA in other countries. *Sugar J.* **2013**, *77*, 20–21.
23. Zhao, D.; Comstock, J.C.; Glaz, B.; Edmé, S.J.; Davidson, R.W.; Gilbert, R.A.; Glynn, N.C.; Sood, S.; Sandhu, H.; McMorkle, K. *et al.* Registration of 'CP 05-1526' sugarcane. *J. Plant Reg.* **2013**, *7*, 305–311.
24. Wang, L.P.; Jackson, P.A.; Lu, X.; Fan, Y.H.; Foreman, J.W.; Chen, X.K.; Deng, H.H.; Fu, C.; Ma, L.; Aitken, K.S. Evaluation of sugarcane × *Saccharum spontaneum* progeny for biomass composition and yield components. *Crop Sci.* **2008**, *48*, 951–961.
25. Asnaghi, C.; Roques, D.; Ruffel, S.; Kaye, C.; Hoarau, J.Y.; Telismart, H.; Girard, J.C.; Raboin, L.M.; Risterucci, A.M.; Grivet, L.; D'Hont, A. Targeted mapping of a sugarcane rust resistance gene *Bru*1 using bulked segregant analysis and AFLP markers. *Theor. Appl. Genet.* **2004**, *108*, 759–764.
26. Virtudazo, E.V.; Nakamura, H.; Kakishima, M. Phylogenetic analysis of sugarcane rusts based on sequences of ITS, 5.8 S rDNA and D1/D2 regions of LSU rDNA. *J. Gen. Plant Pathol.* **2001**, *67*, 28–36.
27. Dixon, L.J.; Castlebury, L.A.; Aime, M.C.; Glynn, N.C.; Comstock J.C. Phylogenetic relationships of sugarcane rust fungi. *Mycol. Progress.* **2010**, *9*, 459–468.
28. Costet, L.; Cunff, L.L.; Royaert, S.; Raboin, L.-M.; Hervouet, C.; Toubi, L.; Telismart, H.; Garsmeur, O.; Rousselle, Y.; Pauquet, J.; Nibouche, S.; Glaszmann, J.-C.; Hoarau, J.-Y.; D'Hont, A. Haplotype structure around *Bru*1 reveals a narrow genetic basis for brown rust resistance in modern sugarcane cultivars. *Theor. Appl. Genet.* **2012**, *125*, 825–836.
29. Glynn, N.C.; Laborde, C.; Davidson, R.W.; Irey, M.S.; Glaz, B.; D'Hont, A.; Comstock, J.C. Utilization of a major brown rust resistance gen in sugarcane breeding. *Mol. Breeding.* **2013**, *31*, 323–331.
30. Asnaghi, C.; Paulet, F.; Kaye, C.; Grivet, L.; Glaszmann, J.C.; D'Hont, A. Application of synteny across the *Poaceae* to determine the map location of a rust resistance gene of sugarcane. *Theor. Appl. Genet.* **2000**, *10*, 962–969.
31. Cunff, L.L.; Garsmeur, O.; Raboin, L.M.; Pauquet, J.; Telismart, H.; Selvi, A.; Grivet, L.; Philippe, R.; Begum, D.; Deu, M.; Costet, L.; Wing, R.; Glaszmann, J.C.; D'Hont, A. Diploid/polyploid syntenic shuttle mapping and haplotype-specific chromosome walking toward a rust resistance gene (*Bru*1) in highly polyploid sugarcane ($2n \sim 12x \sim 115$). *Genetics* **2008**, *180*, 649–660.
32. Raboin, L.-M.; Oliveira, K.M.; Lecunff, L.; Telismart, H.; Roques, D.; Butterfield, M.; Hoarau, J.-Y.; D'Hont, A. Genetic mapping in sugarcane, a high polyploid, using bi-parental progeny: identification of a gene controlling stalk colour and a new rust resistance gene. *Theor. Appl. Genet.* **2006**, *112*, 1382–1391.
33. Hoy, J.W.; Avellaneda, M.C.; Bombecini, J. Variability in *Puccinia melanocephala* pathogenicity and resistance in sugarcane cultivars. *Plant Dis.* **2014**, *98*, 1728–1732.

34. Racedo, J.; Perera, M.F.; Bertani, R.; Funes, C.; Victoria González, V.; Cuenya, M.I.; D'Hont, A.; Welin, B.; Castagnaro, A.P. *Bru*1 gene and potential alternative sources of resistance to sugarcane brown rust disease. *Euphytica.* **2013**, *191*, 429–436.
35. Parco, A.S.; Avellaneda, M.C.; Hale, A.H.; Hoy, J.W.; Kimbeng, C.A.; Pontif, M.J.; Gravois, K.A.; Baisakh, N. Frequency and distribution of the brown rust resistance gene *Bru*1 and implications for the Louisiana sugarcane breeding programme. *Plant Breeding* **2014**, *133*, 654–659.
36. Shanthi, R.M.; Bhagyalakshmi, K.V.; Hemaprabha, G.; Alarmelu, S.; Nagarajan, R. Relative performance of the sugarcane families in early selection stages. *Sugar Tech*. **2008**, *10*, 114–118.
37. Chang, Y.S.; Milligan, S.B. Estimating the potential of sugarcane families to produce elite genotypes using univariate cross prediction methods. *Theor. Appl. Genet*. **1992**, *84*, 662–671.
38. Cox, M.C.; Hogarth, D.M. The effectiveness of family selection in early stages of sugarcane improvement program. In *Focused Plant Improvement: Towards Responsible and Sustainable Agriculture*; Proceedings Tenth Australian Plant Breeding Conference, Gold Coast, Australia, April 1993; Imrie, B.C.; Hacker, J.B. Eds.; Vol. 2, pp. 53–54.
39. McRae, T.A.; Hogarth, D.M.; Foreman, J.F.; Braithwaite, M. Selection sugarcane families in the Burdekin district. In *Focused Plant Improvement: Towards Responsible and Sustainable Agriculture*; Proceedings Tenth Australian Plant Breeding Conference, Gold Coast, Australia, April 1993; Imrie, B.C.; Hacker, J.B. Eds.; Vol.1, pp. 77–82.
40. Tai, P.Y.P.; Miller, J.D.; Glaz, B.; Deren, C.W.; Shine, J.M. Registration of 'CP 78-1628' sugarcane. *Crop Sci*. **1991**, *31*, 236.
41. Deren, C.W.; Glaz, B.; Tai, P.Y.P.; Miller, J.D.; Shine, J.M., Jr.; Registration of 'CP 80-1743' sugarcane. *Crop Sci*. **1991**, *31*, 235–236.
42. Comstock, J.C.; Ferreira, S.A. Sugarcane rust: Factors affecting infection and symptoms development. *Proc. XIX Int. Soc. Sugar Cane Technol. (ISSCT) Congress*. **1986**, *19*, 402–410.
43. Victoria, J.; Moreno, C.; Cassalett, C. Genotype-environment interaction and its effect on sugar cane rust incidence. *Sugar Cane* **1990**, *4*, 13–17.
44. SAS. *SAS System for Windows Release 9.3*; SAS Inst.: Cary, NC, USA, 2010.

13

The Electrochemical Properties of Biochars and How They Affect Soil Redox Properties and Processes

Stephen Joseph [1,2,3,*], Olivier Husson [4], Ellen Ruth Graber [5], Lukas van Zwieten [6,7], Sara Taherymoosavi [2], Torsten Thomas [8], Shaun Nielsen [8], Jun Ye [8], Genxing Pan [3], Chee Chia [2], Paul Munroe [2], Jessica Allen [1], Yun Lin [1], Xiaorong Fan [3] and Scott Donne [1]

[1] Discipline of Chemistry, University of Newcastle, Callaghan NSW 2308, Australia; E-Mails: j.allen@newcastle.edu.au (J.L.); yun.lin@newcastle.edu.au (Y.L.); scott.donne@newcastle.edu.au (S.C.)

[2] School of Materials Science and Engineering, University of NSW, Sydney NSW 2052, Australia; E-Mails: s.taherymoosavi@student.unsw.edu.au (S.T.); c.chia@unsw.edu.au (C.C.); p.munroe@unsw.edu.au (P.M.)

[3] Institute of Resource, Ecosystem and Environment of Agriculture, Nanjing Agricultural University, Nanjing 210095, China; E-Mails: gxpan@njau.edu.cn (P.G.X.); xiaorongfan@njau.edu.cn (X.R.F.)

[4] CIRAD/PERSYST/UPR 115 AIDA and AfricaRice Centre, 01 BP2031 Cotonou, Benin; E-Mail: husson@cirad.fr

[5] Institute of Soil, Water & Environmental Sciences, The Volcani Center, Agricultural Research Organization, P.O. Box 6, Bet Dagan 50250, Israel; E-Mail: ellen.graber@volcani.agri.gov.il

[6] NSW Department of Primary Industries, Bruxner Highway, Wollongbar, 2480 NSW, Australia; E-Mail: lukas.van.zwieten@dpi.nsw.gov.au

[7] Southern Cross Plant Science, Southern Cross University, Lismore, 2477 NSW, Australia

[8] School of Biotechnology and Biomolecular Sciences, University of NSW, Sydney NSW 2052, Australia; E-Mails: t.thomas@unsw.edu.au (T.T.); shaunson26@gmail.com (S.N.); jun.ye2014@gmail.com (J.Y.)

* Author to whom correspondence should be addressed; E-Mail: joey.stephen@gmail.com

Academic Editor: Peter Langridge

Abstract: Biochars are complex heterogeneous materials that consist of mineral phases, amorphous C, graphitic C, and labile organic molecules, many of which can be either electron donors or acceptors when placed in soil. Biochar is a reductant, but its electrical and electrochemical properties are a function of both the temperature of production and the concentration and composition of the various redox active mineral and organic phases present. When biochars are added to soils, they interact with plant roots and root hairs, micro-organisms, soil organic matter, proteins and the nutrient-rich water to form complex organo-mineral-biochar complexes Redox reactions can play an important role in the development of these complexes, and can also result in significant changes in the original C matrix. This paper reviews the redox processes that take place in soil and how they may be affected by the addition of biochar. It reviews the available literature on the redox properties of different biochars. It also reviews how biochar redox properties have been measured and presents new methods and data for determining redox properties of fresh biochars and for biochar/soil systems.

Keywords: biochar; cyclic voltammetry; Pourbaix diagram; electron shuttling

1. Introduction

Biochar is produced from the thermal treatment of biomass in the near absence of air. The utilisation of biochar in agriculture, forestry and land remediation has the potential to not only sequester carbon (C), increase plant yields and enhance soil chemical and physical properties, but also to alter emissions of nitrous oxide (N_2O), affect nutrient leaching and run-off, impact the availability of contaminants in soil and alter soil physical and chemical properties [1]. Biochars are complex heterogeneous materials whose properties depend on feedstock and pyrolysis conditions. They consist of mineral phases, amorphous C, graphitic C, and labile organic molecules, many of which can be either electron donors or acceptors when placed in soil [2,3]. As a result, biochar can be both a reductant and oxidant [3]; its electrical and electrochemical properties are a function of both the temperature of production and the concentration and composition of the various redox active mineral and organic phases present [2–5].

When biochars are added to soils, they interact with plant roots and root hairs, micro-organisms, soil organic matter, proteins and the nutrient-rich soil solution to form organo-mineral-biochar complexes [4]. Redox reactions are hypothesized to play an important role in the development of these complexes, and also to cause significant changes in the original C matrix of the biochar [5].

Electrons are essential reactants in inorganic, organic, and biochemical reactions. While there is a fundamental link between proton transfer and electron transfer, redox reactions, *i.e.*, transfer of electrons, are arguably more important to the chemistry and biochemistry of living systems than proton transfer reactions [5]. In the soil-rhizosphere-plant system, redox conditions play critical roles in cell and plant physiology, microorganism functioning and community structure, soil genesis, and nutrient availability, uptake and transformation, among other processes [5].

This paper reviews the redox processes that take place in soil and how they may be affected by the addition of biochar, a redox active material. It reviews the available literature on the redox properties of different biochars. It also reviews how biochar redox properties have been measured and presents new methods and data for determining redox properties of fresh biochars and for biochar/soil systems.

2. Review of Literature

2.1. An Overview of Redox Processes in Soils with Biochar

Redox processes involving the donation and acceptance of electrons play an important part in soils, such as in nutrient cycling (phosphorous and nitrogen), scavenging of free radicals, formation and destruction of ethylene, methane and nitrous oxide [6,7]. Decomposing organic matter, including biochar, can be regarded as an electron-pump supplying electrons to more oxidised species present in the soil system [8]. However, unlike organic matter, biochar oxidises at much slower rates and has the potential to store electrons [9].

As Husson [10] notes, oxidation and reduction conditions are assessed by measuring the redox potential (Eh;V). Eh is derived by combining the standard free energy change of the generic redox reaction in Equation (1) with the Nerst equation, where F is the Faraday constant, Ox is the oxidized species, Red is the reduced species, R = gas constant, T is temperature in kelvin, *n* = number of electrons, *m* = number of protons exchanged:

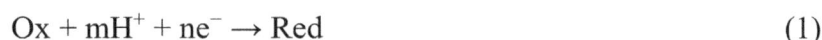

$$\mathrm{Ox} + \mathrm{mH}^{+} + \mathrm{ne}^{-} \rightarrow \mathrm{Red} \quad (1)$$

$$Eh = \mathrm{E}° + \frac{\mathrm{RT} \times \ln[\mathrm{Red}]}{\mathrm{nF} \quad [\mathrm{Ox}]} + \frac{2.303\mathrm{mRT} \times \mathrm{pH}}{\mathrm{nF}} \quad (2)$$

The zero point for the Eh scale (E°) is set by the standard hydrogen electrode (SHE). Equation (2) shows the relationship between Eh and pH. As Eh is correlated to pH, electron activity needs to be compared at a given pH, which can be done by correcting Eh to pH 7 using the regression given by Equation (2) [11,12]. As an example of the importance of this correction, a biochar suspension having an Eh of 500 mV and pH 5 has a higher electron activity than a biochar suspension having an Eh of 400 mV and pH 9. Soil and biochar suspensions can be characterized simultaneously by their Eh and pH using a modified Pourbaix Diagram [4,8]. In the same way, the Eh-pH of a particular biochar suspension can be superimposed on this diagram. Examples are given and techniques are discussed in a latter section. The issue in measurement of the Eh and pH of these diagrams has been discussed in [8,10].

Poise is the resistance to change in Eh when a small amount of oxidant removes electrons from a system or, conversely, a small amount of reductant adds electrons. Poise and Eh are expressions that can be compared to buffer capacity and pH in soils. Poise increases with the total concentration of oxidant plus reductant, and, for a fixed total concentration, it reaches a maximum when the ratio of oxidant to reductant is unity [13]. Since biochars function mainly as reductants, we suggest that, when added to soil, biochar may cause an increase in soil poise [10].

Molecular oxygen is the most common electron-acceptor species, although, in its absence, there are alternative electron acceptors in soil, such as MnOOH (E° = 1.5 V), MnO_2 (E° = 1.2 V), NO_3^- (E° = 0.88 V), and Fe^{3+} (E° = 0.66 V) that can also cause oxidation of organic C [6]. Soil moisture status and soil structure are key factors controlling redox reactions. In well-aerated arable soils, there is a mosaic of anaerobic microsites, in particular in the rhizosphere, resulting from O_2 consumption by both soil organisms and root respiration. Thus, there are two soil redox interfaces; namely (i) the low soil Eh rhizosphere zone, with a high Eh of surrounding well-aerated soil; and (ii) the low Eh rhizosphere region and its interface with the high Eh surfaces of live and growing roots. Similarly, when biochar is added to soil and adsorbs water, there will be internal pores that have no or very low concentrations of O_2 and where the Eh is very low, as well as an area surrounding the biochar where the Eh can be high (in aerated soils [5]).

Redox transformations are most apparent at interfaces where there are two unlike environments and, hence, a driving force for reaction. Typically, these are wetlands, flooded fields and the regions between roots and moist soil in the rhizospheres of many plants [6]. Additions of high concentrations of biochar into the rhizosphere could introduce an environment that contrasts the one that would naturally develop there from the typical soil clays, silt, sand and organic matter components. The redox potential in the immediate area around the biochar particle could change as solutions rich in organic compounds, cations and anions, diffuse in and out of the macro- and meso-pores of the biochar. Joseph *et al.* [5] detailed biotic and abiotic reactions that could take place on the surface and in the pores of the biochar, with regards to both the C matrix and mineral matter.

Research has also highlighted the role that iron minerals play in redox processes in soil [7]. It has been hypothesised that iron minerals in biochar could catalyse a range of redox reactions associated with nutrient cycling, nitrification and denitrification [13]. Iron exists in all types of biomass both in the Fe^{2+} and Fe^{3+} oxidation states and both oxidation states can exist in the biochar usually as an iron oxide (Fe_2O_3, Fe_3O_4, $Fe(OH)_3$). The concentration of different Fe compounds with different oxidation states depends on the pyrolysis conditions (air is often entrained with fuel) as well as complex catalytic processes that take place during pyrolysis between the alkali metals and the organic molecules [5]. Some of these Fe phases have diameters less than 10 nm and are completely surrounded by C, while others exist on the surfaces of pores and can be oxidised by air. Once biochars are placed in soil, Fe compounds can precipitate out on the surface of the biochar or, if they are in a water saturated environment, they can be reduced [4].

The redox processes involving Fe may include:

(1) Electron transfer from organic matter to Fe(III) (hydr) oxides via C oxidation [14].
(2) Reduction of NO_3^- to NO_2^- with the oxidation of Fe^{2+} to Fe^{3+}.
(3) Mineralisation of organic N to NH_4^+ [15] and the oxidation of NH_4^+ to NO_2^- with the consequent reduction of Fe^{3+} to Fe^{2+}.
(4) Oxidation of NH_4^+ to NO_2^- with the consequent reduction of Fe^{3+}, formation and oxidation of FeS minerals in the sulphur (S) cycle [7].
(5) Cycling of S from solid to soluble liquid species driven by oxidation or reduction of Fe species [7].

Biochar with a high content of Fe oxide nanoparticles at the surfaces of pores could significantly increase the rate of reduction of NO_3^- and NO_2^- by lowering the free energy required for the process [13].

Microorganisms can also actively modify and optimize their immediate geochemical surroundings by mobilizing nutrients for uptake and increasing the accessibility of electron acceptors for energy generation to facilitate assimilation and dissimilation [16]. Fe^{3+} educing microorganisms, such as the bacteria *Shewanella* sp., can excrete electrons to increase the availability of Fe minerals. For example, it was reported that small concentrations of biochar stimulate both the rate and extent of microbial reduction of the Fe(III) oxyhydroxide mineral ferrihydrite by *Shewanella oneidensis* MR-1 [17].

2.2. The Electrochemical Properties of Biochars, Summary of Literature

Only a handful of studies have measured redox characteristics of biochars. These are reviewed here. Unfortunately, none have yet explored in a well-controlled manner whether such properties actually play a role in the various effects biochars are reported to have on the soil environment, plant growth and health, gaseous emissions, and the makeup of the rhizosphere microbiome. This is a major knowledge gap.

Ishihara [18] reported that wood charcoal carbonised at <300 °C, 300 °C–800 °C and >800 °C acts as an insulator, a semiconductor and a conductor, respectively. Biochars produced at 600 °C and above are conductors but exhibit a lattice structure that contains a considerable concentration of micropores [19] and stable radicals within the graphitic structure. They contain very low concentrations of water-soluble organic molecules and oxygen-based functional groups. Both graphitic conductors and amorphous C semiconductors produced at temperatures above 600 °C are utilised in electrochemical devices (such as batteries, supercapacitors and microbial fuel cells) and there is now a considerable understanding of the electron transfer processes that take place at the surfaces both in biotic and abiotic redox reactions [20].

Mineral and C phases in biochar have different electrochemical potentials [21]. The C phase has a series of tubular pores (the xylem and phloem of the original biomass structure) that can connect the different mineral phases. These tubular pores are themselves porous at the nanometer scale, and there is also connectivity across the C phases. This porous structure can have similar properties to a semipermeable membrane, whereby two parallel C tubular pores with different concentrations of soluble metals can act as a galvanic cell [22]. Most biochars contain redox active Fe and Mn-based minerals, and many of these redox active minerals exist as nanophase particles [5] often in a number of oxidation states. For example, Fe oxide in biochar can exist as magnetite or hematite. Iron contents as high as 2.3% of the total weight of biochar (biosolids feedstock) has been reported [23], but wood-based biochars tend to have much lower concentrations (0.14%–0.34% w/w). However, biochars produced at lower temperatures (<450 °C) have a much greater concentration of labile organic molecules and surface functional groups.

Graber *et al.* [2] found that aqueous extracts of biochars produced over a range of temperatures from three different feedstocks had substantial reducing capacities. Extracts of two biochars prepared from greenhouse wastes at two different highest heat temperatures (HTTs) were able to solubilize Mn and Fe from different soils over a wide range of pH values, presumably by means of reducing the oxidized metal species. The extract of the lower temperature biochar, having a greater variety and concentration of soluble reducing agents, solubilized more Mn and Fe than the extract of the higher temperature biochar. In the studied systems, the dissolved organic matter (DOM) fraction, in particular, phenolic compounds, was proposed to be responsible for the main part of the reducing capacity, which was on a par with the reducing capacities of various humic and fulvic substances.

Klupfel *et al.* [3] studied the redox properties of 19 wood and grass biochars produced at temperatures ranging from 200 to 700 °C using mediated electrochemical analysis. All the biochars could reversibly accept and donate up to 2 mmol of electrons per gram of biochar, with the high mineral ash grass-based biochars having higher electron exchange capacities (EEC) than the wood-based biochars. Maximum EEC was found for biochars produced at 400 °C. Combined electrochemical, elemental, and spectroscopic analyses of the thermosequence biochars provided evidence that the pool of redox-active moieties was dominated by electron-donating phenolic moieties in the low-HTT biochars, by newly-formed electron accepting quinone moieties in intermediate-HTT biochars, and by electron accepting quinones and possibly condensed aromatics in the high-HTT biochars.

It was also suggested that, due to its conductive properties, biochar solids promote direct interspecies electron transfer (DIET) in co-cultures of *Geobacter metallireducens* with *Geobacter sulfurreducens* or *Methanosarcina barkeri* in a fashion and extent similar to granular activated carbon (GAC), despite being 1000 times less conductive than GAC [9]. The possibility that DIET was promoted by the liquid soluble phase of biochar was rejected because an isolated soluble biochar fraction exhibited less DIET promotion than did the biochar together with the soluble phase. However, this may have been an artifact of the method used for isolating the soluble fraction, which differed from the method used to prepare the biochar itself and could have resulted in a much lower concentration of isolated soluble materials than would have been present in the biochar co-cultures. The authors reported that the reduced biochar was incapable of transferring electrons to Fe(III) citrate, leading them to conclude that electron transport through the solid biochars was unlikely to be due to quinone moieties, but rather, was due to the conductive properties of the solid phase itself. These results contradict those of [3], who found that all 19 studied biochars were able to reversibly donate and accept electrons. It also contradicts the study of [17]. Results from these few studies are insufficient to conclude the mechanisms involved in biochar redox activity, and clearly a gap in knowledge still exists.

2.3. Measurement of the Electrochemical Properties of Fresh Biochars and Soil/Biochar Systems

Different techniques have been used to measure the redox properties of biochars. This section reviews methods that have been published and presents new data using solid-state cyclic voltammetry (SSCV). Graber *et al.* [2] extracted the water-soluble organic fraction of biochars and used the Ferric Reducing Antioxidant Power (FRAP) assay to determine reduction capacity (RC) of the extracted organic molecules. Total phenols in the biochar aqueous extracts and DOM solutions were determined using the Folin Ciocalteu (F-C) assay. $FeSO_4 \cdot H_2O$ standards were also measured by the F-C assay to calculate moles of charge transfer (mol_c), where each mole of $FeSO_4$ represents one mole of charge transfer. Solubilization of Mn and Fe from the soils were compared for biochar aqueous extracts and water as a function of pH, which was controlled with a buffer.

Kluepfel *et al.* [3] finely ground the biochars, removed any O_2 through purging with N_2 at 80 °C, and then ubiquinone (Q10) was added to a solution of ethanol. Biochars were suspended in 0.1 M KCl and 0.1 M phosphate buffer, pH 7, were reduced by addition of borohydride ($NaBH_4$), and then subsequently re-oxidized by O_2 in air. Changes in the biochar redox states were determined by electrochemical analysis of untreated, $NaBH_4$-reduced, and re-oxidized chars. The redox states of the biochars were quantified by mediated electrochemical reduction (MER) and oxidation (MEO) using an electrochemical cell. A glassy carbon cylinder served both as the working electrode (WE) and electrochemical reaction vessel and the counter electrode was a coiled platinum wire separated from the WE compartment by a porous glass frit. The applied redox potentials were measured against Ag/AgCl reference electrodes.

A commonly used technique for measuring the redox characteristics of carbons is solid-state cyclic voltammetry (SSCV) [24]. In this technique, a composite electrode consisting of 10% biochar and 90% SFG6 graphite (Timcal) was prepared through light grinding in a mortar and pestle. This was mixed with a phosphate buffer, equilibrated for 16 h and then transferred to a stainless steel cup cell in which the cylindrical walls are Teflon-lined. This mix was compacted in a hydraulic press to produce a working electrode. A stainless steel inner lining was used as the counter electrode. The cell was then allowed to equilibrate with the electrolyte for 1 h after which the working electrode potential was swept from its open circuit value cathodically to −1.2 V, after which the sweep was reversed to an upper potential limit of 0.2 V, and then back down again to −1.2 V. The potential sweep rate was set at 0.05 mV/s. More details are provided in the supplementary information (S1). In this study, the residual electrode, after cycling, was examined using both a scanning electron microscope (SEM) and a transmission electron microscope (TEM) as per the methods described [5]. In this present study and the supplementary information, SEM imaging was carried out using an Hitachi S3400 fitted with a Bruker Silicon Drift Energy Dispersive X-ray microanalysis system (EDS) for elemental analysis and a JEOL 2100 TEM fitted with an Oxford Instruments EDS detector.

Chen *et al.* [9] determined the electrical conductivities of biochar by two-probe electrical conductance measurements using two gold electrodes separated by a 50 μm non-conductive gap. Biochar was placed between the two gold electrodes to bridge the non-conductive gap. Voltage was then scanned from 0 V to 10.05 V in steps of 0.025 V. For each sample, current was measured 100 s after setting the voltage to allow the exponential decay of the transient ionic current in the gap and to measure steady state electronic current.

Husson *et al.* [25] and [26] note that accurately determining Eh in soils both with and without biochar is difficult, due to:

(i) High variability of soil Eh in space and time: Eh is largely influenced by hydric conditions (water activity), temperature, microbial activity and respiration of living organisms [27,28]. As a consequence, it is difficult to obtain stable measurements, especially in soils with low poising ability (that is, soils with low organic matter and clay content).
(ii) Irreversibility of redox reactions at the surface of the electrodes, which makes it difficult to conduct Eh measurements over long time periods [29].
(iii) Chemical disequilibrium in soils [30].
(iv) Polarisation of and/or leakage from electrodes.

Recently, the possibilities of further complications in Eh measurements were reported [31,32], namely:

(v) The influence of electromagnetic fields on water and living organisms, which can greatly perturb Eh measurements in soil samples through an induced current in the electrode [31].
(vi) The possible role of "structured" interfacial water in cells, affecting electron activity in exclusion zones [32,33].

In both field and bench scale measurements, tandem Eh-pH measurements should be undertaken to help elucidate the role of biochar in affecting redox processes. Husson *et al.* [25] recommends that all measurements be undertaken away from electromagnetic fields (as these can affect the stability of final measurement of the voltage) using very high impedance voltmeters and frequent cleaning of electrodes. Multiple sampling is necessary, as if the electrode touches the biochar in the soil, readings could be much lower than in the overall biochar/soil system. In aerobic soils, stable reproducible Eh measurements need to be carried out at a soil moisture content close to field capacity [25].

For bench scale measurements, soil sample preparation needs to be standardized (especially soil drying and rewetting procedure) and temperature must be maintained at approximately 25 °C [25]. Drying the sample at 35 °C over 3–4 days led to more stable measurements. They found that approximately 10%–15% of water should be added to sandy soils and 25%–35% for clayey soils. When rewetting, water should have an Eh of approximately 400 mV to ensure reproducible results. In flooded soils, the Eh needs to be measured as a function of depth away from electromagnetic fields with frequent cleaning of the electrode and multiple sampling in the experimental plots.

The following section presents new data using the cyclic voltammetry and soil methods described above.

2.4. Electrochemical Properties of Biochar as Measured Using Solid State Cyclic Voltammetry (SSCV), SEM and TEM

Three very different biochars that had been fully characterized and used in field trials were selected. These included:

(1) A wood biochar (Jarrah) produced at 600 °C in a vertical retort [34]. This biochar had a high surface area, high fixed C, high concentration of stable aromatic C and low concentration of functional groups and low ash.
(2) *Acacia saligna* biochar produced at 380 °C in a rotating drum kiln. This biochar has high labile C content and a relative low surface area compared with the Jarrah biochar [35].
(3) Chicken litter high mineral ash biochar produced at 400 °C in a rotating drum kiln [4].

The voltammogram from these biochars was compared to that of the black C particles taken from Terra Preta soils in the Hata Hara region of the Amazon and examined in detailed by [36]. These particles contain high concentrations of mineral matter and an Fe content of approximately 3%.

The voltammetric behaviour of the Saligna and Jarrah biochar electrodes is shown in Figure 1. The initial sweep from the open circuit potential for both the Saligna and Jarrah down to −1.2 V and then back up to +0.2 V shows the electrodes behave like a capacitor, with no real evidence for any faradaic charge transfer processes. However, when the upper potential limit is reached, the Saligna electrode undergoes a substantial activation process at a potential that would be typical of the Eh for many soils, in which case there is significant anodic current generated in the potential range from +0.25 V to −0.1 V. This is followed by significant cathodic current at potentials below −0.7 V. A similar activation process did not occur for the Jarrah biochar.

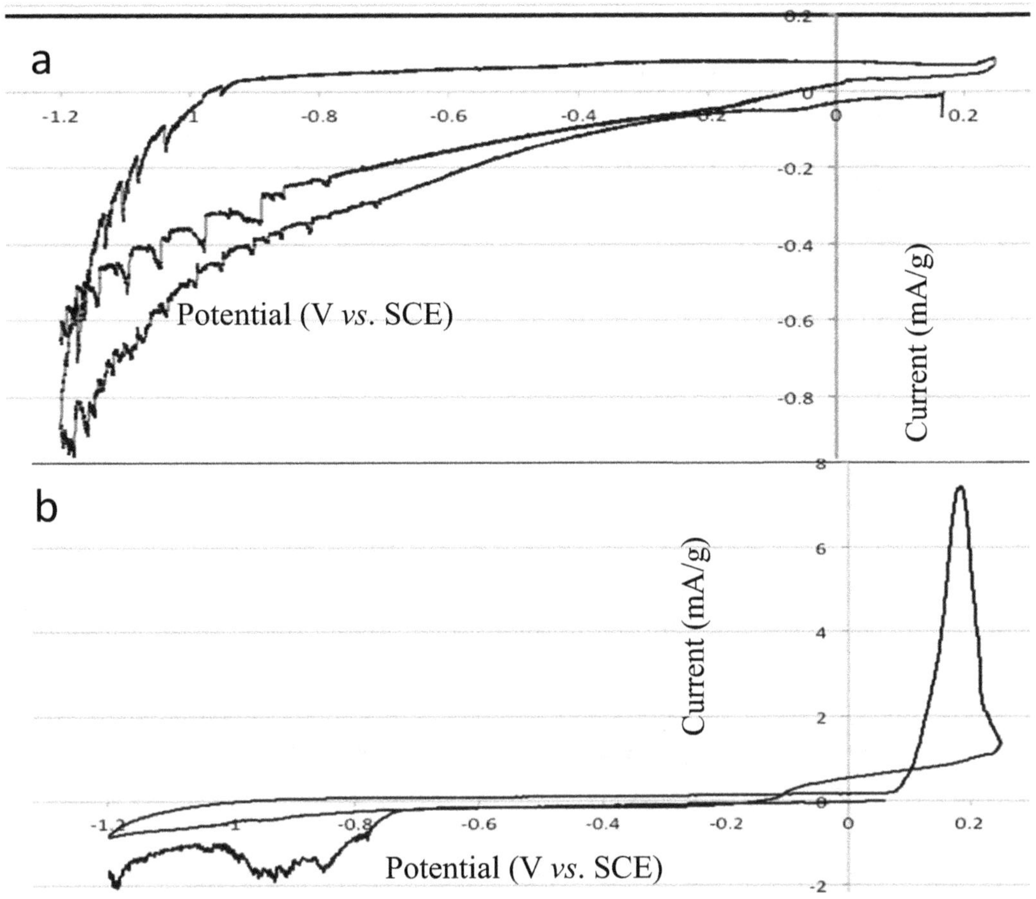

Figure 1. (**a**) Voltammograms of Jarrah biochar produced at 600 °C; and (**b**) Acacia saligna biochar produced at 380 °C.

A similar voltammogram was observed when either a chicken manure biochar (Figure 2a) or Terra Preta soil was cycled (Figure 2b). Nakamura *et al.* [37] and [24] reported that the increase in current would indicate that oxidation of the C to CO_2 takes place preferentially at the nanophase mineral/C interfaces and areas where high concentrations of oxygen functional groups exist [38]. Tulloch *et al.* [24] using cyclic voltammetry has shown that anatase, pyrite, kaolin montmorillinite and alumina increased the rate of oxidation of C in coal. Corrosion experiments carried out on graphite and amorphous C indicate that oxidation of amorphous C surfaces results in an increase in pore surface area [38].

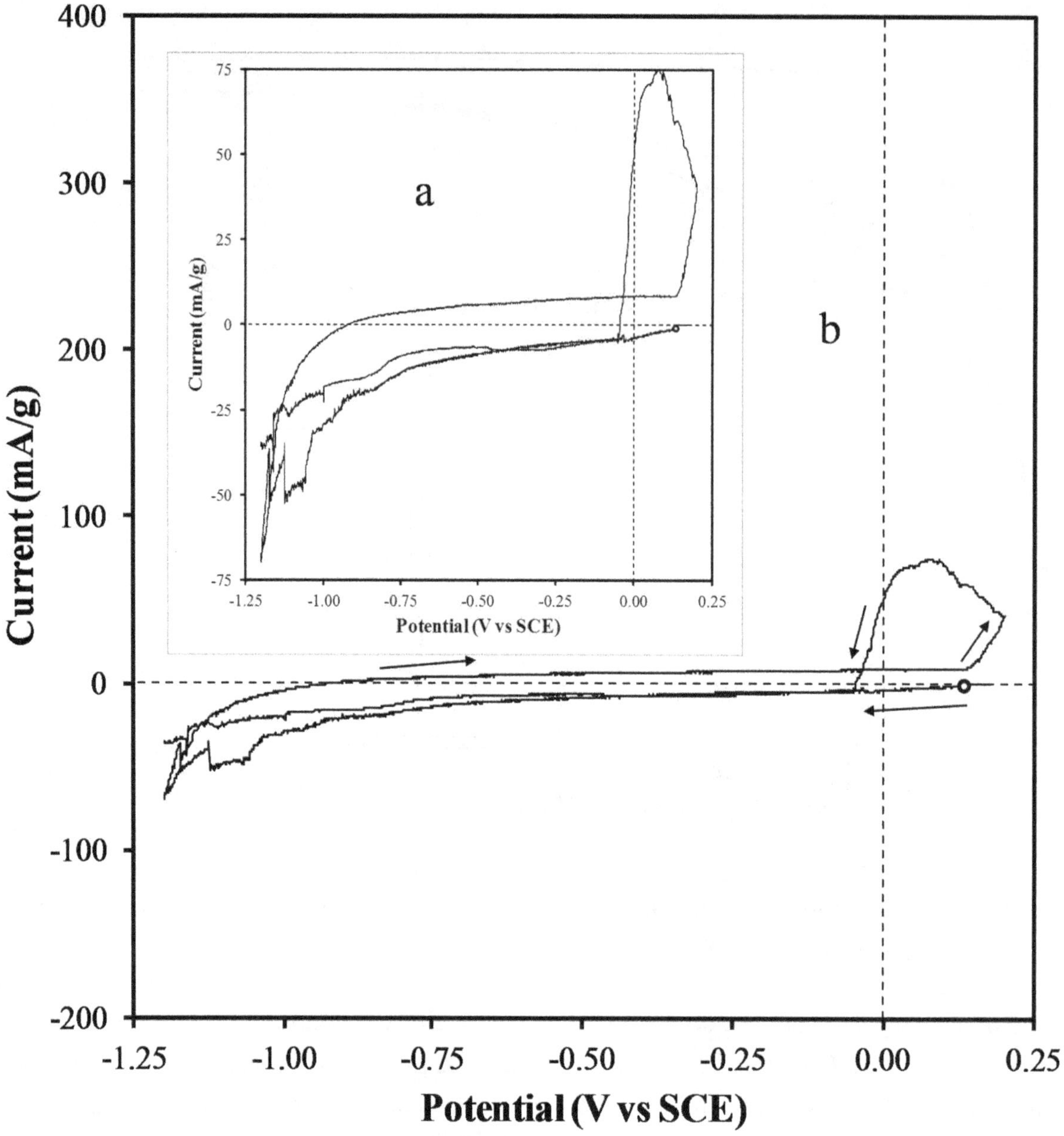

Figure 2. (**a**) Voltammogram of chicken manure biochar produced at 400 °C; (**b**) is that of particles of Terra Preta soils taken from Hata Hara district.

The SEM images and EDS spectra from the cycled *Acacia Saligna* biochar are consistent with the findings above. Significant damage occurred on the surfaces of C matrix (Figure 3a,b) indicated by broken and rounded edges of the walls of the pores and formation of the cracks around an area rich in Fe-based compounds (Figure 3c).

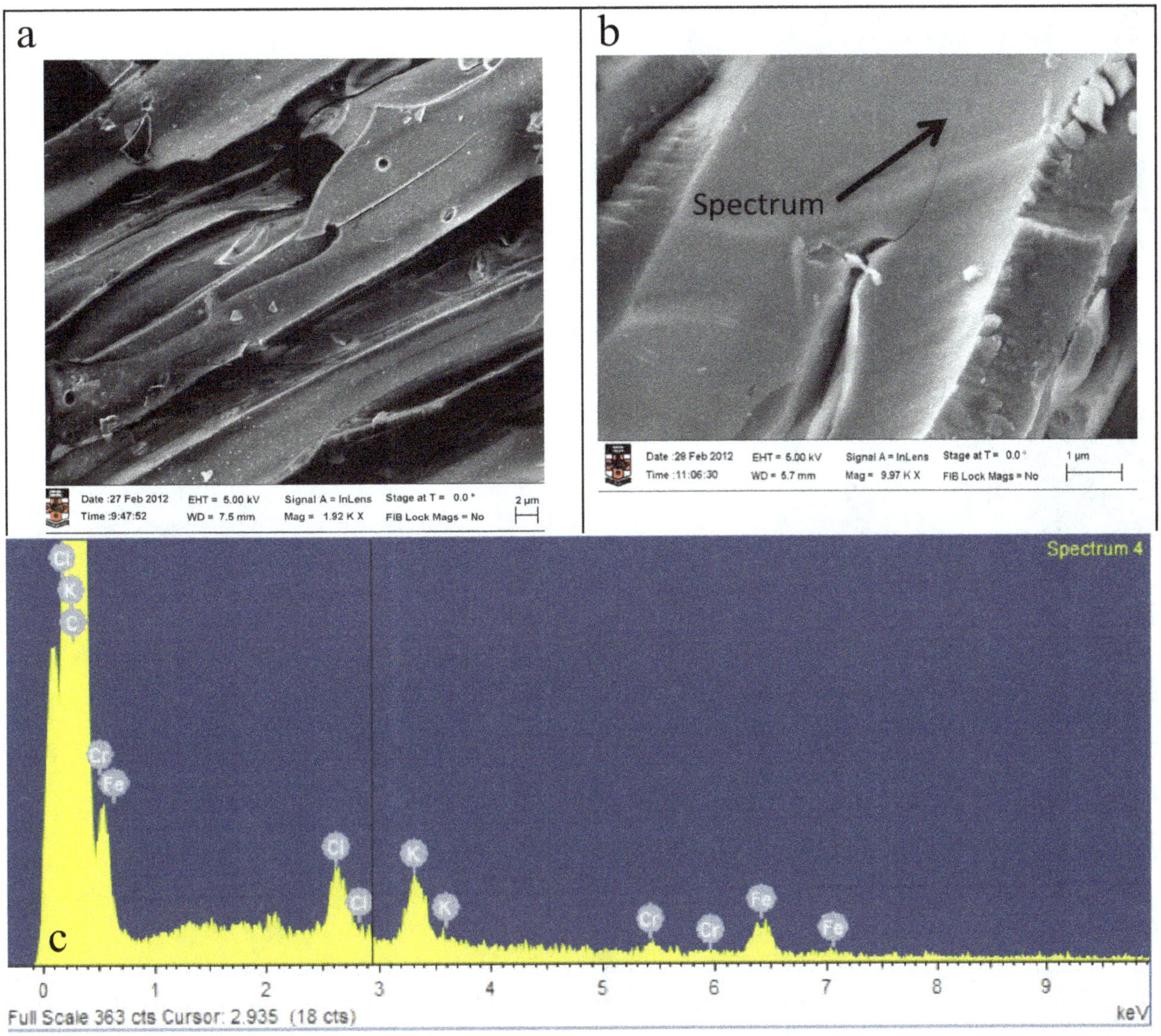

Figure 3. (**a**) SEM image of a surface of a Acacia Saligna biochar where there has been damage on the walls of the biohar (**b**) SEM image of the cracks forming in the carbon matrix of the Acacia Saligna biochar; (**c**) is an EDS analysis of the area shown in (**b**) indicating that there are localised concentrations of Fe-rich and KCl phases on this surface.

The damage noted in these images is similar to that reported by [39]. Formation of cracks in the C matrix could be due to the evolution of gas (probably CO_2), especially as they occur in regions where there is a high concentration high in Fe and O [40]. TEM examination of the surface of the cycled Saligna biochar revealed the formation of a crack at the end of an Fe-rich mineral phase (Figure 4a–c). Figure 5 shows a TEM image of nanophase Fe/C/O rich particles that were separated from the main biochar particle. The nanophase Fe particles are coated with an amorphous C layer. These structures are similar to those observed by [41] on corroded carbon black samples.

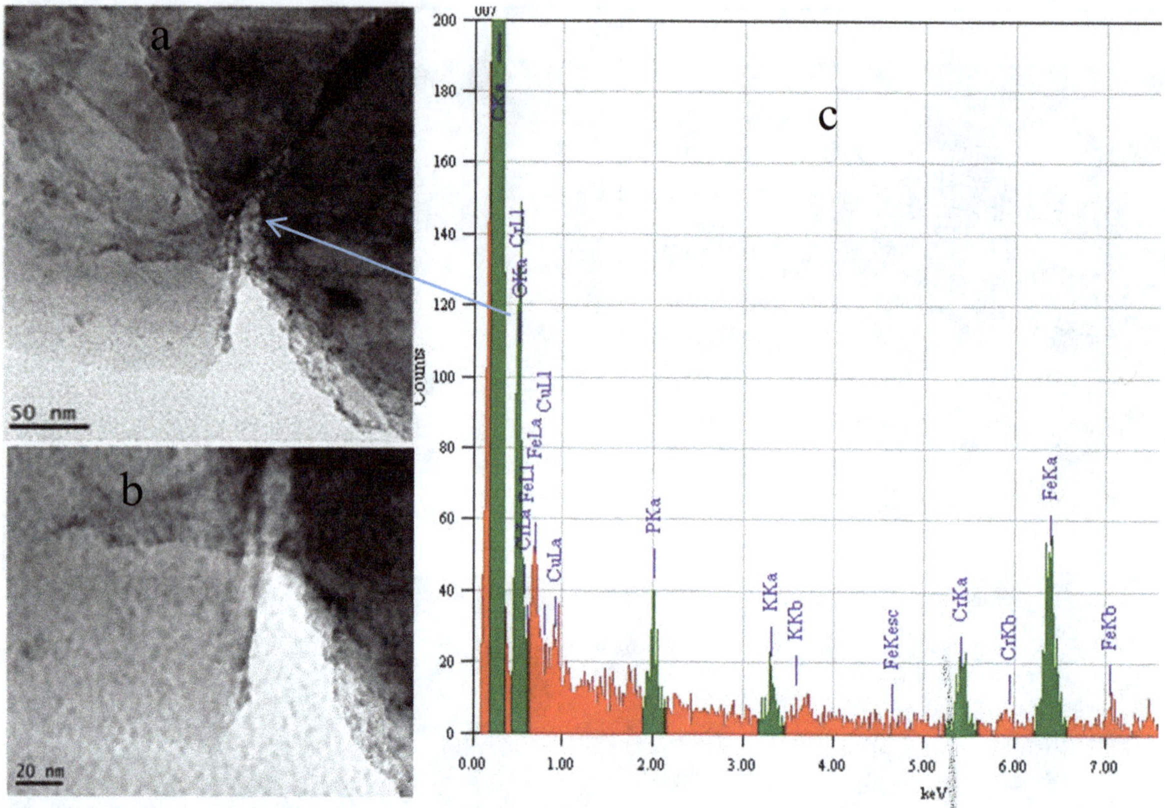

Figure 4. (**a**) TEM image of the oxidized surface of the Saligna biochar where the light coloured region is amorphous carbon and the darker areas are Fe-rich minerals; (**b**) is a magnified TEM image of the crack arrowed in Figure 5a,c; (**c**) is an EDS spectrum taken from the area around the crack.

Similar TEM and SEM images of the Jarrah biochar (Figure S2) with its high concentration of graphitic planes did not reveal similar damage to the surfaces after cycling two times. It is thus apparent that this high temperature wood biochar is much more resistant to oxidation than the lower temperature biochar. These results thus correspond to evidence that higher HTT biochars are more stable in the environment and have substantially longer predicted half-lives than lower HTT biochars [42].

It is possible that once the labile organic C has been oxidized, new oxide layers on more stable graphene sheets could form, and these could protect the surface from further oxidation [43]. These types of processes could account for some of the results from incubation studies where biochar addition causing short-term positive priming of soil organic matter [44] and then longer term negative priming [45]. Whether biotic processes have a larger role than these chemical redox reactions is yet to be determined.

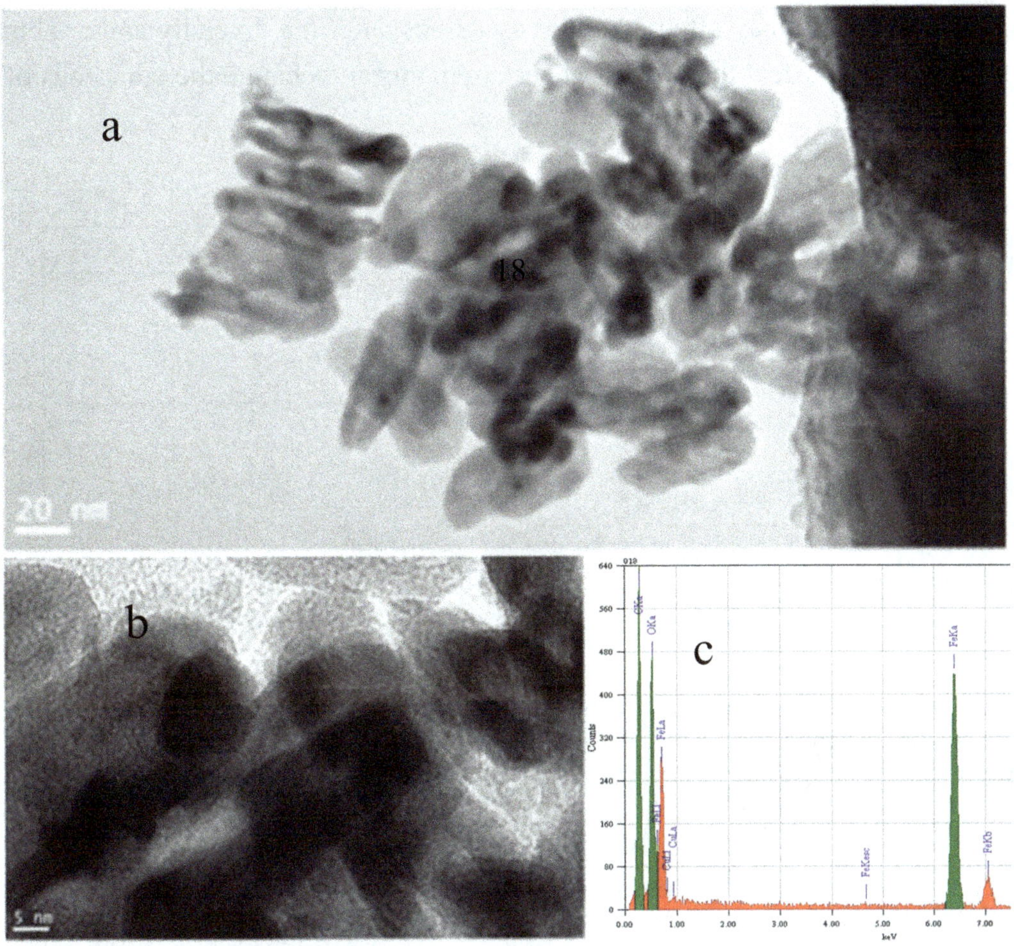

Figure 5. (**a**) shows a TEM image of the surface of biochar after cycling; (**b**) is a higher resolution image of the area shown in (**a**) (18) and (**c**) is an EDS spectrum from area 18 in (**a**).

2.5. Changes in Eh and pH When Biochars are Added to Soil Using the In-Situ *Measurement Technique*

Joseph *et al.* [4,5] and [17] have detailed the potential biotic and abiotic redox reactions that take place when biochar is added to soil. Figure 6 summarises that data on a modified Pourbaix diagram in the changes of Eh (pH corrected) and pH after rice straw biochar (RSB) produced at 400 °C and 625 °C were added to a sandy soil (arenosol, 92% sand, 2% silt, 6% clay) at application rates of 0.5%, 1%, 2% and 5 wt %. Detailed data for this biochar including Eh-corrected for pH are given in supplementary information (Figure S3a–c).

Adding a 1 wt % RSB produced at either 400 °C or 650 °C reduced the soil Eh from 0.55 V to 0.38 V and the pH was increased from 4.6 to 6.0. These conditions are considered to be very favourable for the growth of most plants and for the increase in beneficial micro-organisms [8,10,46]. Adding a small amount (2 wt %) of either temperature RSB reduces the soil Eh from 0.675 V to 0.325 V and the pH was increased from 4.6 to 6.9. Similarly, if the Eh is normalised for a pH of 7 (see supplementary information), the effect of adding 5% biochar does not significantly reduce Eh compared with adding 2%. A similar set of results were obtained for rice husk biochar, although pH was not increased to the same extent as with the straw biochar, due to the high mineral silica content and lower carbonate value

(Figure S4a,4b,4c,4d). A low Eh may affect N and P dynamics, alter the diversity and/or abundance of micro-organisms, and, in turn, affect the growth of certain plants and change the emissions of N_2O and CH_4 [47].

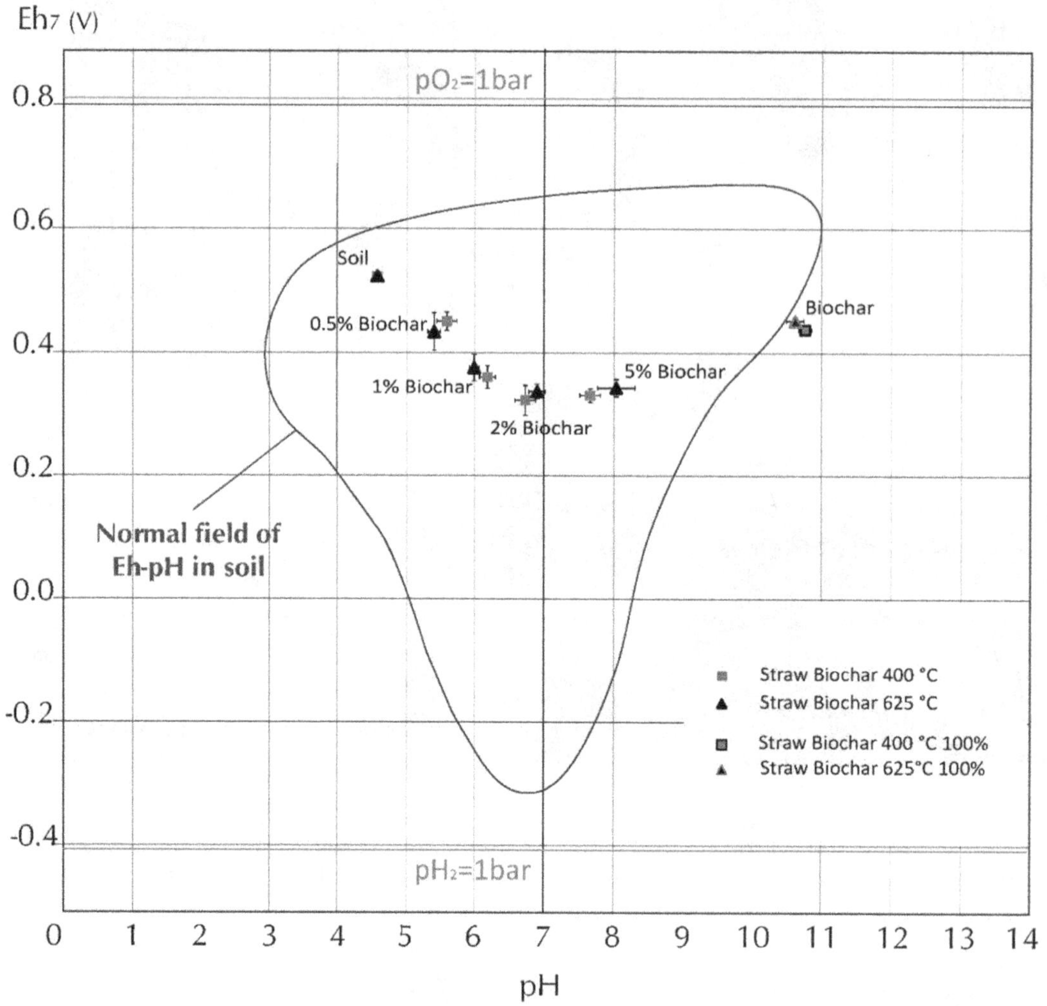

Figure 6. Summary of information on measurement on the change in Eh (corrected for pH) and pH with the addition of 2 rice straw biochars produced at 400 °C and 625 °C.

3. Conclusions and Future Directions

The initial investigations discussed above along with the work of [2–5,10,13,48,49] highlight the importance of redox process affecting the bulk soil properties occurring on the surface and in the vicinity of biochar particles. When biochars are added to soils in significant quantities, we sometimes observe a rapid change in both Eh and pH. The magnitude of the changes that occur for a particular soil/biochar combination are a function of the pyrolysis process conditions (time and maximum temperature), the feedstock, whether there has been any pre or post-treatment of the biochar, and on the soil type and environment. Mineral matter, labile organic molecules and C functional groups at surfaces determine the Eh of a biochar and this, in turn, affects the change in Eh of a soil after the amendment. As [30] noted, redox processes in soil are kinetically hindered and are in disequilibrium with each other. Thus biochar impacts on soil Eh can change over time especially if biochar reacts/interacts with micro-organisms, soil mineral and organic matter, chemical fertilisers and plant roots. As [4] noted, the redox activity of biochars in soils could be a function of the changes in the nature of the organomineral layers that form and/or break down during different crop cycles, changes in environmental conditions and through movement through the soil profile. It is possible that as biochars themselves fragment, new redox active surfaces are exposed, thus resulting in a more complex series of redox reactions.

Key questions that require further research include:

1. Are the redox properties of biochars responsible for some or any of the different effects that biochars have been reported to have in the integrated soil/plant/rhizosphere microbiome system?
2. What are the mutual redox interactions of different biochars in soils of different types?
3. Do biochars increase the poising capacity of soils and why?
4. Are some biochars more effective than others in altering Eh fluctuations, especially in systems (e.g., rice) where flooding and drying cycles occur?
5. Does the penetration of root hairs into the pores of biochar and/or the attachment of roots to the biochar surface change the potential across the plant cell wall and change the take up of specific nutrients? If so, why does this occur?
6. Do Fe and Mn/Oxide particles with diameters less than 20 nm redox active particles on the surfaces of biochar assist in the breakdown of organic matter, increase P availability and reduce the production of greenhouse gases?

Studies to answer such questions have yet to be undertaken, in part because the recognition that biochars are redox active is relatively recent, and, in part, due to the difficulties involved in designing experiments to isolate and test these questions. We propose that one way to do this would be to compare soil system impacts of a given biochar with those of the same biochar doped with redox active components or otherwise altered in its redox activity. Alternatively, the creation of a model system that isolates effects of interest and reduces the number of potential causative factors could be developed and tested.

Acknowledgments

We acknowledge the help of the Electron Microscope and X-ray unit of University of Newcastle and the Electron Microscope Unit of the University of NSW. This work was supported by the grant LP120200418 of the ARC, Renewed Carbon Pty Ltd. and the DAFF Carbon Farming Futures Filling the Research Gap (RG134978).

Author Contributions

All authors have contributed to the production of this paper.

Conflicts of Interest

The authors have no conflict of interest.

References

1. Lehmann, J.; Joseph, S. *Biochar for Environmental Management: Science, Technology and Implementation*; Routledge: Abingdon, UK, 2015.
2. Graber, E.; Tschansky, L.; Cohen, E. Reducing capacity of water extracts of biochars and their solubilization of soil Mn and Fe. *Eur. J. Soil Sci.* **2014**, *65*, 162–172.
3. Klüpfel, L.; Keiluweit, M.; Kleber, M.; Sander, M. Redox properties of plant biomass-derived black carbon (biochar). *Environ. Sci. Technol.* **2014**, *48*, 5601–5611.
4. Joseph, S.D.; Camps-Arbestain, M.; Lin, Y.; Munroe, P.; Chia, C.H.; Hook, J.; van Zweiten, L.; Kimber, S.; Cowie, A.; Singh, B.P.; *et al.* An investigation into the reactions of biochars in soil. *Aust. J. Soil Res.* **2010**, *48*, 501–515.
5. Joseph, S.; Graber, E.R.; Chia, C.; Munroe, P.; Donne, S.; Thomas, T.; Nielsen, S.; Marjo, C.; Rutlidge, H.; Pan, G.X.; *et al.* Shifting paradigms on biochar: Micro/nano-structures and soluble components are responsible for its plant-growth promoting ability. *Carbon Manag.* **2013**, *4*, 323–343.
6. Bartlett, R.J.; James, B.R. Redox chemistry of soils. *Adv. Agron.* **1993**, *50*, 151–208.
7. Li, Y.R.; Yu, S.R.; Strong, J.; Wang, H. Are the biogeochemical cycles of carbon, nitrogen, sulfur, and phosphorus driven by the "FeIII–FeII redox wheel" in dynamic redox environments? *J. Soils Sediments* **2012**, *12*, 683–693.
8. Chesworth, W. Redox, soils, and carbon sequestration. *Edafología* **2004**, *11*, 37–43.
9. Chen, S.; Rotaru, A.; Shrestha, P.; Malvankar, S.; Liu, F.; Fan, W.; Nevin, K.P.; Lovley, D.R. Promoting interspecies electron transfer with biochar. *Sci. Rep.* **2014**, *4*, 5019.
10. Husson, O. Redox potential (Eh) and pH as drivers of soil/plant/microorganism systems: A transdisciplinary overview pointing to integrative opportunities for agronomy. *Plant Soil* **2013**, *362*, 389–417.
11. Pidello, A. Environmental redox potential and redox capacity concept using a simple polarographic experiment. *J. Chem. Educ.* **2003**, *80*, 68–70.
12. Glinski, J.; Stepniewski, W. *Soil Aeration and its Role for Plants*; CRC Press: Boca Raton, FL, USA, 1985.
13. Van Zwieten, L.; Kammann, C.; Cayuela, M.L.; Singh, B.-P.; Joseph, S.; Kimber, S.; Clough T.; Spokas, K. Biochar Effects on Nitrous Oxide and Methane Emissions from Soil. In *Biochar for Environmental Management: Science, Technology and Implementation*; Lehmann, J., Joseph, S., Eds.; Routledge: Abingdon, UK, 2015.
14. Davidson, E.A.; Dail, D.B.; Chorover, J. Iron interference in the quantification of nitrate in soil extracts and its effect on hypothesized abiotic immobilization of nitrate. *Biogeochemistry* **2008**, *90*, 65–73.
15. Yin, X.; Lv, X.; Jiang, M.; Zou, Y. Research progress of the coupling process of Fe and N in wetland soils. *Chin. J. Environ. Sci.* **2010**, *31*, 2254–2259. (In Chinese)
16. Melton, E.; Swanner, E.D.; Behrens, S.; Schmidt, C.; Kappler, A. The interplay of microbially mediated and abiotic reactions in the biogeochemical Fe cycle. *Nat. Rev. Microbiol.* **2014**, *12*, 797–808.

17. Kappler, A.; Wuestner, M.L.; Ruecker, A.; Harter, J.; Halama, M.; Behrens, S. Biochar as an Electron Shuttle between Bacteria and Fe(III) Minerals. *Environ. Sci. Technol.* **2014**, *1*, 339–344.
18. Ishihara, S. Recent trend of advanced carbon materials from wood charcoals. *Mokuzai Gakkai Shi* **1996**, 42, 717–723
19. Bourke, J.; Manley-Harris, M.; Fushimi, C.; Dowaki, K.; Nunoura, T.; Antal, M.J. Do all carbonized charcoals have the same chemical structure? 2. A model of the chemical structure of carbonized charcoal. *Ind Eng Chem Res*. **2007,** 46, 5954–5967
20. Arends, J.; Verstraede, W. 100 years of microbial electricity production:three concepts for the future. *Microbial Biotech*. **2012** , *5*, 333–346.
21. Fabiano, A.; Petter, B.; Madari, E. Biochar: Agronomic and environmental potential in Brazilian savannah soils. *Rev. Bras. Eng. Agríc. Ambient*. **2012,** *16*, 761–768.
22. Suda, F.; Matsuo, T.; Ushioda, D. Transient changes in the power output from the concentration difference cell (dialytic battery) between seawater and river water. *Energy* **2007**, *32*, 165–173.
23. Van Zwieten, L.; Kimber, S.; Morris, S.; Chan, K.Y.; Downie, A.; Rust, J.; Joseph, S.; Cowie, A. Effects of biochar from slow pyrolysis of papermill waste on agronomic performance and soil fertility. *Plant Soil* **2010**, *327*, 235–246.
24. Tulloch, J.; Allen, J.; Wibberley, L.; Donne, S. Influence of selected coal contaminants on graphitic carbon electro-oxidation for application to the direct carbon fuel cell. *J. Power Sources* **2014**, *260*, 140–149.
25. Husson, O.; Huson, B.; Brunet, A.; Babre, D.; Alary, K.; Sarthou, J.; Charpentier, H.; Durand, M.; Benada, J.; Henry, M. Practical improvements in soil redox potential (Eh) measurement for characterisation of soil properties. Application for comparison of conventional and conservation agriculture cropping systems. *Anal. Chim. Acta* **2015**, under review.
26. Fiedler, S.; Vepraskas, M.J.; Richardson, J.L. Soil redox potential: Importance, field measurements, and observations. *Adv. Agron*. **2007**, *94*, 1–54.
27. Thomas, C.R.; Miao, S.L.; Sindhoj, E. Environmental factors affecting temporal and spatial patterns of soil redox potential in Florida Everglades wetlands. *Wetlands* **2009**, *29*, 1133–1145.
28. Benada, J. A non invasive method for redox potential measurement. *Obilnarske Listy* **2009**, *14*, 15–18.
29. Whitfield, A.E.; Ullman, D.E.; German, T.L. Tomato spotted wilt virus glycoprotein G(C) is cleaved at acidic pH. *Virus Res*. **2005**, *110*, 183–186.
30. Grundl, T. A review of current understanding of redox capacity in natural, disequilibrium systems. *Chemosphere* **1994**, *28*, 613–626.
31. Colic, M.; Morse, D. The elusive mechanism of the magnetic "memory" of water. *Colloids Surf. A: Physicochem. Eng. Asp*. **1999**, *154*, 167–174.
32. Del Giudice, E.; Spinetti, P.R.; Tedeschi, A. Water dynamics at the root of metamorphosis in living organisms. *Water* **2010**, *2*, 566–586.
33. Pollack, G.; Clegg, J. Unexpected Linkage between Unstirred Layers, Exclusion Zones, and Water, In *Phase Transitions in Cell Biology*; Pollack, G., Chin, W.C., Eds.; Springer Netherlands: Dordrecht, The Netherlands, 2008; pp. 143–152.

34. Nielsen S.; Minchin T.; Kimber S.; van Zwieten, L.; Caporaso, G.; Gilbert, J.; Munroe, P.; Joseph, S.; Thomas, T. Enhanced biochar causes complex shifts in soil microbial communities. *Agric. Ecosyst. Environ.* **2014**, *191*, 73–82.
35. Chia, C.H.; Singh, B.P.; Joseph, S.; Graber, E.; Munroe, P. Characterization of an enriched biochar. *J. Anal. Appl. Pyrolysis* **2014**, *108*, 26–34.
36. Liang, B.; Lehmann, J.; Solomon, D.; Kinyangi, J.; Grossman, J.; O'Neill, B.; Skjemstad, J.O.; Thies, J.; Luizão, F.J.; Petersen, J.; *et al.* Black carbon increases cation exchange capacity in soils. *Soil Sci. Soc. Am. J.* **2006**, *70*, 1719–1730.
37. Nakamura, M.; Nakanishi, M.; Yamamoto, K. Influence of physical properties of activated carbons on characteristics of electric double-layer capacitors. *J. Power Sources* **1996**, *60*, 225–231.
38. Gallagher, K.G.; Fuller, T.F. Kinetic model of the electrochemical oxidation of graphitic carbon in acidic environments. *Phys. Chem. Phys.* **2009**, *11*, 11557–11567.
39. Siroma, Z.; Ishii, Z.; Yasuda, K.; Miyazaki, Y.; Inaba, M.; Tasaka, A. Imaging of highly oriented pyrolytic graphite corrosion accelerated by Pt particles. *Electrochem. Commun.* **2005**, *7*, 1153–1156.
40. Willsau, J.; Heitbaum, J. The influence of Pt-activation on the corrosion of carbon in gas diffusion electrodes—A dems study. *Electroanal. Chem.* **1984**, *161*, 93.
41. Liu, Z.Y.; Zhang, J.L.; Yu, P.T.; Zhang, J.X.; Makharia, R.; More, K.L.; Stach, E.A. Transmission Electron Microscopy Observation of Corrosion Behaviors of Platinized Carbon Blacks under Thermal and Electrochemical Conditions. *J. Electrochem. Soc.* **2010**, *157*, B906–B913.
42. Crombie, K.; Mašek, O.; Sohi, S.; Brownsort, P.; Cross, A. The effect of pyrolysis conditions on biochar stability as determined by three methods. *GCB Bioenergy* **2013**, *5*, 122–131.
43. Youngmi, Y. Study on the degradation of carbon materials for electrocatalytic applications. Ph.D. Thesis, Technical University of Berline, Berlin, Germany, 2014.
44. Maestrini, B.; Herrmann, A.M.; Nannipieri, P.; Schmidt, M.W.I.; Abiven, S. Ryegrass-derived pyrogenic organic matter changes organic carbon and nitrogen mineralization in a temperate forest soil. *Soil Biol. Biochem.* **2014**, *69*, 291–301.
45. Zimmerman, A.R.; Gao, M.; Ahn, M.Y. Positive and negative carbon mineralization priming effects among a variety of biochar-amended soils. *Soil Biol. Biogeochem.* **2011**, *43*, 1169–1179.
46. Baas Becking, L.G.M.; Kaplan, I.R.; Moore, D. Limits of the natural environment in terms of pH and oxidation-reduction potentials. *J. Geol.* **1960**, *68*, 243–284.
47. Kirk, G. *The Biogeochemistry of Submerged Soils*; John Wiley & Sons: Chichester, UK, 2004; pp. 92–134.
48. Singla, A.; Iwasa, H.; Inubushi, K. Effect of biogas digested slurry based-biochar and digested liquid on N_2O, CO_2 flux and crop yield for three continuous cropping cycles of komatsuna (*Brassica rapa* var. perviridis). *Biol. Fertil. Soils* **2014**, *50*, 1201–1209.

49. Singla, A.; Dubey, S.K.; Singh, A.; Inubushi, K. Effect of biogas digested slurry-based biochar on methane flux and methanogenic archaeal diversity in paddy soil. *Agric. Ecosyst. Environ.* **2014**, *197*, 278–287.

14

Minichromosomes: Vectors for Crop Improvement

Jon P. Cody [†], **Nathan C. Swyers**[†], **Morgan E. McCaw**[†], **Nathaniel D. Graham**[†], **Changzeng Zhao and James A. Birchler** *

Division of Biological Sciences, University of Missouri, 311 Tucker Hall, Columbia, MO 65211-7400, USA; E-Mails: joncody@mail.missouri.edu (J.P.C.); ncs89f@mail.missouri.edu (N.C.S.); mem7b6@mail.missouri.edu (M.E.M.); ndgraham@mail.missouri.edu (N.D.G.); zhaoc@missouri.edu (C.Z.)

† These authors contributed equally to this work.

* Author to whom correspondence should be addressed; E-Mail: BirchlerJ@Missouri.edu

Academic Editor: Gareth Norton

Abstract: Minichromosome technology has the potential to offer a number of possibilities for expanding current biofortification strategies. While conventional genome manipulations rely on random integration of one or a few genes, engineered minichromosomes would enable researchers to concatenate several gene aggregates into a single independent chromosome. These engineered minichromosomes can be rapidly transferred as a unit to other lines through the utilization of doubled haploid breeding. If used in conjunction with other biofortification methods, it may be possible to significantly increase the nutritional value of crops.

Keywords: minichromosomes; biofortification; B chromosomes; telomere truncation; BIBAC; genetic engineering; haploid induction

---

1. Introduction

While efforts to reduce global hunger have been successful, one in nine humans still suffer from malnourishment [1]. Such a statistic is not exclusively dependent on plant nutritional content, but arises from compounded factors in food security, with developing countries being greatly impacted [1].

Many of these developing regions depend on few staple crops for full nutrient intake, which often lack important dietary components, such as essential amino acids, carbohydrates, and minerals [2]. With limited options and access to nutritional supplements, many individuals in these countries suffer from micronutrient malnutrition [1]. Utilization of biofortification strategies may not be the "solve all" answer to global hunger problems. However, these techniques not only offer an opportunity to significantly reduce malnourishment in developing regions, but improve human health worldwide.

Biofortification aims to improve or supplement crop nutritional content through fertilizer application, conventional breeding, and/or genetic engineering [2]. The technique to be used would depend on the specific micronutrient being manipulated; however, complementary utilization of strategies could enhance nutritional output [3]. Inexpensive and simple fertilization strategies are not effective in most scenarios because of several disadvantages including differential effects on crop variants, inability to target edible plant components, and the fact that this approach only modulates the minerals and not the genotype [4]. Conventional breeding is an alternative technique that utilizes favorable characteristics that exist in natural variants and introduces them into commercial lines. The conventional breeding strategy, however, is time consuming and limited by genetic resources that are available [2,4]. Alternatively, genetic engineering/biotechnology, is a powerful tool that can be used to directly manipulate the genetic code of specific crop variants to alter metabolic processes or increase mineral uptake [4].

In 1983, researchers demonstrated that an isolated gene fragment could be transformed into a plant species [5–8]. Since then, biotechnology has been making significant contributions to several fields of scientific study, enabling timely alterations to genetic codes without lengthy introgression processes. Additionally, genetic engineering expands the possibilities that are available with conventional breeding by not relying on natural variations, and allowing increased control over gene expression with diverse promoters that are publically available. Because gene delivery is accomplished through either *Agrobacterium* mediated transformation or particle bombardment, transgene integration is random, which creates a number of limitations with this technology. Notable limitations include, but are not limited to: (1) disruption of endogenous gene function; (2) affected expression from regulatory elements; (3) difficulty of separation from closely associated genes; and (4) inefficient recovery of multiple transgenic events in each successive generation [9].

With many metabolic processes and nutrient accumulation mechanisms requiring multiple gene products, current genetic engineering methods are not an efficient strategy to accomplish the goals of nutrient accumulation in plants. Minichromosome technology, if applied to this problem, offers a unique solution to these limitations through the creation of an autonomous element that works as a platform for transgene stacking and can be transferred efficiently to subsequent generations [10]. Coupled with doubled haploid breeding, transfer of these engineered chromosomes into other varieties could be expedited, allowing rapid analysis of gene aggregates in several crop lines. Here, the basics of minichromosome technology are discussed in the hope of illustrating the potential utility in a number of such applications.

2. Discussion

2.1. Structure Overview

In order for minichromosomes to function properly, they must contain the necessary components required to be successfully propagated during cell division. All eukaryotic chromosomes must contain a centromere for kinetochore formation, origins of replication to maintain proper chromosome numbers during cell division, and telomeres to protect the chromosome ends from degradation. Interestingly, in plants, the telomere is the only component that can be synthesized, as the centromere is epigenetic in nature, and the origin of replication has not been elucidated. As a result, minichromosomes must be created by utilizing endogenous centromeres and telomeres in a process known as the top–down method.

2.2. Centromeres

The centromere is required in order for chromosomes to segregate properly during cell division. The region is responsible for recruitment of kinetochore proteins, and the ultimate attachment of the chromosome to microtubules for proper movement through the cell cycle. While some organisms only require a short sequence to form a centromere, those of plants are much more complex. Plants have regional centromeres, which are composed of satellite repeats, and can vary widely in number between species and even chromosomes [11–13]. CENH3, the histone H3 variant of plants typical of active centromeres, associates with this repeat region [14]; however, studies have shown that the entire region may not interact [14]. Many plant centromeres are additionally interspersed with retrotransposon sequences, in maize known as CRM elements [15]. CRM elements and satellite repeats both interact with CENH3 and are found throughout the centromere region [14,16]. Not all plant centromeres rely on such repeat regions; wheat, for example, does not possess any tandem repeats within its centromere [17].

Despite most centromeres within a genome containing similar sequences, multiple studies have found that new centromeres can form in regions that are unique in sequence structure [18–20]. Additionally, active centromeres that are formed over repeat regions can be inactivated, as was shown by the recovery of structurally dicentric B chromosomes in maize [21]. The realization that both centromere inactivation and de novo formation occur regularly illustrates an epigenetic component to centromere specification rather than a determination by the DNA sequence [22–24]. As a result, simply including centromeric repeats on a potential minichromosome construct is unlikely to induce kinetochore formation when introduced into a cell.

2.3. Origins of Replication

The timely duplication of the genome is an important step in the cell reproduction process. DNA synthesis is strictly regulated by a variety of mechanisms that determine the time and location of replication fork assembly [25]. The number of replication origins in a genome is mostly dependent on chromosome size [25]. In most prokaryotes, there exists only one origin of replication on the circular chromosome. Replication origins have been identified in many prokaryotes and there are even tools for predicting their locations. Eukaryotes, generally, have many origins of replication that assemble at different times, which allows for the replication of the large linear eukaryotic chromosomes.

S. cerevisiae is the only eukaryote where the replication origin has been identified. Neither a specific location nor consensus sequence has been found for origins of replication in any other yeast strains, nor in higher eukaryotes [25]. In metazoans there are many origins of replication, which are not all active in every cell [25]. With regard to minichromosomes, this situation eliminates any concern for specifically identifying replication origin constitution.

2.4. Telomeres

The telomere of the chromosome is a sequence repeat, and accompanying protein complex, that protects the end from damage or chromosome end-to-end fusions. While the majority of the telomere is a heterochromatic region, the extreme chromosome terminus is protected by a single strand overhang that forms a protective loop, known as a G loop [26]. The sequence that confers telomere function is highly conserved, with most plants having a repeat sequence of TTTAGGG that can be extended with the enzyme telomerase [27]. The telomerase is required for telomere extension as DNA polymerase can only extend after a primer template, leading to a loss of sequence at the end of the chromosome during replication. This restriction in synthesis leads to a shorter telomere sequence with each chromosomal replication event. While telomerase has the ability to extend shortened telomeres, telomerase expression varies widely between tissue types, showing little activity in mature tissues [28,29].

For the purposes of minichromosome creation, synthetically produced telomere sequences are introduced with the desired transgene. In most instances, the introduced telomere will be shorter than the endogenous telomere that it replaces. In order to be functional, these repeats must be extended and the end modified to include a G loop. While the exact process that occurs when a synthetic telomere is introduced is unknown, there are studies that suggest that telomere length is monitored, and can be extended in some organisms when the repeat number is too low [30]. Additionally, studies of *Arabidopsis* telomeres show that the repeat number can vary between individual chromosomes, and different ecotypes suggesting that the amount of telomere is being actively regulated [31]. As a result, despite minichromosomes not containing the required amount of telomere upon introduction, the plant contains the machinery required to extend the copy number to the essential number.

2.5. Utilization of B Chromosomes

B chromosomes are supernumerary chromosomes found in many species of plants. B chromosomes are dispensable. They contain no genes essential to the survival of a plant [32], and show little to no effect on fitness except at high copy number [33,34]. B chromosomes are maintained in populations by a drive mechanism. This process consists of nondisjunction of the B chromosome. This results in more B chromosomes in the progeny than in the parents. In maize (*Zea mays*) nondisjunction happens at the second pollen mitosis, which makes the two maize sperm. A plant with one B chromosome produces pollen with sperm containing two B chromosomes or containing none. The sperm with the B chromosomes then preferentially fertilizes the egg cell rather than the polar nuclei. [32]. In rye (*Secale cereale*) nondisjunction occurs at the first pollen mitosis and the two B chromosomes are directed to the generative nucleus resulting in both sperm carrying two B chromosomes [32]. This directed nondisjunction also occurs during the formation of the egg. In maize, two regions are necessary for the nondisjunction of the B centromere. One is located in the proximal euchromatic region, while the other is at the distal end of the long arm [35]. Both act in trans. In rye, only a region on the distal end of the long arm is necessary for nondisjunction.

B chromosomes are useful in the creation of an engineered minichromosome because telomere mediated truncation involves the removal of some or most of at least one chromosome arm. If an A chromosome is truncated, this will generally lead to a detrimental monosomic condition, though rare events of minichromosomes with an A chromosome centromere and otherwise normal complement of chromosomes have been reported [36]. Truncating a B chromosome has no detrimental effect on the phenotype of a plant and therefore does not lower the rate of recovery of truncation events.

B chromosomes do not pair or recombine with A chromosomes [32]. This could allow multiple transgenes to be kept together as a unit instead of requiring intensive breeding strategies to stack multiple transgenes in a single plant. Because B chromosomes do not contain any vital genes, there is no chance of linkage of a transgene to an unfavorable allele or knockout of an important gene.

The accumulation mechanism of the B chromosome can be used to increase the copy number of minichromosomes and increase their dosage [37]. Because the truncation of a B chromosome usually removes the distal end of the long arm, it is necessary to have an unreduced B in the background, which can act in trans to restore the nondisjunction property.

2.6. Alternative Methods for Engineering Minichromosomes

B chromosomes are not found in some agriculturally important crops. In order to create minichromosomes in these crops an A chromosome must be truncated to provide an active centromere. In a euploid individual the truncation of an A chromosome would be detrimental. For that reason telotrisomic lines are a good candidate for telomere truncation. A telotrisomic individual has a normal complement of chromosomes plus a chromosome consisting of one chromosome arm. Telotrisomics can be found by screening the progeny of a triploid plant for trisomic individuals. From trisomics, centromere misdivision events, or splitting, can produce telotrisomics. A complete set of telotrisomics was developed for rice [38].

Aneuploid plants are less vigorous than euploid; generally the less of the genome for which a plant is aneuploid, the less negative the phenotype. Optimally, a minichromosomes should contain as little of the A genome as possible. Telotrisomics are good candidates for creating minichromosomes because they require only one chromosome arm truncation to remove much of the genic region from the A chromosome for which they are trisomic and relieve the aneuploid phenotype.

The effects of aneuploidy can also be lessened in a higher ploidy background. In maize, a truncated A chromosome was rescued in a spontaneous tetraploidy event [39]. Tetraploid plants have also been used for A chromosome truncation in *Arabidopsis thaliana* [40] and barley (*Hordeum vulgare*) [41]. The truncated chromosome can then be transferred to a diploid background by successive crosses of the tetrapliod by a diploid to produce a triploid and again by a diploid to recover the truncated minichromosome as an extra entity in a diploid background [39].

2.7. Stability in Plants

Minichromosomes derived from the B chromosome of maize are heritable, but transmitted at varying frequencies. Many minichromosomes fail to pair by metaphase 1 of meiosis. Other irregularities such as early sister chromatid separation and lagging chromosomes during division have also been observed. These meiotic abnormalities appear to be correlated with minichromosome size [42]. Nondisjunction can also occur in somatic tissues, especially at high minichromosome copy number. When mitotic metaphase cells were examined in maize root meristems, B chromosome and minichromosome copy number varied between cells when high numbers were present [37].

Stable transmission of minichromosomes can be achieved in multiple ways. Firstly, larger B minichromosomes tend to pair better and behave more normally in meiosis than smaller B-derived minichromosomes. A minichromosome derived from a truncation of only the distal tip of a B chromosome will behave like an A chromosome if there are two copies to allow pairing and there are no normal B chromosomes present [42]. To improve transmission of smaller, less stable minichromosomes or minichromosomes in a species without B chromosomes, a gametophytic selection mechanism has been proposed. Cytoplasmic male sterile (cms) plants have a mitochondrial mutation which causes pollen abortion. A nuclear restorer of fertility (Rf) gene prevents pollen abortion in cms plants [43]. Inserting Rf onto a minichromosome would allow any pollen grains with a minichromosome to develop, while any pollen grains without a minichromosome would abort [44].

2.8. Transgene Expression

Random integration of genetic fragments can lead to varying levels of expression, due to possible silencing events that occur at different loci throughout the host genome. This phenomenon, coupled with repressed transcription of repeated or homologous sequences, is an obstacle in the field of genetic engineering [45,46]. Artificial minichromosome technology offers an autonomous platform for sequential transgene integration events at a specific locus that has been demonstrated to successfully express genes at detectable levels. However, utilization of repeated genetic elements needs to be avoided when designing minichromosomes due to possible homologous gene silencing events [46]. While still in preliminary stages of development, the use of synthetic promoters will be a valuable tool in avoiding sequence repeats [47]. As both artificial minichromosomes and synthetic promoter technologies progress in parallel, complementary use of both systems will work synergistically to expand limits of genetic engineering and avoid possible silencing events that occur from sequence homology.

2.9. Creating Large Minichromosomes with BIBACs

Genetic engineering has, so far, focused mostly on introducing single or a few genes into crops. While these crops have been successful, more complex traits such as improved nutrition will rely on stacking multiple genes [48]. Thus far, introduction of multiple genes has relied on complex breeding to combine separately introduced genes, or co-introducing multiple genes during one transformation. As a result, the ability to introduce multiple traits on one T-DNA is preferred.

Vectors capable of carrying large (>300 kb) DNA fragments, known as bacterial artificial chromosome (BAC) vectors, were first created for use in *E. coli* [49]. Shortly after, the vectors were modified to have the ability to be maintained in *Agrobacterium tumefaciens* as transformation vectors [50]. The result, known as binary bacterial artificial chromosome (BIBAC) vectors, have been shown to transfer large intact DNA fragments in a variety of different plant species [50–52]. By including telomere sequences near the right border of a BIBAC vector, it should be possible to create minichromosomes while introducing large DNA fragments. These minichromosomes would allow multiple traits to be introduced at a single locus, and would only require a single selection for the entire transgene array.

2.10. In Vivo Modification of Minichromosomes

As previously stated, traditional genome manipulation tools, such as *Agrobacterium* mediated transformation and biolistic bombardment, impose a number of notable limitations. Introducing multiple traits into a single background will require several rounds of transformation, which can be labor intensive and time consuming. Through the utilization of genome targeting mechanisms, minichromosome technology offers a novel solution to these constraints. Precise targeting is derived from a number of site-specific recombination systems that add, remove, or modify specific genetic elements, allowing the assembly of large gene aggregates in a regulated manner.

A diverse number of site-specific recombination systems are publically available for research purposes [53,54]. While unique in sequence recognition, all systems follow the same basic steps: (1) Expressed recombinase enzymes bind to respective sites; (2) Bound enzymes form a synaptic complex, with sites assembled in a parallel orientation; (3) Recombinases catalyze the crossover and fusion of genetic material between recognized cassettes; (4) Disassembly of the synaptic complex [53]. The outcome of the recombination reaction is dependent on location and orientation of the recognition sites. Integration, inversion, and excision of specific genetic elements are all possible, so careful planning of experimental design needs to be carried out to ensure anticipated outcomes. It should be noted that all enzymes fall under a particular sub-category of recombinases that reflect the nature of the system [54]. These categories differ in a number of characteristics, including size and directionality [54]. Such a collection of recombinases can be exploited for the purpose of large-scale genetic manipulations, specifically the creation of artificial minichromosomes.

Regulation of recombinase expression is of most importance in assembling gene aggregates on minichromosomes. Several strategies to control the timing and duration of recombinase activity have been demonstrated [55]. With sexually propagating plants, backgrounds that are actively expressing a recombinase enzyme can be bred with lines containing a minichromosome that is to be modified. The recombinase gene could simply be crossed out in the next generation, leaving a background that only contains a modified minichromosome. Organisms that reproduce vegetatively, or require lengthy germination cycles, could favorably use tactics that transiently express the recombinase enzyme. An example of one of these methods is the auto-excision strategy. Under the control of an inducible promoter, recombinases can be activated by a number of factors, including heat-shock, chemical, or developmental cues [55]. If flanked by respective recombination sites, induction of an inducible promoter will lead to expression of the recombinase, followed by its removal via intra-molecular recombination [54,55].

Taken together, minichromosome construction will require strategic utilization of several site-specific recombination systems. Each respective recognition site is assembled in a specific orientation to allow integration of gene fragments and removal of selectable markers in a successive manner [56]. The proposed transgene stacking strategy exploits recombinase directionality, which enables recombination site and selectable marker recycling with each round of integration [56]. Before this process can be initiated, however, the minichromosome generated must contain a specific sequence that is to be acted upon by a well-characterized recombination system for the purpose of integration. A background that contains a minichromosome, and actively expresses a recombinase, will be introduced to a gene fragment that contains gene(s) of interest, recombination sites, and a selectable marker. The expressed recombinase will form a synaptic complex between its respective recognition sites on the minichromosome and the gene fragment, initiating strand exchange and integration. Transgene orientation upon insertion is predictable and unidirectional, due to the nature of the specific recombination system used. The selectable marker, now bordered by recombination cassettes, is used to identify a successful integration event. Depending on the method of reproduction, a different site-specific recombination system, for the purpose of excision, will be activated through genetic crosses or transient expression. This recombinase will identify the flanking recognition sequences and remove the selectable marker, restoring a single recombination site to be used in the next round of modifications.

2.11. Transfer of Minichromosomes with Haploid Induction

It may be possible to transfer minichromosomes to multiple lines rapidly through haploid induction. If a haploid inducing line containing a minichromosome is used to produce haploid embryos it should be possible to recover haploids with one or more minichromosomes as follows. In maize, high haploid induction rate lines have been derived from a line called Stock 6 [57], and its derivatives. When used as a pollen donor Stock 6 derived haploid inducer lines produce maternal haploids at a high rate. Many groups have noted the transfer of markers from the inducer line to the maternal haploids [58,59], and the transfer of complete B chromosomes has also been reported [59]. Minichromosomes introduced into the inducer line should be able to be transferred in an analogous fashion.

Once a haploid plant with a minichromosome has been obtained, its ploidy can be doubled by colchicine or high-pressure nitrous oxide gas [60]. These treatments produce diploid sectors in the plant which are fertile and the plant can be self-pollinated. This results in a doubled haploid line, which is completely homozygous. By this method, once a minichromosome is back-crossed into a haploid inducing line, it can be transferred to new lines with just two crosses: One to generate the haploid and one for self-pollination. The resultant progeny will be completely homozygous which is basically not possible if the minichromosome had been back-crossed into the line using a typical introgression scheme.

3. Conclusions

As the population of the world continues to increase, there is a growing need to find innovative ways to utilize the static amount of arable land. Biofortification of crops through conventional breeding and fertilizer application are two strategies to make headway toward improving nutritional value, but genetic engineering offers another strategy for addressing global food security. Gene stacking could allow for complex traits and pathways to be expressed with minimal selective breeding. Minichromosomes could provide a stable and heritable platform for this gene stacking. Because a bottom up approach toward creating an artificial chromosome is not yet possible, the top–down strategy for minichromosome creation is a viable option. B chromosomes, which are present or could be introduced in some crop species, offer a platform for creation of minichromosomes with the advantage that they lack essential genes. Other options as outlined above are also available. Minichromosomes have been shown to reliably express transgenes and to be transmitted from generation to generation. There is potential to make large scale additions to minichromosomes, which could allow for the introduction of multiple genes at once instead of through several transformations. There are now a variety of recombination systems, as well as genome editing technologies, which can be utilized to edit an existing minichromosome and make minichromosomes a truly custom platform. The ability to transfer minichromosomes through haploid breeding would allow for the rapid introduction of minichromosomes to many inbred lines, including those not amenable to transformation, without the need for generations of introgression. For these reasons, minichromosomes could provide a tool for improved biofortification of a variety of crop species.

Acknowledgments

Research on this topic was funded by National Science Foundation grant IOS-1339198.

Conflicts of Interest

The authors declare no conflict of interest.

References

1. FAO; IFAD; WFP. Strengthening the enabling environment for food security and nutrition. In *The State of Food Insecurity in the World 2014*; FAO: Rome, Italy, 2014.
2. Hirschi, K.D. Nutrient biofortification of food crops. *Annu. Rev. Nutr.* **2009**, *29*, 401–421.
3. Bruulsema, T.W.; Heffer, P.; Welch, R.M.; Cakmak, I.; Moran, K. *Fertilizing Crops to Improve Human Health: A Scientific Review*; International Plant Nutrition Institute: Norcross, GA, USA, 2012.
4. Zhu, C.; Naqvi, S.; Gomez-Galera, S.; Pelacho, A.M.; Capell, T.; Christou, P. Transgenic strategies for the nutritional enhancement of plants. *Trends Plant Sci.* **2007**, *12*, 548–555.
5. Bevan, M.W.; Flavell, R.B.; Chilton, M. A chimaeric antibiotic resistance gene as a selectable marker for plant cell transformation. *Nature* **1983**, *304*, 184–187.
6. Herrera-Estrella, L.; Depicker, A.; Montagu, V.M.; Schell, J. Expression of chimaeric genes transferred into plant cells using a Ti-plasmid-derived vector. *Nature* **1983**, *303*, 209–213.

7. Fraley, R.T.; Rogers, S.G.; Horsch, R.B.; Sanders, P.R.; Flick, J.S.; Adams, S.P.; Bittner, M.L.; Brand, L.A.; Fink, C.L.; Fry, J.S.; *et al.* Expression of bacterial genes in plant cells. *Proc. Natl. Acad. Sci. USA* **1983**, *80*, 4803–4807.
8. Murai, N.; Kemp, J.D.; Sutton, D.W.; Murray, M.G.; Slightom, J.L.; Merlo, D.; Reichert, N.A.; Sengupta-Gopalan, C.; Stock, C.A.; Barker, R.F.; *et al.* Phaseolin gene from bean is expressed after transfer to sunflower via tumor-inducing plasmid vectors. *Science* **1983**, *222*, 476–482.
9. Yu, W.; Han, F.; Birchler, J.A. Engineered minichromosomes in plants. *Curr. Opin. Biotechnol.* **2007**, *18*, 425–431.
10. Gaeta, R.T.; Masonbrink, R.E.; Krishnaswamy, L.; Zhao, C.; Birchler, J.A. Synthetic chromosome platforms in plants. *Annu. Rev. Plant Biol.* **2012**, *63*, 307–330.
11. Kanizay, L.; Dawe, R. Centromeres: Long intergenic spaces with adaptive features. *Funct. Integr. Genomics* **2009**, *9*, 287–292.
12. Burrack, L.S.; Berman, J. Neocentromeres and epigenetically inherited features of centromeres. *Chromosome Res.* **2012**, *20*, 607–619.
13. Kato, A.; Lamb, J.C.; Birchler, J.A. Chromosome painting using repetitive DNA sequences as probes for somatic chromosome identification in maize. *Proc. Natl. Acad. Sci. USA* **2004**, *101*, 13554–13559.
14. Zhong, C.X.; Marshall, J.B.; Topp, C.; Mroczek, R.; Kato, A.; Nagaki, K.; Birchler, J.A.; Jiang, J.; Dawe, R.K. Centromeric retroelements and satellites interact with maize kinetochore protein CENH3. *Plant Cell* **2002**, *14*, 2825–2836.
15. Wolfgruber, T.K.; Sharma, A.; Schneider, K.L.; Albert, P.S.; Koo, D.; Shi, J.; Gao, Z.; Han, F.; Lee, H.; Xu, R.; *et al.* Maize centromere structure and evolution: Sequence analysis of centromeres 2 and 5 reveals dynamic Loci shaped primarily by retrotransposons. *PLoS Genet.* **2009**, *5*, e1000743.
16. Jin, W.; Melo, J.R.; Nagaki, K.; Talbert, P.B.; Henikoff, S.; Dawe, R.K.; Jiang, J. Maize centromeres: Organization and functional adaptation in the genetic background of oat. *Plant Cell* **2004**, *16*, 571–581.
17. Liu, Z.; Yue, W.; Li, D.; Wang, R.; Kong, X.; Lu, K.; Wang, G.; Dong, Y.; Jin, W.; Zhang, X. Structure and dynamics of retrotransposons at wheat centromeres and pericentromeres. *Chromosoma* **2008**, *117*, 445–456.
18. Nasuda, S.; Hudakova, S.; Schubert, I.; Houben, A.; Endo, T. Stable barley chromosomes without centromeric repeats. *Proc. Natl. Acad. Sci. USA* **2005**, *102*, 9842–9847.
19. Gong, Z.; Yu, H.; Huang, J.; Yi, C.; Gu, M. Unstable transmission of rice chromosomes without functional centromeric repeats in asexual propagation. *Chromosome Res.* **2009**, *17*, 863–872.
20. Fu, S.; Lv, Z.; Gao, Z.; Wu, H.; Pang, J.; Zhang, B.; Dong, Q.; Guo, X.; Wang, X.; Birchler, J.A.; *et al.* De novo centromere formation on a chromosome fragment in maize. *Proc. Natl. Acad. Sci. USA* **2013**, *110*, 6033–6036.
21. Han, F.; Lamb, J.C.; Birchler, J.A. High frequency of centromere inactivation resulting in stable dicentric chromosomes of maize. *Proc. Natl. Acad. Sci. USA* **2006**, *103*, 3238–3243.
22. Han, F.; Gao, Z.; Birchler, J.A. Reactivation of an inactive centromere reveals epigenetic and structural components for centromere specification in maize. *Plant Cell* **2009**, *21*, 1929–1939.

23. Birchler, J.A.; Gao, Z.; Sharma, A.; Presting, G.G.; Han, F. Epigenetic aspects of centromere function in plants. *Curr. Opin. Plant Biol.* **2011**, *14*, 217–222.
24. Henikoff, S.; Furuyama, T. Epigenetic inheritance of centromeres. *Cold Spring Harb. Symp. Quant. Biol.* **2010**, *75*, 51–60.
25. Leonard, A.C.; Mechali, M. DNA replication origins. *Cold Spring Harb. Perspect. Biol.* **2013**, *5*, doi:10.1101/cshperspect.a010116.
26. Makarov, V.L.; Hirose, Y.; Langmore, J. Long g tails at both ends of human chromosomes suggest a C strand degradation mechanism for telomere shortening. *Cell* **2001**, *88*, 657–666.
27. Fajkus, J.; Fulnecková, J.; Hulanova, M.; Berkova, K.; Ríha, K.; Matyásek, R. Plant cells express telomerase activity upon transfer to callus culture, without extensively changing telomere lengths. *Mol. Gen. Genet.* **1998**, *260*, 470–474.
28. Fitzgerald, M.S.; McKnight, T.D.; Shippen, D.E. Characterization and developmental patterns of telomerase expression in plants. *Proc. Natl. Acad. Sci. USA* **1996**, *93*, 14422–14427.
29. Kilian, A.; Heller, K.; Kleinhofs, A. Development patterns of telomerase activity in barley and maize. *Plant Mol. Biol.* **1998**, *37*, 621–628.
30. Hemann, M.T.; Strong, M.A.; Hao, L.; Greider, C.W. The Shortest Telomere, Not Average Telomere Length, Is Critical for Cell Viability and Chromosome Stability. *Cell* 2001, 107. 67–77.
31. Shakirov, E.V.; Shippen, D.E. Length Regulation and dynamics of individual telomere tracts in wild-type arabidopsis. *Plant Cell* **2004**, *16*, 1959–1967.
32. Jones, N.; Houben, A. B chromosomes in plants: Escapees from the A chromosome genome. *Trends Plant Sci.* **2003**, *8*, 417–423.
33. Randolph, L.F. Genetic characteristics of the B chromosomes in maize. *Genetics* **1941**, *26*, 608–631.
34. Staub, R.W. Leaf striping correlated with the presence of B chromosomes in maize. *J. Hered.* **1987**, *78*, 71–74.
35. Roman, H. Mitotic nondisjunction in the case of interchanges involving the B-type chromosome in maize. *Genetics* **1947**, *32*, 391–409.
36. Gaeta, R.T.; Masonbrink, R.E.; Zhao, C.; Sanyal, A.; Krishnaswamy, L.; Birchler, J.A. *In vivo* modification of a maize engineered minichromosome. *Chromosoma* **2013**, *122*, 221–232.
37. Masonbrink, R.E.; Birchler, J.A. Accumulation of multiple copies of maize minichromosomes. *Cytogenet. Genome Res.* **2012**, *137*, 50–59.
38. Cheng, Z.; Yan, H.; Yu, H.; Tang, S.; Jiang, J.; Gu, M.; Zhu, L. Development and applications of a complete set of rice telotrisomics. *Genetics* **2001**, *157*, 361–368.
39. Yu, W.; Han, F.; Gao, Z.; Vega, J.M.; Birchler, J.A. Construction and behavior of engineered minichromosomes in maize. *Proc. Natl. Acad. Sci. USA* **2007**, *104*, 8924–8929.
40. Teo, C.H.; Ma, L.; Kapusi, E.; Hensel, G.; Kumlehn, J.; Schubert, I.; Houben, A.; Mette, M.F. Induction of telomere-mediated chromosomal truncation and stability of truncated chromosomes in Arabidopsis thaliana. *Plant J.* **2011**, *68*, 28–39.
41. Kapusi, E.; Ma, L.; Teo, C.H.; Hensel, G.; Himmelbach, A.; Schubert, I.; Mette, M.; Kumlehn, J.; Houben, A. Telomere-mediated truncation of barley chromosomes. *Chromosoma* **2012**, *121*, 181–190.
42. Han, F.; Gao, Z.; Yu, W.; Birchler, J.A. Minichromosome analysis of chromosome pairing, disjunction, and sister chromatid cohesion in maize. *Plant Cell* **2007**, *19*, 3853–3863.

43. Chase, C.D.; Gabay-Laughnan, S. Cytoplasmic male sterility and fertility restoration by nuclear genes. In *Molecular Biology and Biotechnology of Plant Organelles*; Daniell, H., Chase, C.D., Eds.; Springer: Dordrecht, The Netherlands, 2004; Volume 22, pp. 593–621.
44. Birchler, J.; Krishnaswamy, L.; Gaeta, R.; Masonbrink, R.; Zhao, C. Engineered Minichromosomes in Plants. *CRC Crit. Rev. Plant. Sci.* **2010**, *29*, 135–147.
45. Ye, F.; Signer, E.R. RIGS (repeat-induced gene silencing) in Arabidopsis is transcriptional and alters chromatin configuration. *Proc. Natl. Acad. Sci. USA* **1996**, *93*, 10881–10886.
46. Matzke, A.J.; Matzke, M.A. Position effects and epigenetic silencing of plant transgenes. *Curr. Opin. Plant Biol.* **1998**, *1*, 142–148.
47. Liu, W.; Yuan, J.S.; Stewart, C. Advanced genetic tools for plant biotechnology. *Nat. Rev. Genet.* **2013**, *14*, 781–793.
48. Halpin, C. Gene stacking in transgenic plants—The challenge for 21st century plant biotechnology. *Plant Biotechnol. J.* **2005**, *3*, 141–155.
49. Shizuya, H.; Birren, B.; Kim, U.; Mancino, V.; Slepak, T.; Tachiiri, Y.; Simon, M. Cloning and stable maintenance of 300-kilobase-pair fragments of human DNA in Escherichia coli using an F-factor-based vector. *Proc. Natl. Acad. Sci. USA* **1992**, *89*, 8794–8797.
50. Hamilton, C.M.; Frary, A.; Lewis, C.; Tanksley, S.D. Stable transfer of intact high molecular weight DNA into plant chromosomes. *Proc. Natl. Acad. Sci. USA* **1996**, *93*, 9975–9979.
51. Vega, J.M.; Yu, W.; Han, F.; Kato, A.; Peters, E.M.; Zhang, Z.S.; Birchler, J.A. Agrobacterium-mediated transformation of maize (Zea mays) with Cre-lox site specific recombination cassettes in BIBAC vectors. *Plant Mol. Biol.* **2008**, *66*, 587–598.
52. Hamilton, C.M.; Frary, A.; Xu, Y.; Tanksley, S.D.; Zhang, H. Construction of tomato genomic DNA libraries in a binary-BAC (BIBAC) vector. *Plant J.* **1999**, *18*, 223–229.
53. Grindley, N.D.; Whiteson, K.L.; Rice, P.A. Mechanisms of site-specific recombination. *Annu. Rev. Biochem.* **2006**, *75*, 567–605.
54. Wang, Y.; Yau, Y.; Perkins-Balding, D.; Thomson, J.G. Recombinase technology: Applications and possibilities. *Plant Cell Rep.* **2011**, *30*, 267–285.
55. Gidoni, D.; Srivastava, V.; Carmi, N. Site-specific excisional recombination strategies for elimination of undesirable transgenes from crop plants. *In Vitro Cell. Dev. Biol. Plant* **2008**, *44*, 457–467.
56. Ow, D.W. 2004 SIVB congress symposium proceeding: Transgene management via multiple site-specific recombination systems. *In Vitro Cell. Dev. Biol. Plant* **2005**, *41*, 213–219.
57. Coe, E.H. A line of maize with high haploid frequency. *Am. Natl.* **1959**, *93*, 381–382.
58. Zhang, Z.; Qiu, F.; Liu, Y.; Ma, K.; Li, Z.; Xu, S. Chromosome elimination and *in vivo* haploid production induced by stock 6-derived inducer line in maize (*Zea mays* L.). *Plant Cell Rep.* **2008**, *27*, 1851–1860.
59. Zhao, X.; Xu, X.; Xie, H.; Chen, S.; Jin, W. Fertilization and uniparental chromosome elimination during crosses with maize haploid inducers. *Plant Physiol.* **2013**, *163*, 721–731.

60. Kato, A. Chromosome doubling of haploid maize seedlings using nitrous oxide gas at the flower primordial stage. *Plant Breed.* **2002**, *121*, 370–377.

Permissions

All chapters in this book were first published in Agronomy, by MDPI; hereby published with permission under the Creative Commons Attribution License or equivalent. Every chapter published in this book has been scrutinized by our experts. Their significance has been extensively debated. The topics covered herein carry significant findings which will fuel the growth of the discipline. They may even be implemented as practical applications or may be referred to as a beginning point for another development.

The contributors of this book come from diverse backgrounds, making this book a truly international effort. This book will bring forth new frontiers with its revolutionizing research information and detailed analysis of the nascent developments around the world.

We would like to thank all the contributing authors for lending their expertise to make the book truly unique. They have played a crucial role in the development of this book. Without their invaluable contributions this book wouldn't have been possible. They have made vital efforts to compile up to date information on the varied aspects of this subject to make this book a valuable addition to the collection of many professionals and students.

This book was conceptualized with the vision of imparting up-to-date information and advanced data in this field. To ensure the same, a matchless editorial board was set up. Every individual on the board went through rigorous rounds of assessment to prove their worth. After which they invested a large part of their time researching and compiling the most relevant data for our readers.

The editorial board has been involved in producing this book since its inception. They have spent rigorous hours researching and exploring the diverse topics which have resulted in the successful publishing of this book. They have passed on their knowledge of decades through this book. To expedite this challenging task, the publisher supported the team at every step. A small team of assistant editors was also appointed to further simplify the editing procedure and attain best results for the readers.

Apart from the editorial board, the designing team has also invested a significant amount of their time in understanding the subject and creating the most relevant covers. They scrutinized every image to scout for the most suitable representation of the subject and create an appropriate cover for the book.

The publishing team has been an ardent support to the editorial, designing and production team. Their endless efforts to recruit the best for this project, has resulted in the accomplishment of this book. They are a veteran in the field of academics and their pool of knowledge is as vast as their experience in printing. Their expertise and guidance has proved useful at every step. Their uncompromising quality standards have made this book an exceptional effort. Their encouragement from time to time has been an inspiration for everyone.

The publisher and the editorial board hope that this book will prove to be a valuable piece of knowledge for researchers, students, practitioners and scholars across the globe.

List of Contributors

James Harrison, Karen A. Moore, Konrad Paszkiewicz, Thomas Jones, Murray R. Grant and David J. Studholme
College of Life and Environmental Sciences, University of Exeter, Geoffrey Pope Building, Stocker Road, Exeter EX4 4QD, UK

Daniel Ambacheew and Sadik Muzemil
Southern Agricultural Research Institution (SARI). P.O. Box. 06, Hawassa, Ethiopia

Clyde D. Boyette and Kenneth C. Stetina
USDA-ARS, Biological Control of Pests Research Unit, Stoneville, MS, 38776, USA

Robert E. Hoagland
USDA-ARS, Crop Production Systems Research Unit, Stoneville, MS, 38776, USA

Cynthia L. Sagers
Department of Biological Sciences, University of Arkansas, Fayetteville, AR 72701 USA

Jason P. Londo
United States Department of Agriculture, Agricultural Research Service, Geneva, NY 14456 USA

Nonnie Bautista
Plant Biology Division, Institute of Biological Sciences, University of the Philippines Los Baños, Laguna 4031, Philippines

Edward Henry Lee and Lidia S. Watrud
US Environmental Protection Agency, Western Ecology Division, National Health and Environmental Effects Research Lab, Corvallis, OR 97333, USA

George King
CSS-Dynamac, Corvallis, OR 97333, USA

Hardy Schulz and Bruno Glaser
Soil Biogeochemistry, Martin-Luther-University Halle-Wittenberg, Von-Seckendorff-Platz 3, Halle 06120, Germany

Gerald Dunst
Sonnenerde, Oberwarterstraße 100, Riedlingsdorf A-7422, Austria

Malinda S. Thilakarathna and Manish N. Raizada
Department of Plant Agriculture, University of Guelph, 50 Stone Road East, Guelph, ON N1G 2W1, Canada

Franz-W. Badeck and Fulvia Rizza
CRA-GPG—Council for Agricultural Research and Economics, Genomics Research Centre, Via San Protaso 302, Fiorenzuola d'Arda (PC) 29017, Italy

Garry Rosewarne
CSIRO Agriculture Flagship, Clunies Ross St, Black Mountain, Canberra 2601, Australia
Department of Environment and Primary Industries, 110 Natimuk Rd, Horsham Vic 3400, Australia

David Bonnett
CSIRO Agriculture Flagship, Clunies Ross St, Black Mountain, Canberra 2601, Australia
International Maize and Wheat Improvement Centre, Apdo. 06600 Mexico Distrito Federal, Postal 6-6-41, Mexico

Greg Rebetzke, Paul Lonergan and Philip J. Larkin
CSIRO Agriculture Flagship, Clunies Ross St, Black Mountain, Canberra 2601, Australia

Emanuel Lekakis and Aristotelis Papadopoulos
ELGO-DIMITRA, Soil Science Institute of Thessaloniki, Thermi 57001, Greece

Vassilis Aschonitis
Department of Life Sciences and Biotechnology, University of Ferrara, Via L.Borsari 46, Ferrara 44121, Italy

Athina Pavlatou-Ve and Vassilis Antonopoulos
Department of Hydraulics, Soil Science & Agricultural Engineering, School of Agriculture, Aristotle University of Thessaloniki, Thessaloniki 54124, Greece

Hussien Mohammed Beshir
Department of Plant Sciences, University of Saskatchewan, 51 Campus Drive, Saskatoon, SK S7N 5A8, Canada
School of Plant and Horticultural Sciences, Hawassa University, Hawassa, Ethiopia

Frances L. Walley
Department of Soil Science, University of Saskatchewan, 51 Campus Drive, Saskatoon, SK S7N 5A8, Canada

Rosalind Bueckert and Bunyamin Tar'an
Department of Plant Sciences, University of Saskatchewan, 51 Campus Drive, Saskatoon, SK S7N 5A8, Canada

Thomas Panagopoulos
Research Center of Spatial and Organizational Dynamics (CIEO), University of Algarve, Campus Gambelas, Faro 8005-139, Portugal

Jorge de Jesus
Ben-Gurion University of the Negev, Beer Sheva 84105, Israel

Jiftah Ben-Asher
Katif research center for coastal deserts development, Ministry of Science Sedot Negev Academic Campus, Sedot 86200, Israel

Gurpreet Kaur
Department of Soil, Environmental and Atmospheric Sciences, University of Missouri, Columbia, MO 65211, USA

Kelly A. Nelson
Division of Plant Sciences, University of Missouri, Novelty, MO 63460, USA

Duli Zhao, Jack C. Comstock, Per McCord and Sushma Sood
USDA-Agricultural Research Service, Sugarcane Field Station, Canal Point, FL 33438, USA

R. Wayne Davidson and Miguel Baltazar
Florida Sugar Cane League, Inc., Clewiston, FL 33440, USA

Stephen Joseph
Discipline of Chemistry, University of Newcastle, Callaghan NSW 2308, Australia
School of Materials Science and Engineering, University of NSW, Sydney NSW 2052, Australia
Institute of Resource, Ecosystem and Environment of Agriculture, Nanjing Agricultural University, Nanjing 210095, China

Olivier Husson
CIRAD/PERSYST/UPR 115 AIDA and AfricaRice Centre, 01 BP2031 Cotonou, Benin

Ellen Ruth Graber
Institute of Soil, Water & Environmental Sciences, The Volcani Center, Agricultural Research Organization, P.O. Box 6, Bet Dagan 50250, Israel

Lukas van Zwieten
NSW Department of Primary Industries, Bruxner Highway, Wollongbar, 2480 NSW, Australia
Southern Cross Plant Science, Southern Cross University, Lismore, 2477 NSW, Australia

Sara Taherymoosavi, Chee Chia and Paul Munroe
School of Materials Science and Engineering, University of NSW, Sydney NSW 2052, Australia

Torsten Thomas, Shaun Nielsen and Jun Ye
School of Biotechnology and Biomolecular Sciences, University of NSW, Sydney NSW 2052, Australia

Genxing Pan and Xiaorong Fan
Institute of Resource, Ecosystem and Environment of Agriculture, Nanjing Agricultural University, Nanjing 210095, China

Jessica Allen, Yun Lin and Scott Donne
Discipline of Chemistry, University of Newcastle, Callaghan NSW 2308, Australia

Jon P. Cody, Nathan C. Swyers, Morgan E. McCaw, Nathaniel D. Graham, Changzeng Zhao and James A. Birchler
Division of Biological Sciences, University of Missouri, 311 Tucker Hall, Columbia, MO 65211-7400, USA

www.ingramcontent.com/pod-product-compliance
Lightning Source LLC
Chambersburg PA
CBHW080459200326
41458CB00012B/4029